THE FUNGI OF SOUTHEAST ENGLAND

THE FUNGI OF SOUTHEAST ENGLAND

by

R.W.G. DENNIS

The Royal Botanic Gardens, Kew

First published 1995

ISBN 0 947643 80 X

Address of author:

Dr R.W.G. Dennis, Royal Botanic Gardens, Kew, Richmond, Surrey TW9 3AE,
U.K.

General Editor: J.M. Lock

Cover design by Media Resources, Royal Botanic Gardens, Kew

Typeset at the Royal Botanic Gardens, Kew, by
Pam Arnold, Christine Beard, Dominica Costello
Margaret Newman, Pam Rosen
and Helen Ward

Printed and bound in Great Britain by Whitstable Litho Ltd., Whitstable, Kent.

CONTENTS

Contents continued

The Fungi of Southeast England

The present catalogue was originally undertaken some 40 years ago to provide a comprehensive picture of the mycota in a southern lowland area as a standard with which to compare that of a northern largely highland area like the Hebrides. The four south-eastern counties were chosen largely because they are clearly delimited by water on three sides, by the Channel, Straits of Dover and the lower Thames with its estuary. They include a comprehensive range of habitats from salt marsh and freshwater swamp to ancient broad-leaved forest, chalk pasture and sandy heath. The proximity to London and the long tradition of mycological research at Kew in Surrey has led to its fungi being sampled over a long period by specialists in nearly every group from phycomycetes to gasteromycetes so that the bias in most areas in favour of foray records is largely overcome. The presence of the Ministry of Agriculture's Veterinary laboratory at Weybridge makes it possible to include animal mycoses. These may have originated in any part of the country but since the hosts are by their nature mobile and subject to dispersal in normal trading practice it has seemed acceptable to include the fungi as ?Surrey. The western boundary of the area is an arbitrary administrative one of no biological or geographical importance and there have been problems over the acceptance or assignment of some records. Thus, Virginia Water lies for the greater part within Berkshire but part is in Surrey and collections so labelled have been accepted for the latter. The town of Farnham is in Surrey but is the postal address of the Forestry Commission's laboratory in Alice Holt forest, just across the Hampshire border. Similarly, Haslemere is in Surrey and it has been a popular centre for fungus forays but it lies in the extreme southwest corner of the county and at least one favourite collecting site, Waggoners Wells, lies just within Hampshire. Other local collecting grounds are divided about equally between Surrey and West Sussex so, unless the locality is precisely stated, a "Haslemere" record is ambiguous. Tunbridge Wells is in Kent but adjacent to the East Sussex border and collections so labelled are likely to have originated within the latter, where High Rocks, Broadwater Forest, Frant and the like were favourite hunting grounds for A.A. Pearson before he moved house to Hindhead. Even such a prolific station as Bedgebury Pinetum has rather perversely been sited right on the Sussex border so, though it provided many Kent records, it can hardly be held to be representative of the county.

Nevertheless these technical problems in assigning individual collections to their appropriate division should not be taken very seriously. Fungi are not like perennial flowering plants for which the precise station can be indicated to a metre so that presence or absence on one side of a boundary ditch may decide inclusion in a county flora. Fungi may be present at a site for a very long period, even centuries in the case of, for example fairy rings, but their presence can only be demonstrated when they fruit, usually at most for a week or two a year and not necessarily every year. Recording them is a matter of chance, it is improbable most have a restricted distribution and it is unlikely that quibbles about the precise locality of a collection have any relevance to a species' presence in a narrowly defined area. On the contrary, it is axiomatic that virtually all species reported from adjacent counties but not yet within those which form the basis of the present catalogue have merely temporarily escaped observation there. For this reason care has been taken to indicate under each genus additional species recorded from nearby counties. These are not included in the enumerations but they must certainly be taken into consideration in forming a balanced assessment of the southeastern mycota.

Species concepts in fungi have differed widely, both over a period of time and between groups of fungi. In the last century agaricales were identified by the look, sniff, nibble and spit method and species concepts were wide; today, not only are much finer

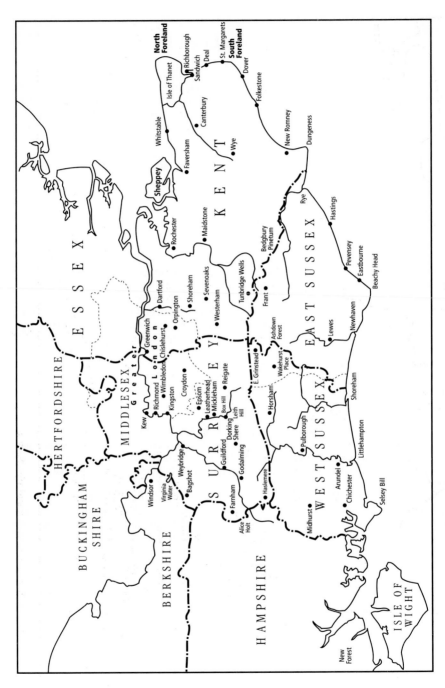

Map 1. Southeast England. Watsonian vice-county boundaries are marked ‑‑‑‑‑‑‑‑‑‑ as is the boundary of Greater London

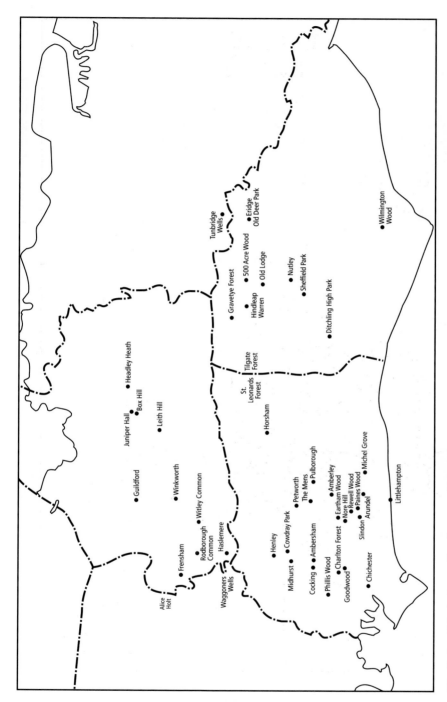

Map 2. Localities in Surrey, East Sussex and West Sussex

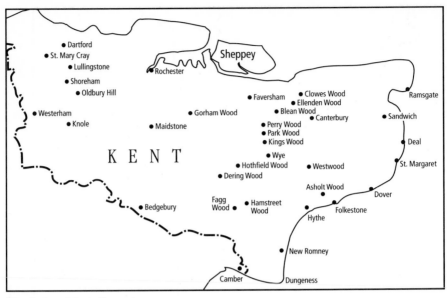

Map 3. Localities in Kent

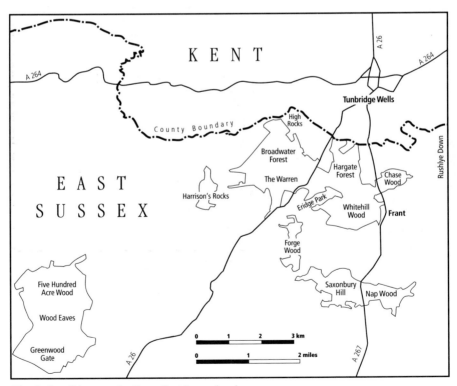

Map 4. Localities near the Kent - East Sussex border

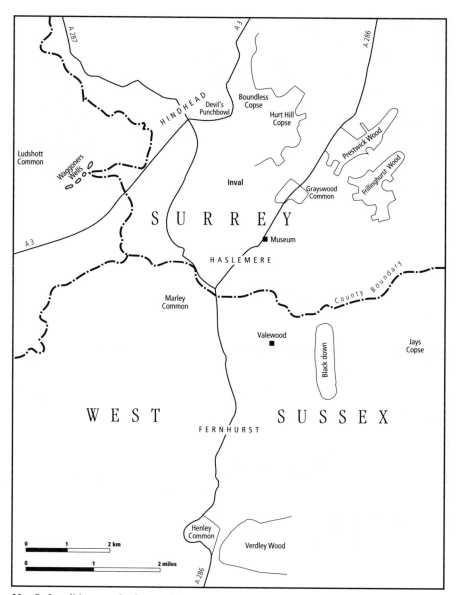

Map 5. Localities near the Surrey - West Sussex border

morphological characters taken into account but if two mycelia derived from similar carpophores fail to diploidise one another there is a tendency to assign them to different species. Amongst obligate parasites specific concepts have fluctuated perhaps even more widely and care has been taken in the lists of Uredinales, Erysiphales and Peronosporales to list host species separately in case *formae speciales* are involved. An even worse problem for the enumerator of fungi is the existence of pleomorphism. Many, perhaps most, ascomycetes include an accessory spore form or anamorph in their life cycle and in many species this is more readily collected than the ascal frutification, the teleomorph. Anamorphs are classifiable in a parallel system to that of the teleomorphs, the fungi imperfecti. Hence it is important not to count an ascomycete twice over in any enumeration of species in an area. In the present catalogue anamorphs are listed under their teleomorphs when this is known so a record of the latter does not signify that it has been actually seen or collected here. For example, *Polythrincium trifolii* Kunze is extremely abundant and ubiquitous on leaves of *Trifolium repens* but no material is available of the known teleomorph *Mycosphaerella killiani* Petrak though the fungus is catalogued under the latter name. There remain. however, very many fungi imperfecti for which a teleomorph is not known. Either the connection between the two states of the life cycle has not yet been demonstrated or the teleomorph is seldom formed or even obsolete. The form-genera of imperfecti are listed alphabetically at the end of the catalogue under the traditional categories "Coelomycetes" with discrete carpophores and "Hyphales" without such but the distinction between the two is no longer clear cut and sometimes largely traditional. Teleomorphs are catalogued under orders within the three classical divisions of fungi: Basidiomycotina, Ascomycotina and Phycomycotera. Myxomycota, bacteria and Actinomycetes are not included. Within these fundamental categories the classification of fungi has undergone revolutionary changes in the last half century. The basis for recognising and circumscribing families is comparatively new and while those currently employed in the system of Basidiomycotina may be fairly stable and widely accepted those of Ascomycotina are not; even the assignment of some genera to an appropriate order is debatable. In the following tabular summary the families are those currently recognised in the herbarium of the Royal Botanic Gardens. It should be recognised, however, that the traditional "Gasteromycetes" are an unnatural assemblage of unrelated families, some of which show close affinities with families of Agaricales. A glance at the statistics in Table 1 which lists the currently recognised families of Basidiomycotina suggests there may be a difference of opinion among specialists regarding the circumscription of families. In Agaricales 1674 species have been accommodated in 17 families, approximately 98.5 per family, whereas only 485 Aphyllophorales require 45 families, 17 of them monogeneric, approximately 11.2 per family. The latter figure is comparable with that found in phanerogamic floras, for example in the Flora of Sussex about 1400 species of phanerogams and vascular cryptogams are distributed through 95 families, approximately 14.7 species to a family.

A botanist, aware that 95 families suffice to accommodate the vascular plants of Sussex, may feel sceptical that more than double the number is needed in a system of Eumycetes alone, viz 86 families of Basidiomycotina and another 105 for Ascomycotina of the same area. Experience may indeed show that taxonomic mycologists have indulged their fissiparous propensity to excess in the postwar years and that sober reflection may lead to consolidation or amalgamation of some genera and families. Nevertheless the large number of higher taxa is an expression of reality on two counts. First, in spite of their small stature, there is vastly greater diversity in fundamental structure among fungi than among flowering plants. Secondly, whereas the British flora is only a minute and unrepresentative fraction of the world's plants, fungi as a whole tend to have very wide geographical ranges, girdling the globe in the temperate or tropical zones or in both. Hence the British mycota is far more representative of the world's fungi than in the case of flowering plants. It is also salutary to remember that the present catalogue includes the names of nearly 1300 taxa of Fungi Imperfecti and, though one can make informed guesses at the likely teleomorph of many, one cannot feel certain that if and when they are all to be accommodated, and all but a few in which the hyphae have clamp connections belong to genera of ascomycotina, the newfound teleomorphs representing undiscovered ascomycotina, will not require additional families to house them.

The massive preponderance of Agaricales in the basidiomycotina (excluding the obligatory parasites, the smuts and rusts) is clear. To the 1674 Agaricales one should add 13 so-called gasteromycetes of agaricoid affinity or 1687 taxa against 485 Aphyllophorales plus 49 nonagaricoid gasteromycetes and 64 heterobasidae, approximately 1700 to 600 or 73.9% to 26.1%. The slightly higher figure 75.8% Agaricales given in Dennis 1973 applied to homobasidial nongastroid taxa only. The distinction between a homobasidium and a heterobasidium is a little less clear cut today and the insistance on the agaricoid affinities of some gastroid families makes it preferable to include all but rusts and smuts in the comparison with agaricales. Ectomycorrhizal fungi are mainly agaricales, where they amount to 35.3% of the species but only 19% of the genera, mainly in the families Amanitaceae, Boletaceae, Cantharellaceae, Cortinariaceae, Gomphidiaceae and Russulaceae, but the ectomycorrhizal habit is adopted also by a few genera of operculate ascomycotina. Obviously it is the large families Amanitaceae, Boletaceae, Cortinariaceae and Russulaceae which are of primary importance in forestry, especially in the afforestation of previously treeless land. The endomycorrhizal habit predominant amongst herbaceous angiosperms seems, however, to be the sphere of the Mucorales, especially the family Endogonaceae and the genus *Glomus* and its segregates.

A few fungi, mainly hyphales, are involved in animal mycoses, many more in plant diseases. In the latter field it is necessary to distinguish between the obligate parasites, the Uredinales, Ustilaginales, Exobasidiaceae, Taphrinales, Erysiphales, Peronosporales where survival of the parasite depends on survival of the host, and facultative parasites with which parasitism is often an opportunist matter and death of the host may be a positive advantage. Not many plant parasites are of economic significance, most have no more than a disfiguring effect on the host though this may greatly reduce the saleability of a plant product, but a few pose a real threat to agriculture even, rarely, amounting to the failure of staple crops. Modern plant pathology is involved rather with virus infections, nutritional deficiences, even bacterial diseases than with fungi.

Compared with botany and zoology, mycology is a modern discipline. Only a few bizarre fungal carpophores, usually gastromycetes, attracted the attention of 16th and 17th century writers. Inevitably one of the first was *Phallus*, the subject of Jonge's dissertation "Phallus ex fungorum genere in Hollandiae sabulatis passim crescentis" 1564, followed by *Cyathus* "Fungus calyciformis seminifer" (Camerarius 1689), *Clathrus ruber* "Fungus lupi crepitus vulgi, efflorescens (Colonna 1616), *Myriostoma* "Fungus pulverulentus coli instar perforatus cum volva stellata" (Ray 1704), *Langermannia* "Lupi crepitus sive Fungus ovatus" (Parkinson 1640), *Lycoperdon perlatum* "Fungus pulverulentus Crepitus Lupi dictus major, pediculo longiore ventricoso" (Ray 1704). Dillenius (1718) recognised a genus *Peziza* "vel ut Plinius habet Pezica" which probably included some fungi to which the validated name is still applied and also *Morchella* "Fungi pileati et pediculo donati, scrobiculati". Apart from the morphologically odd looking, early writers were primarily interested in edible fungi, including truffles, but Vaillant (1727) for the first time described *Amanita phalloides* "Fungus phalloides anulatus sordide virescens et patulus", though it was sixty years before Paulet demonstrated its toxicity. Micheli (1729), for the first time provided the outline of a mycological classification, with a number of generic names still in use, and also demonstrated that fungi were not the product of spontaneous generation but autonomous organisms reproduced by spores. It was this uncertainty regarding their origin, enshrined in myths like the origin of *Polyporus tuberaster* from fossilized lynx urine or *Agaricalus arvensis* from urine or semen of a stallion, that made it difficult for botanists to take them seriously. As to parasitic fungi the tendency to dismiss them as exanthemata, a product of the diseased plant rather than agents of the disease was only effectively refuted by the observations of Berkeley and Montagne in the wake of Potato Blight in 1845. Even after the adoption of binomial nomenclature and the sexual system of botanical classification it was difficult for botanists to cope with "plants" lacking the obvious sexual organs on which the Linnaean classification was founded. To Linnaeus himself fungi were incomprehensible, partly because he rejected microscopy and chose to return to the ideas of Dillenius rather than adopt the approach pioneered by Micheli. "Fungus ex doctrine Dilleniana potius dividere quam ex Micheliana nobis visum suit, qui oculis *armatis* non adsueti, suimus." Hence, for Linnaeus fungi were chaos in which it was impossible for botanists to recognise species and varieties. His great authority stifled mycological research throughout the 18th century until its end, when it was effectively reborn in the works of Bulliard, Persoon and Sowerby, systematised in 1821 by E.M. Fries.

Thus people have only been seriously collecting and recording fungi in our area for 200 years and the catalogue is far from complete. Every year at Kew we receive from local sources fungi we do not recognise, though admittedly these are predominatly microfungi. For some a name is eventually found though this can be a tedious business since they may have been described anywhere in Europe or North America, occasionally even in the southern hemisphere. There still remain a number for which no name is found and which are presumably new to science. It also seems probable that the local mycota has not remained stable for two centuries during which there have been great social and economic changes affecting the agriculture, landscape and vegetation or even slight unverifiable climatic modification. To demonstrate extinction of micro-organisms which fruit erratically and are only collected by chance is almost impossible; a species may appear to be absent for a century and then reappear. Has it been reintroduced, lying dormant or simply overlooked? But at least one fungus, with large very distinctive carpophores, *Myriostoma coliforme* (With.) Corda, once present in the area and indeed first found here, seems to have been lost. Ray received it in 1695 from "the lane from Crayford to Bexley Common", perhaps now the A207 in a totally built up area. Apparently it was widespread though uncommon from Kent to Norfolk at the end of the 18th century but it slowly died out and has not been reported from Britain since 25th September 1890, when Plowright received specimens from "a hedgebank in the village of Hillington, Norfolk". It may have been a steppe species, always at the edge of its range in England, but the reason for its extinction is not clear: one may suspect more intensive agriculture with ploughing of the habitat, as well as sealing under concrete or tarmac, without any necessity for speculation about climatic changes.

One can feel more confident about additions to our mycota if they have large conspicuous carpophores unlikely to have been missed by 19th century collectors or when they are the agents of hitherto unknown plant diseases. There has been a succession of the latter over the last century and a half. First came the best known invader, with the most dramatic and permanent consequences, *Phytophthora infestans* (Mont.) de Bary, introduced directly or indirectly from some part of Latin America where tuber-forming species of *Solanum* are native.

As it is very narrowly host-limited and the host is not native, *Puccinia vincae* Berk. must actually have arrived even earlier for Berkeley had it from Suffolk by 1836. Though widespread it is not common, perhaps because of hyperparasitism by *Tuberculina persicina* (Ditmar) Sacc. *Puccinia asparagi* DC. was at Swancombe, Kent by 1865. *P. malvacearum* Mont., native to Chile, was in Spain by 1869 but did not reach England until 1873, since when it has caused ubiquitous disease on *Althaea rosea* as well as native *Malva* spp. *Uromyces dianthi* (Pers.) Niessl was imported about 1890 and *Cronartium rubicola* Fisch. had arrived in Britain by 1892, probably from Asia though it was in Estonia by 1854. *P. chrysanthemi* Roze came in 1895 and *Synchytrium endobioticum* (Schilb.) Perc. probably in the following year, though it did not arouse alarm until 1902. This was potentially almost as disastrous as *Phytophthora infestans* but the existence of total immunity (actually hypersensitivity) to the disease in some potato varieties made it possible for it to be completely eliminated. Its importance lay in the consequent necessity for strict inspection, rogueing and certification of potato seed crops to ensure varietal purity. Development of the inspection and certification system in Scotland made it possible to control tuber-borne potato virus diseases once they had been recognised later in the century. *Sphaerotheca mors-uvae* (Schw.) Berk. & Curt. reached northern Ireland in July 1900, presumably from North America, and was in southeast England by 1906 at latest. *Cumminsiella mirabilissima* (Peck) Nannf., also from North America, was in Scotland by 1922, Northumberland by 1930 and spread inexorably southwards to become coextensive with *Mahonia aquifolium*. By 1927 *Ceratocystis ulmi* had reached the London area and it has subsequently changed the landscape of lowland England by destroying most hedgerow elms. *Puccinia antirrhini* Diet. & Holw. appeared in Kent in 1933 and was at first destructive to *Antirrhinum majus* throughout the country, until resistant varieties were selected. *P. opizii* Bub. appeared on lettuce in Norfolk in 1951 alternating with carices but apparently died out, there is material on *Carex muricata* from Sussex only. *Puccinia lagenophorae* Cooke is an Australian rust on compositae which appeared at Dungeness in August 1961 and has spread progressivley across the country to the remotest Hebrides on *Senecio squalidus* and *S. vulgaris*. *P. pelargonii-zonalis* Doidge arrived on the Sussex coast at Eastbourne in 1965 having crossed France from an introduction in Provence and seems to have survived at least in Surrey. *P. oxalidis* Diet. & Ellis was in Surrey in 1973 and I saw it in the Hebrides in 1986.

It will be observed that the introduction of strict plant quarantine legislation has made no difference to the steady stream of immigrant fungi. Not only will travellers always circumvent customs control by secreting plant cuttings in baggage or pockets, and rapid air travel means the plant material will easily survive any distance, but the English channel is not wide enough to impede wind transport of fungal spores from Europe.

Uredinales are usually the best documented fungi but another small group of incomers is easily identified, those associated with glasshouses, "Stove aliens" like *Amanita nauceosa* (Wakef.) Reid, *Conocybe intrusa* (Peck) Sing., *Lactocollybia angiospermarum* Sing., sundry *Lepiota* and *Leucocoprinus* species, *Mycena osmundicola* Lange, many of which are associated with Botanic Gardens. Some, however, have been around in their restricted artificial environments for a very long time. *Leucocoprinus birnbaumii* (Corda) Sing. is among the earliest recorded and figured English fungi, as "Agaricus totus luteus, stipite subbulboso annulato, pileo obtusoconico piloso squamoso, velo araneoso, substantia sicca", amongst the bark in the pine-stove belonging to *J. Caygill*, Esq., at Sha, near Halifax, in August, 1785. (Bolton 1788). For Sowerby (1796) it, with *L. cepaestipes* (Sow.) Pat., was "not uncommon in bark-beds about London", but the collection he cited and figured was from Wormley Bury, Cheshunt, Hertfordshire. The species is still plentiful in the plant houses at Kew and occurs from time to time in domestic circumstances with potted plants such as *Monstera deliciosa*. With such a respectable pedigree this is surely a British fungus yet it finds no place in the 1960 Check List of British Agarics and Boleti , presumably because it is an alien. It seems impossible to adopt so xenophobic an attitude which must always involve subjective distinctions and all species found to have been recorded from the area have been included in this catalogue.

Though one suspects the stove aliens to have been introduced by human agency, albeit inadvertently, a few apparent recent immigrants among the agaricales seem likely to have made landfall unassisted in this way. Two, with large, conspicuous and distinctive carpophores which could not have been overlooked by 19th century collectors have appeared in southern England in postwar years. *Omphalotus olearius* (DC.) Sing. was first observed in Sussex in September 1967, nearby again in 1975 on a stump of *Quercus* and subsequently in Hampshire. It was already known in Normandy so an extension of range across the Channel is not surprising. *Amanita ovoidea* (Bull.) Quél. had been reported from England twice before, in Somerset October 1880 and in Norfolk, on the site of a military camp, in 1920. There was no voucher material for either record and they have been rejected. There are now two recent collections from the west of our area, Wiltshire

July 1973 and Isle of Wight October 1990, both fully studied and supported by voucher material. The type localities of the species are la Forêt de Fontainebleau 48°23'N and les bois de Malesherbes 48°18'N. Bulliard remarked that this was a fungus of the southern provinces, where it was known as Oronge blanche; if it is followed here in due course by l'Oronge propre, *A. caesarea* (Scop.) Pers., the classical "Boletus", one will, indeed, have to take climatic change seriously.

One fungus of uncertain status is the unmistakeable *Clathrus ruber* Pers. This is certainly native throughout the western mediterranean countries and up the west coast of France to Bretagne, probably as far as Granville. It is associated usually with woodland especially of *Pinus* or *Quercus* spp., apparently as a saprophyte on decaying roots or buried wood. In England the first records are from the Isle of Wight 1844 and south Devon 1848, but it is apparently now well established in Scilly and there are scattered records along the south coast to Kent. Though a few of these refer to copses, pine woods or damp pasture the majority are from gardens though, surprisingly, not from horticultural institutions or botanic gardens. Is it a native at the limit of its range, clinging to a few sheltered places with a southern exposure or a casual introduction that sometimes persists for a few years in similar situations? Either interpretation of the evidence is tenable and in catalogueing the species as "probable alien" I have been influenced by the lack of records prior to 1844. Undoubtedly alien is our other species, *C. archeri* (Berk.) Dring. This is native to Tasmania, New South Wales, New Zealand and perhaps South Africa. It appeared in a cornish garden in 1945 and in Sussex in 1953 but it has since established itself amongst native vegetation and spread into Kent so it must be accepted as naturalised. Some form of human activity must have been involved in the move from the southern to the northern hemispheres but not necessarily so in the invasion of England for the fungus has been well established in eastern France since 1920, originally at a site where rotten meat had been buried. Inevitably one suspects introduction with stores of some kind originating in Australia and distributed either to troops at the end of the 1914-18 war or during relief work in its aftermath. Maire (1930) commented "presque toutes les stations où il a été rencontré sont situées sur l'emplacement de campements des troupes américaines". If there is significance in this observation one must assume they were fed on Australian rations for the fungus is not known in North America. More recently it has spread through south Germany, Switzerland, into Austria and Italy, also south to the Pays basque.

More recently another southern phalloid, *Ileodictyon cibarium* Tul., native to New Zealand, Australia and Chile appeared in a garden in Middlesex in December 1963 and in one in Surrey in January 1971. As yet this fungus is not known in Europe so the introduction to England from the antipodes seems to have been direct. The same must be true of the ascogenous *Pilocotylis pila* Berk. not yet in the southeastern counties but introduced from New Zealand to the midlands by 1973 and evidently locally established there.

Tables 1 Basidiomycotina and 2 Ascomycotina present a taxonomic analysis indicating the number of genera and species in each family of the Eumycotina represented in south-east England. It will be seen that many genera in Table 2 cannot yet be assigned with confidence to any currently accepted family. There follows an alphabetical catalogue of fungi as recorded from the area in published literature supplemented by collections and data in the herbarium of the Royal Botanic Gardens, Kew. Distribution between the counties is indicated by numerals: 1 = Surrey 2 = West Sussex 3 = East Sussex 4 = Kent. The boundary between 2 and 3 is that between the watsonian vicecounties 13 & 14, near the prewar boundary between West and East Sussex. The current boundary between the counties so called lies for the greater part further east. The significance in the present context is that collections from Wakehurst Place are entered under 3 though the estate lies within the modern administrative county West Sussex.

Table 1 Families of Basidiomycotina

	Genera	Species		Genera	Species
Agaricaceae	8	145	Podoscyphaceae	3	3
Amanitaceae	2	29	Pterulaceae	1	2
Bolbitiaceae	4	58	Ramariaceae	1	15
Boletaceae	8	59	Schizophyllaceae	1	1
Cantharellaceae	2	10	Sistotremataceae	7	26
Coprinaceae	5	137	Sparassidaceae	1	3
Cortinariaceae	10	335	Sporodiolaceae	5	6
Crepidotaceae	6	25	Steccherinaceae	3	4
Entolomataceae	7	93	Stereaceae	3	9
Gomphidiaceae	2	4	Thelephoraceae	5	37
Hygrophoraceae	3	74	Tubulicrinaceae	2	5
Paxillaceae	4	9	Tulasnellaceae	1	13
Pluteaceae	2	48	Typhulaceae	2	11
Polyporaceae	6	17	Xenasmataceae	1	4
Russulaceae	2	160			
Strophariaceae	8	79	Astraeaceae	1	1
Tricholomataceae	47	392	Clathraceae	4	5
			Geastraceae	1	9
Aleurodiscaceae	2	3	Hydnangiaceae	1	1
Amylocorticiaceae	1	3	Hymenogastraceae	2	7
Aphelariaceae	-	-	Lycoperdaceae	6	18
Atheliaceae	9	19	Melanogastraceae	1	3
Auriscalpiaceae	2	5	Nidulariaceae	4	5
Bankeraceae	1	4	Phallaceae	2	2
Botryobasidiaceae	2	8	Rhizopogonaceae	1	2
Carcinomycetaceae	1	2	Sclerodermataceae	1	5
Cejpomycetaceae	1	1	Sphaerobolaceae	1	1
Ceratobasidiaceae	3	5	Stephanosporaceae	1	1
Clavariaceae	3	29	Tulostomataceae	2	2
Clavariadelphaceae	1	5			
Clavicoronaceae	1	1	Graphiolaceae	1	1
Clavulinaceae	2	4	Tilletiaceae	4	19
Coniophoraceae	6	16	Ustilaginaceae	7	23
Coriolaceae	40	81	Chaconiaceae	1	1
Corticiaceae	6	6	Coleosporaceae	1	1
Cyphellaceae	6	6	Cronartiaceae	1	2
Epitheliaceae	-	-	Melampsoraceae	1	8
Exobasidiaceae	1	4	Phragmidiaceae	4	12
Fistulinaceae	1	1	Pucciniaceae	7	132
Ganodermataceae	1	6	Pucciniastraceae	4	11
Gleocystidinellaceae	5	7	Sphaerophragmiaceae	1	2
Hericiaceae	2	7	Uropyxidaceae	1	2
Hydnaceae	1	2			
Hymenochaetaceae	5	24	Dacrymycetaceae	5	15
Lachnocladiaceae	3	4	Auriculariaceae	6	10
Lindtneriaceae	2	3	Ecchynaceae	1	1
Meruliaceae	13	33	Tremellaceae	11	37
Peniophoraceae	1	18	Sirobasidiaceae	1	1

Table 2 Families of Ascomycotina

	Genera	Species		Genera	Species
Taphrinaceae	1	12	Amphisphaeriaceae	10	19
Protomycetaceae	3	4	Halosphaeriaceae	3	3
			Trichosphaeriaceae	12	38
Ascobolaceae	5	26	Xylariaceae	15	63
Balsaminaceae	1	1	Sphaeriales		
Geneaceae	1	2	unassigned	8	9
Helvellaceae	4	16			
Humariaceae	25	77	Chaetomiaceae	2	14
Monascaceae	1	2	Lasiosphaeriaceae	12	41
Morchellaceae	4	7	Melanosporaceae	3	10
Pezizaceae	8	53	Sordariaceae	3	11
Sarcosomataceae	4	5	Sordariales		
Terfeziaceae	2	2	unassigned	2	2
Thelobolaceae	3	9	Nitschkiaceae	4	8
Tuberaceae	1	8	Diatrypaceae	5	19
			Gnomoniaceae	11	23
Ascocorticiaceae	1	1	Melanconidaceae	6	15
Dermateaceae	27	95	Melogrammataceae	1	1
Geoglossaceae	5	11	Psaeudovalsaceae	3	5
Helotiaceae	38	105	Valsaceae	9	64
Hemiphacidiaceae	1	1	Diaporthales		
Hyaloscyphaceae	23	82	unassigned	7	14
Orbiliaceae	1	10	Ophiostomataceae	1	6
Sclerotiniaceae	9	35	Elaphomycetaceae	1	3
			Erysiphaceae	8	72
Mycocaliciaceae	1	1			
Caliciaceae	1	1	Cephalothecaceae	1	2
			Pseudeurotiaceae	4	5
Ascodichaeniaceae	1	1	Trichocomaceae	10	19
Leptopeltidaceae	1	1	Eurotiales		
Phacidiaceae	1	2	unassigned	2	2
Rhytismataceae	15	31	Microascaceae	4	7
			Pithoascaceae	1	1
Agyriaceae	1	1	Ascosphaeriaceae	2	2
Arthoniaceae	1	2	Gymnoascaceae	7	20
Lecanorales			Onygenaceae	3	4
unassigned	1	3	Saccharomycetaceae	5	8
Opegraphaceae	1	1	Laboulbeniaceae	6	7
Seuratiaceae	1	1	Verrucariales	1	1
			Lecanidiaceae	1	1
Ostropaceae	1	1			
Stictidaceae	2	6	Asterinaceae	1	1
Clavicipitaceae	5	13	Aulographaceae	1	1
Hypomycetaceae	4	8	Botryosphaeriaceae	4	12
Hypocreaceae	11	63	Capnodiaceae	1	1
Phyllachoraceae	2	6	Dacampiaceae	2	2
Polystigmataceae	1	1	Didymosphaeriaceae	1	4

Table 2 Continued

	Genera	Species		Genera	Species
Dothideaceae	7	52	Phaeosphaeriaceae	8	37
Dothioraceae	3	3	Phaeotrichaceae	5	17
Elsinoaceae	1	3	Pleomassariaceae	3	5
Fenestellaceae	2	4	Pleosporaceae	13	25
Herpotrichiellaceae	2	8	Pyrenophoraceae	2	8
Hysteriaceae	4	7	Schizothyriaceae	1	1
Lophiostomataceae	3	12	Testudinaceae	1	1
Massarinaceae	1	2	Tubeufiaceae	2	3
Melanommataceae	3	8	Venturiaceae	6	20
Meliolaceae	1	1	Dothidiales		
Metacapnodiaceae	1	1	unassigned	5	5
Micropeltidaceae	1	2			
Microthyriaceae	2	13	Trypetheliaceae	2	3
Mytilinidiaceae	3	3			

British Mycological Society Forays in South-east England.

Haslemere, Surrey	25–30 Sept. 1905	Trans. Brit. Myc. Soc.	2:101–111
Haslemere, Surrey and West Sussex	22–27 Sept. 1913	Trans. Brit. Myc. Soc.	4:207–223
Haslemere, Surrey and West Sussex	13–16 May 1921	Trans. Brit. Myc. Soc.	7:221–225
Arundel, West Sussex	21–25 May 1926	Trans. Brit. Myc. Soc.	12:1–5
Littlehampton, West Sussex	1–6 Oct. 1928	Trans. Brit. Myc. Soc.	14:185–193
Horsham, West and East Sussex	29 May–2 June 1931	Trans. Brit. Myc. Soc.	17:1–4
Haslemere, Surrey and West Sussex	19–24 Sept. 1932	Trans. Brit. Myc. Soc.	18:7–17
Tunbridge Wells, Kent and mainly E. Sussex	22–26 May 1936	Trans. Brit. Myc. Soc.	22:1–4
Arundel, West Sussex	19–20 May 1939	Trans. Brit. Myc. Soc.	24:1–3
Juniper Hall, Surrey	29 May–1 June 1953	Trans. Brit. Myc. Soc.	37:183
Midhurst, West Sussex	21–26 May 1959	Brit. Myc. Soc. News Bull.	13:4–6
Egham, Surrey (Virginia Water only)	5–12 Sept. 1961	Brit. Myc. Soc. News Bull.	18:3–5
Chichester, West Sussex	1–5 Sept. 1967	Bull. Brit. Myc. Soc.	2:61–66
Nutley, East Sussex	2–9 Sept. 1970	Bull. Brit. Myc. Soc.	5:48–52
Guildford, Surrey and West Sussex	5–12 Sept. 1979	Bull. Brit. Myc. Soc.	15:19–29
Wye, Kent	16–23 Sept. 1981	Bull. Brit. Myc. Soc.	17:16–26
Pulborough, West Sussex	24–30 May 1985	Bull. Brit. Myc. Soc.	20:82–88

BASIDIOMYCETES

AGARICALES

Agaricus L.

Terrestrial saprophytes. The "species" are distinguished by few, even subjective, characters and as such are liable to almost indefinite proliferation of names.

	1	2	3	4
Agaricus abruptibulbus Peck	1	2		4
A. altipes (Møll.) Pilat	1			
A. arvensis Schaeff.	1	2	3	4
A. augustus Fr.	1	2	3	4
A. bernardii Quél.	1	2		4
A. bisporus (Lange) Imbach	1	2		4
A. bitorquis (Quél.) Bon	1	2		4
A. bohusii M. Bon	1			
A. bresadolianus Bohus	1			
A. campestris L.	1	2	3	4
A. comtulus Fr.	1	2	3	4
A. cupreobrunneus (Schaeff. & Steer) Pilat	1			
A. depauperatus (Møll.) Pilat				4
A. devoniensis Orton		2		4
A. dulcidulus Schaeff.	1	2		
A. essettei M.Bon	1			
A. excellens (Møll.) Møll.	1			4
A. fissuratus (Møll.) Møll.	1			4
A. fuscofibrillosus (Møll.) Pilat	1		3	
A. gennadii (Chatin & Boud.) Orton			3	
A. haemorrhoidarius Schulzer	1	2		4
A. langei (Møll.) Møll.	1	2		4
A. lanipes (Møll. & Schaeff.) Sing.	1			4
A. leucotrichus (Møll.) Moll.	1			
A. luteolorufescens Orton			3	
A. luteomaculatus (Møll.) Møll.	1			
A. lutosus (Møll.) Møll.	1			4
A. macrocarpus (Møll.) Møll.	1			
A. macrosporus (Møll. & Schaeff.) Pilat	1	2	3	4
A. maleolens Møll.	1			
A. nivescens (Møll.) Møll.	1			4
A. nivescens var *parkensis* (Møll.) Møll.	1			
A. phaeolepidotus (Møll.) Møll.	1			
A. placomyces Peck	1	2	3	4
A. porphyrizon Orton	1	2		4
A. porphyrocephalus Møll.	1			4
A. purpurellus (Møll.) Møll.	1			4
A. romagnesii Wasser	1			
A. rusiophyllus Lasch	1			
A. semotus Fr.	1	2	3	4
A. silvaticus Schaeff.	1	2	3	4
A. silvicola (Vott.) Peck	1	2		4
A. spissicaulis (Møll.) Møll.	1		3	4
A. squamuliferus (Møll.) Pilat	1	2		
A. stramineus Krombh.	1			
A. subfloccosus (Lange) Pilat	1			
A. subperonatus (Lange) Sing.	1	2		4
A. vaporarius (Vill.) Moser	1	2	3	4

A. variegans Møll.	1		3 4
A. vinosobrunneus Orton	1		
A. xanthodermus Genev.	1	2	4

Agrocybe Fayod

Saprophytes, mostly terrestrial but A. *cylindracea* is lignicolous and a good edible fungus.

Agrocybe arvalis (Fr.) Sing.	1		3
A. cylindracea (DC.) Maire (A. aegerita (Brig.) Sing.)	1	2	3 4
A. dura (Bolt.) Sing. (A. molesta (Lasch) Sing.)	1	2	3 4
A. erebia (Fr.) Kühner	1	2	3 4
A. paludosa (Lange) Kühner & Romagn.	1	2	4
A. praecox (Pers.) Fayod	1	2	3 4
A. pusiola (Fr.) Metrod	1		4
A. putaminum (Maire) Sing.	1		
A. semiorbicularis (Bull.) Fayod	1	2	3 4
A. tabacina (DC.) Konrad & Maubl.	1		
A. temulenta (Fr.) Orton	1		
A. vervacti (Fr.) Maire	1	2	3 4

Amanita Pers.

Mycorrhizal with trees and shrubs but not narrowly host-limited. The genus includes both popular edible and notoriously lethal fungi, the latter notably A. *pantherina*, A. *phalloides* and A. *virosa*, A. *ovoidea* Quél is present in the Isle of Wight and other British species include A. *friabilis* (Karst.) Bas and A. *lividopallescens* Boud.

Amanita argentea Huijsman	1	2	4
A. battarae Boud. (A. umbrinolutea Secr nom. illeg.)	1	2	
A. citrina (Schaeff.) Gray	1	2	3 4
A. citrina var. alba (Gill.) Gilb.	1	2	3 4
A. crocea (Quél.) Kühner & Romagn.	1		4
A. echinocephala (Vitt.) Quél.	1	2	4
A. eliae Quél.		2	3 4
A. excelsa (Fr.) Kummer incl. A. spissa (Fr.) Kummer	1	2	3 4
A. franchetii (Boud.) Fayod (A. aspera auct. non Pers.)	1	2	3 4
A. fulva (Schaeff.) Pers.	1	2	3 4
A. gemmata (Fr.) Gill.	1	2	4
A. inopinata Reid & Bas	1		4
A. muscaria (L.) Hook.	1	2	3 4
A. nauseosa (Wakef.) Reid Glasshouse alien	1		
A. pantherina (DC.) Krombh.	1	2	3 4
A. phalloides (Vaill.) Link	1	2	3 4
A. phalloides var. verna (Bull.) Maire	1	2	
A. porphyria (A. & S.) Mlady	1	2	3 4
A. rubescens (Pers.) Gray	1	2	3 4
A. rubescens var. annulosulphurea Gill.	1		3 4
A. singeri Bas	1		
A. strangulata (Fr.) Roze (A. ceciliae (B. & Br.) Bas)	1	2	3 4
A. strobiliformis (Paul.) Bertil. (A. solitaria (Bull.) Secr.)	1	2	3 4
A. submembranacea (Bon) Gröger	1	2	4
A. vaginata (Bull.) Vitt.	1	2	3 4
A. virosa (Fr.) Bertil.	1	2	4
A. vittadinii (Moretti) Vitt.		2	

Armillaria (Fr.) Kummer

Facultative parasites of woody plants. A. *mellea* may be a collective species and a number of its races have received specific names but they are not readily separable in the field and are not listed here.

Armillaria mellea (Vahl) Kummer	1	2	3	4
A. tabescens (Scop.) Emel	1	2		4

Asterophora Ditmar

Hyperparasites of *Russula* spp. *Nyctalis* Fr. is a synonym preferred by some authors.

Asterophora lycoperdoides (Bull.) Gray (*Nyctalis asterophora* Fr.)	1	2	3	4
A. parasitica (Bull.) Sing.	1	2	3	4

Baeospora Sing.

Saprophytic on fallen cones of *Pinus*, many old records of *Collybia conigena* (Pers.) Kummer may belong here, distinguished from *Strobilurus* spp. by autumnal fruiting.

Baeospora myosura (Fr.) Sing.	1	2	3	4

Bolbitius Fr.

Saprophytic on dung or manured soil. Some authors distinguish *B. titubans* (Bull.) Fr. with a similar distribution.

Bolbitius vitellinus (Pers.) Fr.	1	2	3	4

Boletus Dill

Here inclusive of *Chalciporus* Bat., *Pulveroboletus* Murr, *Suillus* Mich. and *Xerocomus* Quél. Mycorrhizal with woody plants.

Boletus aereus Bull.	1			
B. (Suillus) aeruginascens Opat (*B. viscidus* Fr.)	1	2	3	4
B. appendiculatus Schaeff.	1	2	3	4
B. armeniacus Quél.	1			
B. badius Fr.	1	2	3	4
B. (Suillus) bovinus L.	1	2	3	4
B. calopus Fr.	1	2	3	4
B. chrysenteron Bull.	1	2	3	4
B. citrinovirens Watling	1			
B. edulis Bull.	1	2	3	4
B. fechtneri Vel.	1	2		4
B. (Suillus) fluryii Huijsman				4
B. fragrans Vitt. Dubious record	1			
B. gentilis Quél. (*B. cramesinus* Secr. nom. illeg.)	1	2		
B. (Suillus) granulatus L.	1	2	3	4
B. (Suillus) grevillei Klotzsch (*B. elegans* Schum.)	1	2	3	4
B. hemichrysus (B. & C.) Sing. (*B. sulphureus* Fr.)		2		
B. impolitus Fr.	1	2	3	
B. junquilleus (Quél.) Boud.		2		
B. lanatus Rostk.	1	2		4
B. lignicola Kallenb.	1			
B. luridiformis Rostk. (*B. erythropus* (Fr.) Secr. non Pers.)	1	2	3	4
B. luridus Schaeff.	1	2	3	4
B. (Suillus) luteus L.	1	2	3	4
B. parasiticus Bull.	1	2	3	4
B. pinicola Venturi	1	2		
B. (Chalciporus) piperatus Bull.	1	2	3	4
B. (Suillus) placidus Bon.				4
B. porosporus (Imler) Watling	1			
B. pruinatus Fr. & Hok	1	2	3	4
B. pulverulentus Opat.	1	2	3	
B. purpureus Pers.	1	2		4
B. queletii Schulzer	1	2		4
B. radicans Pers. (*B. albidus* Rocq.)	1	2	3	4
B. regius Krombh.	1			

B. reticulatus Schaeff. (*B. aestivalis* Fr.)	1	2	3	
B. rubellus Krombh. (*B. versicolor* Rostk.)	1	2	3	4
B. rubinus W.G. Smith	1	2	3	4
B. spadiceus Fr. Dubious record	1			
B. subtomentosus L.	1	2	3	4
B. tridentinus Bres.	1	2		4
B. (Suillus) variegatus Sow.	1	2	3	4

Calocybe Kühner
Saprophytes: also in England *C. cerina* (Pers.) Donk

Calocybe carnea (Bull.) Kühner	1	2	3	4
C. constricta (Fr.) Sing. (*Tricholoma leucocephalum* Fr.) Quél.)	1	2	3	4
C. gambosum (Fr.) Sing.	1	2	3	4
C. ionides (Bull.) Kühner	1	2	3	4
C. obscurissima (Pearson)	1			4
C. persicolor (Fr.) Sing.	1			

Calyptella Quél.
Saprophytes, especially on herbaceous stems.

Calyptella capula (Holmsk.) Quél.	1	2	3	4
C. laeta (Fr.) W.B. Cooke	1			

Campanella Henn.
Saprophytes with pleurotoid habit but reduced anastomosing lamellae, mainly tropical.

Campanella caesia Romagn. (*C. europaea* Sing.)	4

Cantharellula Sing.
Terrestrial saprophytes, here including *Pseudoclitocybe* (Sing.) Sing. and *Pseudoomphalina* (Sing.) Sing., to the latter of which the missing *C. graveolens* (Petersen) Moser has been referred.

Cantharellula (*Pseudoclitocybe*) *cyathiformis* (Bull.) Sing.	1	2	3	4
C. (*Ps.*) *expallens* (Pers.) Orton	1	2	3	
C. obbata (Fr.) Bousset	1			
C. umbonata (Gmelin) Sing.	1		3	

Cantharellus Adanson
Mycorrhizal with forest trees; *C. cibarius* is a familiar edible fungus of considerable economic importance though not cultivated artificially as such.

Cantharellus amethysteus (Quél.) Sacc.		2		4
C. cibarius Fr.	1	2	3	4
C. cinereus (Pers.) Fr.	1	2		4
C. ferruginascens Orton	1			4
C. friesii Quél. Dubious record				4
C. infundibuliformis (Scop.) Fr.	1	2	3	4
C. melanoxeros Desm.				4
C. tubaeformis (Bull.) Fr.	1	2	3	4

Cellypha Donk
Saprophytes

Cellypha berkeleyi (Massee) W.B. Cooke	1
C. griseopallida (Weinm.) W.B. Cooke	1
C. goldbachii Weinm.) Donk	1

Chaetocalathus Sing.
Saprophyte. *C. craterellus* (Dur. & Lév.) Sing. in Devon

Chamaemyces Earle
Terrestrial saprophyte

Chamaemyces fracidus (Fr.) Donk (*Drosella irrorata* (Quél.) Kühn. & Maire)   1   2   3   4

Cheimonophyllum Sing.
Saprophyte. British species *C. candidissimum* (B. & C.) Sing., probably overlooked

Chromocyphella de Toni & Levi
Saprophytes, on wood or moss.

Chromocyphella galeata (Schum.) W.B. Cooke   1

Chroogomphus (Sing.) Miller
Mycorrhizal with conifers

Chroogomphus rutilus (Fr.) Miller   1   2   3   4

Claudopus (W.G. Smith) Gillet
Saprophytes, entolomoid fungi with pleurotoid habit.

Claudopus byssisedus (Pers.) Gill   1
C. depluens (Batsch) Gill.   1

Clitocybe (Fr.) Kummer
Saprophytes, mainly terrestrial. The species are often ill-defined. In addition to those listed *C. agrestis* Harmaja, *C. alexandri* (Gill.) Konrad, *C. barbularum* (Romagn.) Orton, *C. paropsis* (Fr.) Karst. *C. scyphoides* (Fr.) Orton, *C. subsinopica* Harmaja & *C. truncicola* (Peck) Sacc. might be expected to occur.

	1	2	3	4
Clitocybe americana Bigelow	1			4
C. angustissima (Lasch) Kummer		2		
C. borealis (Lange & Skifte) Orton & Watling	1			
C. brumalis (Fr.) Quél.	1		3	4
C. candicans (Pers.) Kummer	1		3	4
C. clavipes (Pers.) Kummer	1	2	3	4
C. clusiliformis Kühner & Romagn.		2		
C. (Pseudoomphalina) compressipes (Peck) Sacc.	1			
C. costata Kühner & Romagn. (*C. incilis* auct.)	1			
C. dealbata (Sow.) Kummer	1	2	3	4
C. depauperata (Lange) Orton	1			
C. diatreta (Fr.) Kummer	1			4
C. dicolor (Pers.) Lange	1	2	3	4
C. ditopus (Fr.) Gill.	1	2	3	4
C. ericetorum (Bull.) Quél.	1	2		
C. flaccida (Sow.) Kummer(*C. inversa* (Scop.) Quél.)	1	2	3	4
C. fragrans (Sow.) Kummer	1	2	3	4
C. fritilliformis (Lasch) Gill.	1			
C. geotropa (Bull.) Quél.	1	2	3	4
C. geotropa var. *maxima* (F.) Nuesch	1	2		
C. gibba (Fr.) Kummer (*C. infundibuliformis* (Schaeff.) Quél.)	1	2	3	4
C. gilva (Pers.) Kummer (*C. splendens* (Pers.) Konr.)	1			4
C. houghtonii (Berk. & Br.) Dennis	1	2		4
C. hydrogramma (Bull.) Kummer	1	2	3	
C. incomis (Karst.) Orton	1	2		4
C. inornata (Sow.) Gill.	1		3	
C. langei Sing.	1			

	1	2	3	4
C. metachroa (Fr.) Kummer	1	2	3	4
C. nebularis (Batsch) Kummer	1	2	3	4
C. nebularis var. *alba* Bat.	1			4
C. obsoleta (Batsch) Quél.	1	2	3	4
C. odora (Bull.) Kummer	1	2	3	4
C. orbiformis (Fr.) Gill.	1			
C. osmophora Gilbert	1			
C. phyllophila (Fr.) Kummer	1	2		4
C. pseudoclusilis (Joss. & Konr.) Orton	1			
C. quercina Pearson (cf. *C. fritilliformis*)	1			
C. rivulosa (Pers.) Kummer	1	2	3	4
C. sericella Kühner & Romagn.				4
C. sinopica (Fr.) Kummer		2		4
C. sinopicoides Peck	1	2		
C. squamulosa (Pers.) Fr.	1	2		4
C. squamulosoides Orton	1			
C. striatula (Kühner) Orton	1		3	
C. suaveolens (Schum.) Kummer	1	2		
C. subalutacea (Batsch) Kummer				4
C. tenuissima Romagn.	1			
C. tornata (Fr.) Kummer	1			
C. trullaeformis (Fr.) Karst.			3	4
C. tuba (Fr.) Gill.	1		3	
C. umbilicata (Schaeff.) Kummer	1			
C. vermicularis (Fr.) Quél.		2		4
C. vibecina (Fr.) Quél. Old records may refer to *C. dicolor* or *C. langei*	1	2		4

Clitocybula (Sing.) Sing.
   Saprophyte: *C. lacerata* (Lasch) Sing. is recorded from Dorset.

Clitopilus (Fr.) Kummer
   Saprophytes, *C. omphaliformis* Joss. is in the west midlands.

	1	2	3	4
Clitopilus cretatus (B. & Br.) Sacc.	1	2		4
C. hobsonii (B. & Br.) Orton	1	2		4
C. passeckerianus (Pilat) Sing.	1	2		4
C. pinsitus (Fr.) Joss.	1			
C. prunulus (Scop.) Kummer	1	2	3	4

Collybia (Fr.) Kummer
   Saprophytes, especially on forest debris. Includes *Pseudobaeospora* Sing.

	1	2	3	4
Collybia acervata (Fr.) Kummer	1			4
C. alkalinovirens Sing. (*C. obscura* Favre)				4
C. butyracea (Bull.) Kummer	1	2	3	4
C. cirrhata (Schum.) Kummer	1	2	3	4
C. confluens (Pers.) Kummer	1	2	3	4
C. cookei (Bres.) J.D. Arnold	1	2	3	4
C. distorta (Fr.) Quél.	1			4
C. dryophila (Bull.) Kummer	1	2	3	4
C. dryophila var. *aquosa* (Bull.) Quél.	1			
C. erythropus (Pers.) Kummer (*C. kuehneriana* Sing.)	1	2	3	4
C. fuscopurpurea (Pers.) Kummer (*C. konradii* Sing.)	1	2		4
C. fusipes (Bull.) Quél.	1	2	3	4
C. luteifolia Gill.	1	2		4
C. luxurians Peck Glasshouse alien	1			
C. maculata (A. & S.) Kummer	1	2	3	4
C. nephelodes (B. & Br.) Sacc. Glasshouse alien	1			
C. ocellata (Fr.) Kummer	1			
C. ocior (Pers.) Vilgalys & Miller	1			

C. peronata (Bolt.) Kummer	1	2	3	4
C. (*Pseudobaeospora*) *pillodii* Quél.	1			
C. prolixa (Hornemann) Gill.	1	2	3	
C. racemosa (Pers.) Quél.	1			
C. succinea (Fr.) Quél.		2	3	4
C. tergina (Fr.) Lundell	1			4
C. tuberosa (Bull.) Kummer	1	2	3	4

A record of *C. exsculpta* (Fr.) Gillet from 4 has been rejected for the present.

Conocybe Fayod

Saprophytes, mainly terrestrial. *C. utriformis* Orton is on record from Middlesex and many other species such as *C. pseudopilosella* Kühner, *C. sienophylla* (B. & Br.) Sing are to be looked for in the area.

Conocybe ambigua Watling	1			
C. anthracophila (Maire & Kühner) Sing.				4
C. antipus (Lasch) Fayod Dubious record	1			
C. aporos Kits van Waveren	1	2		4
C. appendiculata Lange & Kühner	1	2		4
C. arrhenii (Fr.) Kits van Waveren (*C. togularis* pp.)	1	2	3	4
C. blattaria (Fr.) Kühner	1	2		4
C. brunnea Lange & Kühner	1	2		
C. brunneola Kühner	1	2		4
C. dentatomarginata Watling				4
C. dumetorum (Vel.) Svrček (*C. laricina* (Kühner) Kühner)	1			
C. dunensis T.J. Wallace				4
C. exannulata Kühner	1			4
C. farinacea Watling	1			
C. filaris (Fr.) Kühner	1	2	3	4
C. fuscomarginata (Murr.) Sing.	1			
C. hadrocystis (v.Wav.) Watling	1			
C. inocybeoides Watling	1			
C. intrusa (Peck) Sing. Glasshouse alien	1			
C. Kuehneriana Sing. (*C. ochracea* pp.)	1	2	3	4
C. lactea (Lange) Metrod	1	2	3	4
C. leucopus Kühner		2		
C. macrocephala Kühner	1			
C. magnicapitata Orton	1			
C. mairei Kühner	1	2		4
C. mesospora Kühner	1	2		4
C. moseri Watling (*C. plumbeicincta* pp.)	1	2		
C. murinacea Watling	1			
C. percincta Orton	1	2		4
C. pilosella (Peck) Kühner	1	2		4
C. plicatella (Peck) Kühner	1			4
C. pubescens (Gill.) Kühner	1	2		
C. pygmaeoaffinis (Fr.) Kühner	1		3	4
C. rickeniana Orton	1	2		
C. rickenii (Schaeff.) Kühner	1	2	3	4
(*C. siliginea* pp.)				
C. rugosa (Peck) Watling	1			4
C. semiglobata Kühner Many old records as *C. tenera* belong here	1	2		
C. siliginea (Fr.) Kühner	1		3	
C. sordida (Kühner) Kühner & Watling	1			
C. striaepes (Cooke) Lundell	1			
C. subovalis Kühner	1	2		4
C. subpubescens Orton	1			
C. tenera (*Schaeff.*) Fayod	1			
C. vestita (Fr.) Kummer	1			

Coprinus Pers.

Saprophytes, mainly coprophilous, a few anthracophilous or lignicolous, *C. poliomallus* Romagn. has been recorded from Middlesex.

Coprinus acuminatus (Romagn.) Orton	1	2		4
C. amphithallus Lange & Smith		2		
C. angulatus Peck (*C. boudieri* Quél.)	1	2	3	4
C. argenteus Orton	1			
C. atramentarius (Bull.) Fr.	1	2	3	4
C. auricomus Pat.	1	2		4
C. bellulus Uljé	1			
C. bisporiger Orton	1	2		
C. bisporus Lange	1			
C. brunneofibrillosus Dennis	1			
C. callinus M. Lange & A.H. Smith	1	2		4
C. cinereofloccosus Orton	1	2		4
C. cinereus (Schaeff.) Gray	1	2	3	4
C. cinnamomeotinctus Orton	1			
C. comatus (Müll.) Gray	1	2	3	4
C. congregatus (Bull.) Fr.	1			4
C. cordisporus Gibbs (*C. patouillardii* auct.)	1	2		
C. cortinatus Lange	1	2		
C. cothurnatus Godey	1			
C. curtis Kalchbr.	1			
C. disseminatus (Pers.) Gray	1	2	3	4
C. domesticus (Bolt.) Gray	1	2		4
C. echinosporus Buller	1			4
C. ellisii Orton	1	2	3	
C. ephemeroides (Bull.) Fr.	1			
C. ephemerus (Bull.) Fr.	1	2		4
C. episcopalis Orton	1	2		
C. erythrocephalus (Lév.) Fr.	1	2		4
C. extinctorius (Bull.) Fr.				4
C. flocculosus DC	1			
C. friesii Quél.	1	2		
C. galericuliformis Losa		2		4
C. gonophyllus Quél.	1			
C. griseofoetidus Orton	1			
C. heptemerus M. Lange & A.H. Smith	1			
C. hercules Ulji & Bas	1			
C. heterosetulosus Locquin		2		
C. hexagonosporus Joss.	1			
C. hiascens (Fr.) Quél.	1	2	3	4
C. impatiens (Fr.) Quél.	1	2		
C. insignis Peck	1			
C. kuehneri Ulje & Bas	1			
C. laanii Kits van Waveren				4
C. lagopides Karst. (*C. funariarum* Metr.)	1	2		4
C. lagopus (Fr.) Fr.	1	2	3	4
C. latisporus Orton	1			
C. leiocephalus Orton	1	2		4
C. macrocephalus (Berk.) Berk.	1	2		4
C. micaceus (Bull.) Fr.	1	2	3	4
C. miser Karst.	1			
C. narcoticus (Batsch) Fr.	1		3	
C. niveus (Pers.) Fr.	1	2	3	4
C. nudiceps Orton	1			
C. pellucidus Karst.	1		3	
C. phlyctidosporus Romagn.	1			
C. picaceus (Bull.) Gray	1	2		4

C. plagioporus Romagn.	1			
C. plicatilis (Fr.) Fr.	1	2	3	4
C. radians (Desm.) Fr.	1	2	3	4
C. radiatus (Bolt.) Gray	1	2	3	4
C. rhombisporus Orton	1			
C. saccharomyces Orton		2		
C. saichiae Reid	1			
C. sclerocystidiosus M. Lange & A.H. Smith		2		
C. semitalis Orton	1	2		4
C. silvaticus Peck (C. tardus (Karst.) Karst.)	1	2		
C. stanglianus Enderle et al.	1			
C. stercoreus Fr. (C. velox auct.)	1	2	3	4
C. sterquilinus (Fr.) Fr.	1			4
C. subdisseminatus M. Lange (formerly listed as C. hemerobius Fr.)		2		4
C. subimpatiens M. Lange & A.H. Smith	1			
C. subpurpureus Smith	1			
C. tigrinellus Boud.	1	2		
C. trisporus Kemp & Watling				4
C. truncorum (Schaeff.) Fr.	1	2		4
C. tuberosus Quél. (C. stercorarius auct.)	1	2	3	
C. urticaecola (Berk. & Br.) Buller	1	2		
C. xanthothrix Romagn.	1	2		

In view of their unspecialised requirements it seems very unlikely the further 20 or so British species already recognised are not also lurking here awaiting discovery.

Cortinarius Roussel

Probably all mycorrhizal with trees and shrubs. In view of the very large number of species involved it will be convenient to follow tradition and list them under the subgenera especially as there is a tendency to raise some of these to generic rank.

Subgenus Myxacium Fr.

Cortinarius causticus Fr.	1	2	3	4
C. collinitus (Sow.) Fr.	1	2		4
C. croceocaeruleus (Pers.) Fr.	1	2		4
C. crystallinus Fr.	1	2		
C. delibutus Fr.	1	2	3	4
C. elatior Fr.	1	2	3	4
C. emollitus Fr.		2		
C. livido-ochraceus (Berk.) Berk.	1			
C. mucosus (Bull.) Kickx	1	2		
C. ochroleucus (Schaeff.) Fr.	1	2	3	4
C. pinicola Orton	1	2	3	4
C. pluvius (Fr.) Fr.		2		
C. pseudosalor Lange	1	2	3	4
C. stillatitius Fr.	1		3	
C. trivialis Lange	1	2		4
C. vibratilis Lange	1	2	3	

Subgenus Phlegmacium Fr.

In addition C. langei Henry & C. saginus Fr. have been reported from Essex and C. xanthophyllus (Cooke) Orton from Hampshire. Records of C. cyanopus Fr. and C. glaucopus Fr. have been omitted pending confirmation.

C. amarescens Moser		1			
C. amoenolens Henry	Old records as C. glaucopus may belong here	1	2	3	4
C. aureoturbinatus Lange			2		4
C. balteatocumatilis Henry		1			4
C. balteatus (Fr.) Fr.	Needs confirmation	1	2		

23

		1	2	3	4
C. caerulescens (Schaeff.) Fr.		1	2		4
C. caesiocyaneus Britz.		1			4
C. calochrous (Pers.) Fr.		1	2		
C. caroviolaceus Orton		1			
C. cedretorum Maire		1			
C. citrinus Lange		1	2		
C. crassus Fr.					4
C. crocolitus Quél.		1			4
C. cumatilis Fr.	Needs confirmation		2		
C. durus Orton	Old records as *C. claricolor* may belong here also		2		
C. elegantior (Fr.) Fr.		1	2		
C. fulgens (A. & S.) Fr.	Needs confirmation		2	3	4
C. fulmineus (Fr.) Fr.		1	2		4
C. infractus (Pers.) Fr.		1	2		
C. largus Fr.		1		3	4
C. magicus Eichhorn		1			
C. mairei (Moser) Orton			2		
C. melliolens J. Schaeff.		1			
C. multiformis (Fr.) Fr.		1	2	3	4
C. nemorensis (Fr.) Lange		1		3	4
C. olidus Lange	Including *C. cephalixus* auct.	1			
C. orichalceus (Batsch) Fr.	Needs confirmation		2		
C. osmophorus Orton		1			
C. parherpeticus Henry			2		
C. parvus Henry		1	2		
C. porphyropus (A. & S.) Fr.	Confirmation desirable	1	2	3	
C. pseudosulphureus Henry		1			
C. purpurascens (Fr.) Fr.		1	2		4
C. rickenianus Maire			2		
C. rufoalbus Kühner		1			
C. rufo-olivaceus (Pers.) Fr.		1	2		4
C. sebaceus Fr.	Doubtful record	1			
C. sodagnites Henry		1	2		
C. splendens Henry		1			4
C. subarquatus (Moser) Moser		1			
C. subpurpurascens (Batsch) Kickx			2		
C. subturbinatus Henry		1			
C. triumphans Fr.	Old records may have been of *C. crocolitus*	1	2	3	4
C. varius (Schaeff.) Fr.		1			
C. xanthocephalus Orton			2		4
C. xanthochrous Orton		1			
C. xantho-ochraceus Orton		1			

Subgenus *Sericeocybe* Orton

		1	2	3	4
C. alboviolaceus (Pers.) Fr.		1	2	3	4
C. anomalus (Fr.) Fr.		1	2	3	4
C. argentatus (Pers.) Fr.	Probably a misdetermination	1			
C. azureus Fr.			2		
C. caninus (Fr.) Fr.		1		3	4
C. epsomensis Orton		1			
C. fuliginosus Orton					4
C. hillieri Henry					4
C. lepidopus Cooke		1	2	3	4
C. malachius (Fr.) Fr.		1	2		4
C. myrtillinus Fr.			2	3	4
C. opimus Fr.		1			
C. pearsonii Orton		1	2	3	4
C. simulatus Orton			2		

C. spilomeus (Fr.) Fr.	1	2	3	
C. suillus Fr.	1	2		
C. tabularis (Bull.) Fr.	1	2	3	4
C. turgidus Fr.	1			4

Subgenus Cortinarius (= Inoloma Fr.)

In addition C. speciosissimus Kühner & Romagn. & C. venetus Fr. are listed from Hampshire.

C. bolaris (Pers.) Fr.	1	2	3	4
C. callisteus (Fr.) Fr.	1			
C. cotoneus Fr.	1	2		4
C. orellanoides Henry	1			
C. orellanus Fr.	1		3	
C. penicillatus Fr. Doubtful record	1			
C. pholideus (Fr.) Fr.	1	2	3	4
C. psammocephalus (Bull.) Fr.	1		3	4
C. rubicundulus (Rea) Pearson		2	3	
C. tophaceus Fr.		2		
C. violaceus (L.) Fr.	1	2		4

Subgenus Dermocybe Fr.

C. anthracinus (Fr.) Fr.	1		3	
C. cinnabarinus Fr.	1	2	3	4
C. cinnamomeobadius Henry	1			4
C. cinnamomeoluteus Orton	1	2		4
C. cinnamomeus (L.) Fr. s. str.	1		3	
C. croceoconus Fr.	1	2		
C. croceofolius Peck	1	2		
C. malicorius Fr.	1	2		
C. olivaceofuscus Kühner	1			
C. phoeniceus (Bull.) Maire	1	2	3	4
C. puniceus Orton	1			4
C. raphanoides (Pers.) Fr. (C. subnotatus Fr.)	1			
C. sanguineus (Wulf.) Fr.	1	2	3	4
C. semisanguineus (Fr.) Gill.	1	2	3	4
C. sphagneti Orton	1			
C. uliginosus Berk. (C. queletii Bat.)	1	2		4

Subgenus Hydrocybe Fr./Telamonia Fr.

Most species are ill-defined, all old records are unreliable and the names subject to revision.

C. acutus (Pers.) Fr.	1	2	3	4
C. acutostriatulus Henry	1		3	
C. alnetorum Vel.	1			4
C. armillatus (Fr.) Fr. (C. haematochelis (Bull.) Fr.)	1	2	3	4
C. aureomarginatus Pearson	1			
C. balaustinus Fr.	1	2		
C. basililaceus Pearson	1			
C. basiroseus Pearson	1			
C. betuletorum Moser	1			
C. bicolor Cooke	1	2	3	4
C. biformis Fr.	1	2		
C. bovinus Fr.				4
C. brunneus (Pers.) Fr.				4
C. bulbosus (Sow.) Fr. Confirmation desirable			3	
C. bulliardii (Pers.) Fr. Confirmation desirable	1			
C. candellaris Fr. Confirmation desirable	1			
C. castaneus (Bull.) Fr. Confirmation desirable	1	2	3	4
C. damascenus Fr.	1	2		

25

		1	2	3	4
C. decipiens (Pers.) Fr.	Confirmation desirable	1	2	3	4
C. decumbens (Pers.) Fr.	Confirmation desirable	1			
C. depressus (Weinm.) Fr.	Confirmation desirable	1			
C. dolabratus Fr.	Confirmation desirable	1	2		
C. duracinus Fr.		1		3	4
C. erythrinus (Fr.) Fr. ss. Lange		1	2	3	4
C. evernius (Fr.) Fr.	Confirmation desirable		2		
C. fasciatus (Scop.) Fr. ss. Lange		1			4
C. firmus (Weinm.) Fr.	Confirmation desirable		2	3	
C. flexipes (Pers.) Fr.		1	2		
C. fuscopallens Fr.		1			
C. gentilis (Fr.) Fr.		1	2	3	4
C. glandicolor (Fr.) Fr.		1		3	
C. helvelloides (Fr.) Fr.		1	2	3	4
C. helvolus Fr.		1			
C. hemitrichus (Pers.) Fr.		1	2		4
C. hinnuleus Fr.		1	2	3	4
C. hoeftii (Weinm.) Fr.			2		
C. holophaeus Lange			2		
C. iliopodius (Bull.) Fr.	Confirmation desirable	1			4
C. illuminus Fr.	Confirmation desirable		2		
C. imbutus Fr.	Confirmation desirable		2		4
C. impennis Fr.	Confirmation desirable		2		
C. incisus (Pers.) Fr.	Confirmation desirable	1	2	3	
C. injucundus (Weinm.) Fr.	Confirmation desirable	1			
C. isabellinus (Batsch) Fr.	Confirmation desirable	1			
C. jubarinus Fr.	Confirmation desirable	1			
C. junghuhnii Fr.			2		4
C. krombholzii Fr.	Confirmation desirable	1			
C. leucopus (Bull.) Fr.	Confirmation desirable	1	2		4
C. lilacinopusillus Orton		1			
C. melleopallens (Fr.) Lange	Confirmation desirable	1			
C. milvinus Fr.	Confirmation desirable				4
C. obtusus (Fr.) Fr.		1	2	3	4
C. paleaceus Fr.		1	2	3	4
C. paleiferus Svrček		1			4
C. periscelis Fr.	Confirmation desirable		2		
C. pulchellus Lange		1			
C. reedii (Berk.) Berk.					4
C. rigens (Pers.) Fr.		1	2	3	4
C. rigidus (Scop.) Fr.		1	2	3	4
C. rubricosus (Fr.) Fr.		1			
C. safranopes Henry		1			
C. salicicola Orton		1			4
C. saniosus (Fr.) Fr.		1	2		4
C. saturninus (Fr.) Fr.		1	2		4
C. scandens Fr.		1	2	3	4
C. sciophyllus Fr. ss. Lange			2		
C. scutulatus (Fr.) Fr.	Confirmation desirable	1	2		
C. sertipes Kühner		1			
C. stemmatus Fr.		1			
C. subbalaustinus Henry		1		3	4
C. subferrugineus (Batsch) Fr.		1	2	3	
C. tabacinus Orton		1			
C. tortuosus (Fr.) Fr.	Confirmation desirable	1			
C. torvus (Fr.) Fr.		1	2	3	4
C. triformis Fr.		1	2		
C. umbrinolens Orton		1			
C. uraceus Fr.			2		4
C. violilamellatus Pearson		1			

Craterellus Pers.
Saprophytes. Includes *Pseudocraterellus* Corner.

Craterellus cornucopioides (L.) Fr.	1	2	3	4
C. (Ps.) sinuosus (Fr.) Fr. (*C. pusillus* (Fr.) Fr.)	1	2		4

Crepidotus (Fr.) Kummer
Saprophytes, especially on herbaceous and woody debris. See also *Simocybe*.

Crepidotus amygdalosporus Kühner	1	2		4
C. applanatus (Pers.) Kummer	1	2		
C. autochthonus Lange	1			4
C. calolepis (Fr.) Karst.	1	2		
C. cesatii (Rab.) Sacc.	1	2	3	
C. herbarum (Peck) Sacc. (*Pleurotellus graminicola* Fayod)	1	2	3	4
C. inhonestus Karst.	1		3	
C. luteolus (Lamb.) Sacc.	1	2	3	4
C. mollis (Schaeff.) Kummer	1	2	3	4
C. pubescens Bres.	1			4
C. subsphaerosporus (Lange) Kühner & Romagn.	1			
C. subtilis Orton	1			4
C. variabilis (Pers.) Kummer	1	2	3	4

Crinipellis Pat.
Saprophytes, especially on dead Gramineae.

Crinipellis stipitarius (Fr.) Pat.	1	2	3	4

Cyphellopsis Donk
Lignicolous saprophytes, formerly referred to *Solenia* Pers.

Cyphellopsis anomala (Pers.) Donk	1	2	3	4
C. confusa (Bres.) Reid	1	2	3	

Cystoderma Fayod
Terrestrial saprophytes.

Cystoderma amianthinum (Scop.) Fayod	1	2	3	4
C. carcharias (Pers.) Fayod	1	2		4
C. cinnabarinum (A. & S.) Fayod	1		3	
C. granulosum (Batsch) Fayod	1	2	3	4

Deconica (W.G. Smith) Karst. see *Psilocybe*

Delicatula Fayod see *Mycena*

Dermoloma (Lange) Sing.
Terrestrial saprophytes.

Dermoloma atrocinereum (Pers.) Orton	1			4
D. cuneifolium (Fr.) Sing.	1	2	3	4
D. fuscobrunneum Orton	1			
D. pseudocuneifolium Orton	1			

Drosella Maire see *Chamaemyces*

Eccilia (Fr.) Kummer
Terrestrial saprophytes, better united with *Entoloma*.

Eccilia cancrina (Fr.) Ricken (*Entoloma neglectum* (Lasch) Moser)	1

E. nigrella (Pers.) Gill.	1	2	
E. pallens Maire		2	
E. paludicola Orton	1	2	
E. parkensis (Fr.) Quél.			3

E. rhodocyclix (Lasch) Kummer	1	2		4
E. sericeonitida Orton	1	2	3	4

Entoloma (Fr.) Kummer
Terrestrial saprophytes.

Entoloma ameides (Berk. & Br.) Sacc.	1	2		4
E. aprile (Britz.) Sacc.	1	2		
E. clypeatum (L.) Kummer	1	2	3	4
E. costatum (Fr.) Kummer Confirmation desirable	1			4
E. erophilum (Fr.) Karst. (E. plebejum (Kalchbr.) Noordelos)	1	2		
E. excentricum Bres.		2		
E. fuscomarginatum Orton	1			
E. helodes (Pers.) Kummer	1	2		
E. jubatum (Fr.) Karst.	1	2		4
E. madidum (Fr.) Gill. (E. bloxami (B. & Br.) Sacc.)	1	2		4
E. myrmecophilum (Romagn.) Moser	1			
E. nidorosum (Fr.) Quél.	1	2	3	4
E. niphoides (Romagn.) Orton		2		4
E. nitidum Quél.			3	
E. percandidum Noordeloos (Rhodophyllus omphaliformis Romagn.)				4
E. porphyrophaeum (Fr.) Karst.	1	2	3	4
E. prunuloides (Fr.) Quél. (E. repandum auct.)	1	2		4
E. rhodopolium (Fr.) Kummer	1	2	3	4
E. saundersii (Fr.) Sacc.	1			
E. scabiosum (Fr.) Quél.	1			
E. sepium (Noulet & Dassiet) Richon & Roze	1			
E. sericatum (Britz.) Sacc.	1	2		
E. sinuatum (Bull.) Kummer (E. lividum (Bull.) Quél.)	1	2	3	4
E. sordidulum (Kühn. & Romagn.) Orton	1			
E. sphagneti Naveau	1			
E. subradiatum (Kühner & Romagn.) Moser	1			
E. turbidum (Fr.) Quél.	1	2	3	4

Episphaeria Donk
Lignicolous saprophyte.

Episphaeria fraxinicola (Berk. & Br.) Donk	1

Fayodia Kühner
Saprophytes. The type species, *Fayodia bisphaerigera* (Lange) Kühner has been recorded from Berkshire; *Myxomphalia* Hora is commonly regarded as no more than a subgenus.

Flagelloscypha Donk
Saprophytes on plant debris.

Flagelloscypha faginea (Lib.) W.B. Cooke	1
F. minutissima (Burt) Donk	1
F. pilatii Agerer	1

Flammulaster Earle
Lignicolous saprophytes. *Flocculina* Orton is a synonym.

Flammulaster carpophila (Fr.) Earle	1	2
F. denticulata Orton	1	

	1	2	3	4
F. denticulata Orton	1			
F. ferrugineus (Maire) Watling				4
F. granulosus (Lange) Watling	1	2		4
F. limulatus (Fr.) Watling	1	2		4
F. siparius (Fr.) Watling	1		3	4
F. subincarnatus (Joss. & Kühner) Watling	1			

Flammulina Karst.

Lignicolous saprophyte or weak facultative parasite of fruit trees.

	1	2	3	4
Flammulina velutipes (Curt.) Sing.	1	2	3	4

Galerina Earle

Saprophytes, terrestrial or less often lignicolous. Additional species reported from Essex include *G. camerina* (Fr.) Kühner, *G. heimansii* Reijnd. & *G. pseudocamerina* Sing.

	1	2	3	4
Galerina ampullaceocystis Orton		2	3	
G. badipes (Fr.) Kühner	1			4
G. calyptrata Orton			3	4
G. cedretorum (Maire) Sing.	1			
G. heterocystis (Atk.) Sing. & Smith (*G. clavata* (Vel.) Kühner)	1	2		4
G. hypnorum (Schrank) Kühner	1	2	3	4
G. laevis (Pers.) Sing. (*G. graminea* (Vel.) Kühner)	1	2		4
G. luteofulva Orton				4
G. mniophila (Lasch) Kühner	1	2		4
G. mycenoides (Fr.) Kühner	1	2	3	
G. nana (Petri) Kühner			3	4
G. paludosa (Fr.) Kühner	1	2	3	4
G. permixta (Orton) Pegler & Young (*Naucoria permixta* Orton)	1			4
G. phillipsii Reid	1			
G. praticola (Moll.) Orton	1	2		4
G. pumila (Pers.) Sing. (*G. mycenopsis* (Fr.) Kühner)	1	2	3	4
G. salicicola Orton	1			
G. sideroides (Bull.) Kühner	1			
G. sphagnorum (Pers.) Kühner	1	2	3	4
G. stylifera (Atk.) Smith & Sing.				4
G. subbadipes Huijsman		2		
G. tibiicystis (Atk.) Kühner			3	
G. triscopa (Fr.) Kühner	1			
G. uncialis (Britz.) Kühner	1	2		
G. unicolor (Vahl) Sing. (*G. marginata* (Batsch) Kühner)	1	2	3	4
G. vittaeformis (Fr.) Moser (*G. rubiginosa* auct.)	1	2	3	4

Geopetalum Pat.

Anthracophilous. *Faerberia* Pouzar is a synonym.

	1	2	3	4
Geopetalum carbonarium (A. & S.) Pat.	1	2	3	

Gomphidius Fr.

Mycorrhizal with conifers, see also *Chroogomphus*.

	1	2	3	4
Gomphidius glutinosus (Schaeff.) Fr.	1	2	3	4
G. maculatus (Scop.) Fr. (*G. gracilis* B. & Br.)	1	2	3	4
G. roseus (Fr.) Karst.	1		3	4

Gomphus Pers.

There is an unverified old record of the very rare *Gomphus clavatus* (Pers.) Gray from Essex.

Gymnopilus Karst.
Saprophytes, mostly lignicolous.

	1	2	3	4
Gymnopilus bellulus (Peck) Murr.	1			
G. decipiens (W.G. Smith) Orton	1	2	3	
G. flavus (Bres.) Sing.	1	2		4
G. fulgens (Favre & Maire) Sing.	1		3	
G. hybridus (Fr.) Sing.	1	2		
G. junonius (Fr.) Orton (*G. spectabilis* (Fr.) Sing.)	1	2	3	4
G. odini (Fr.) Kühner & Romagn.	1			
G. penetrans (Fr.) Murr.	1	2	3	4
G. sapineus (Fr.) Maire Records prior to 1953 applied to the preceding	1	2		4
G. stabilis (Weinm.) Kühner & Romagn.				4

Gyrodon Opat
See *Uloporus* Quél.

Gyroporus Quél.
Mycorrhizal

	1	2	3	4
Gyroporus castaneus (Bull.) Quél.	1	2	3	4
G. cyanescens (Bull.) Quél.	1			4

Haasiella Kotlaba & Pouzar
Saprophytes, united with the predominantly tropical genus *Gerronema* Sing. by Singer.

	1
Haasiella venustissima (Fr.) Kotlaba & Pouzar	1

Hebeloma (Fr.) Kummer
Mycorrhizal. The genus is readily recognisable, species within it much less so and most records must be regarded with considerable reserve though there seems no reason to doubt the probable presence in the area of those listed here.

	1	2	3	4
Hebeloma alpinum (Favre) Bruchet				4
H. anthracophilum Maire	1	2	3	4
H. calyptrosporum Bruchet	1			
H. colossum Huijsman		2		
H. crustuliniforme (Bull.) Quèl.	1	2	3	4
H. edurum Metrod	1	2		4
H. fastibile (Pers.) Kummer	1	2		4
H. funariophyllum Moser	1			
H. helodes Favre		2		
H. hiemale Bres.	1			
H. leucosarx Orton	1			4
H. longicaudum (Pers.) Kummer	1	2	3	4
H. mesophaeum (Pers.) Quél.	1	2	3	4
H. pumilum Lange	1			
H. pusillum Lange	1			
H. radicosum (Bull.) Ricken	1	2		4
H. sacchariolens Quél. (*H. nauceosum* (Cooke) Sacc.)	1	2	3	4
H. sinapizans (Paulet) Gill.	1	2	3	4
H. sinuosum (Fr.) Quél.	1	2		4
H. spoliatum (Fr.) Karst. (*H. radicatum* (Cooke) Maire)		2		
H. strophosum (Fr.) Sacc.	1	2		4
H. testaceum (Batsch) Quél.	1	2	3	4
H. truncatum (Schaeff.) Kummer		2		
H. versipelle (Fr.) Gill.	1	2		

Hebelomina Maire
?Mycorrhizal.

	1	2	3	4
Hebelomina neerlandica Huijsman	1			

Hemimycena (Sing.) Sing. see *Mycena*

Hohenbuehelia Schulzer
Saprophytes, lignicolous or terrestrial.

	1	2	3	4
Hohenbuehelia algida (Fr.) Sing.	1			
H. atrocaerulea (Fr.) Sing.	1	2	3	4
H. culmicola Bon				4
H. geogenia (DC.) Sing.	1	2		4
H. geogenia var. *queletii* Kühner				4
H. mastrucata (Fr.) Sing.	1	2		4
H. petaloides (Bull.) Schulzer	1	2		4
H. reniformis (Meyer) Sing.	1			

Hygrophoropsis (Schroet.) Maire
Saprophytes, terrestrial or lignicolous.

	1	2	3	4
Hygrophoropsis albida (Fr.) Maire	1	2		
H. aurantiaca (Wulf.) Maire	1	2	3	4
H. aurantiaca var. *pallida* (Cooke) Kühner & Romagn.	1			4
H. aurantiaca var. *rufa* Reid		2		
H. fuscosquamula Orton	1			4

Hygrophorus Fr.
The species are listed under the subgenera as there is a tendency to raise these to generic rank.

Subgenus *Hygrophorus* (= *Limacium* Fr.)
Mycorrhizal with forest trees.

		1	2	3	4
Hygrophorus agathosmus (Fr.) Fr. (*H. cerasinus* (Berk.) Fr.)		1		3	
H. arbustivus Fr.		1	2	3	4
H. bresadolae Quél.			2		
H. chrysaspis Metrod		1	2		4
H. chrysodon (Batsch) Fr.		1	2		
H. cossus (Sow.) Fr.		1	2	3	4
H. dichrous Kühner & Romagn. (*H. persoonii* Arnolds)		1	2		4
H. discoideus (Pers.) Fr.		1	2		4
H. eburneus (Bull.) Fr.		1	2	3	4
H. hypothejus (Fr.) Fr.		1	2	3	4
H. lucorum (Kalchbr.) Henn.		1	2	3	4
H. mesotephrus Berk. & Br.		1	2		
H. nemoreus (Pers.) Fr. (*H. leporinus* Fr.)		1			4
H. olivaceoalbus (Fr.) Fr.	Old records relate to *H. dichrous*		2		
H. penarius Fr.	Confirmation required		2		
H. pudorinus (Fr.) Fr.				3	
H. quercetorum Orton			2		
H. russula (Schaeff.) Quél.	Confirmation required	1			
H. unicolor Grög. (*H. leucophaeus* auct.)		1	2		4

Subgenus *Cuphophyllus* Donk
Terrestrial saprophytes *H. subviolaceus* Peck occurs in the Isle of Wight.

	1	2	3	4
H. cinereus (Pers.) Quél.		2		4
H. colemannianus Blox.				4
H. lacmus (Schum.) Kalchbr.				4

	1	2	3	4
H. niveus (Scop.) Fr.	1	2	3	4
H. ortonii (Bon) Dennis (*H. Berkeleyi* Orton non Sacc.)	1			
H. pratensis (Pers.) Fr.	1	2	3	4
H. russocoriaceus Berk. & Miller	1	2		4
H. subradiatus (Schum.) Fr.	1	2	3	4
H. virgineus (Wulf.) Fr.	1	2	3	4

Subgenus *Hygrocybe* Fr.
  Terrestrial saprophytes

	1	2	3	4
H. aurantiosplendens (Haller) Orton	1	2		
H. aurantius (Murr.) Murr.				4
H. calyptraeformis Berk. & Br.	1	2	3	4
H. ceraceus (Wulf.) Fr.	1	2	3	4
H. chlorophanus (Fr.) Fr.	1	2	3	4
H. citrinus Rea				4
H. coccineocrenatus Orton	1			
H. coccineus (Schaeff.) Fr.	1	2	3	4
H. conicoides Orton		2		
H. conicus (Scop.) Fr.	1	2	3	4
H. flavescens (Kauffm.) A.H. Smith & Hesler	1	2	3	4
H. fornicatus Fr.	1	2		4
H. glutinipes (Lange) Orton	1	2	3	4
H. insipidus (Lange) Lundell	1	2		4
H. intermedius Pass.	1			4
H. konradii (Haller) Orton	1			
H. laetus (Pers.) Fr. (*H. houghtonii* Berk.)	1	2	3	4
H. langei Dodge (*H. constans* Lange, *H. persistens* (Britz.) Sing.)	1	2		4
H. lepida Arnolds (*H. cantharellus* (Schw.) Murr.)	1			
H. marchii Bres. (*H. reidii* Kühner)	1	2		4
H. metapodius (Fr.) Fr. Needs confirmation	1			
H. miniatus (Fr.) Fr.	1	2	3	4
H. mollis (Berk. & Br.) Kauffm.	1			
H. nigrescens (Quél.) Quél.	1	2	3	4
H. nitiosus Blytt			3	
H. nitratus (Pers.) Fr.	1	2	3	
H. obrusseus (Fr.) Fr.	1	2		4
H. ovinus (Bull.) Fr.	1			
H. psittacinus (Schaeff.) Fr.	1	2	3	4
H. puniceus (Fr.) Fr.	1	2	3	4
H. quietus Kühner		2	3	4
H. reai R. Maire	1	2		
H. sciophanus (Fr.) Fr.	1	2		
H. splendidissimus Orton		2		4
H. strangulatus Orton	1	2	3	4
H. subglobisporus Orton	1	2		
H. subminutulus (Murr.) Orton	1	2		4
H. substrangulatus Orton	1			
H. turundus (Fr.) Fr.	1	2		4
H. unguinosus (Fr.) Fr.	1	2	3	4
H. vitellinus Fr.	1	2		4

Hygrotrama Sing.
  Terrestrial saprophytes, including *Camarophyllopsis* Horak.

	1	2	3	4
Hygrotrama atropunctum (Pers.) Sing.		2		
H. foetens (Phill.) Sing.	1			
H. hymenocephalum (Smith & Hesler) Sing.	1			
H. schulzeri (Bres.) Sing.		2		

Hypholoma (Fr.) Kummer
Saprophytes, lignicolous or terrestrial especially on peaty soils.

Hypholoma capnoides (Fr.) Kummer	1	2	3	4
H. elongatum (Pers.) Ricken	1	2	3	4
H. epixanthum (Fr.) Quél.	1			4
H. ericaeoides Orton	1	2	3	4
H. ericaeum (Pers.) Kühner	1	2	3	4
H. fasciculare (Huds.) Kummer	1	2	3	4
H. marginatum (Pers.) Schroet.) (*H. dispersum* (Fr.) Quél.)	1	2	3	
H. polytrichi (Fr.) Ricken	1			
H. radicosum Lange	1	2	3	
H. subericaeum (Fr.) Kühner	1	2	3	4
H. sublateritium (Fr.) Quél.	1	2	3	4
H. sublateritium var. *squamosum* (Cooke) Massee				4
H. udum (Pers.) Quél.	1	2	3	4

Hypsizygus Sing.
Lignicolous saprophyte.

Hypsizygus ulmarius (Bull.) Redhead (*Pleurotus ulmarius* (Bull.) Kummer)	1	2	3	4

Inocybe (Fr.) Fr.
Mycorrhizal with woody plants, includes *Astrosporina* Schroet., *Clypeus* (Britz.) Fayod. *Inocybe calamistrata* (Fr.) Gill. is reported from Hampshire & *I. tenebrosa* Quél. from Berkshire. Records of *I. auricoma* (Batsch) Lange are omitted as unverifiable and confused.

Inocybe abjecta (Karst.) Sacc.	Doubtful record				4
I. (*Clypeus*) *acuta* Boud.		1			
I. agardhii (Lund) Orton		1	2		4
I. (*C.*) *asterospora* Quél.		1	2	3	4
I. (*C.*) *boltonii* Heim		1	2	3	
I. bongardii (Weinm.) Quél.		1	2		4
I. (*C.*) *brevispora* Huijsman		1	2		4
I. brunneoatra (Heim) Orton		1	2		
I. (*C.*) *calospora* Quél.		1		3	
I. cervicolor (Pers.) Quél.	Needs confirmation	1	2		4
I. cincinnata (Fr.) Quél. (*I. phaeocomis* (Pers.) Kuyper)		1	2		
I. cincinnatula Kühner					4
I. cookei Bres.		1	2	3	4
I. corydalina Quél.		1	2	3	4
I. (*C.*) *curtipes* Karst. (*I. lanuginella* auct.)		1	2	3	4
I. descissa (Fr.) Quél.	1932 Foray record rejected	1		3	
I. dulcamara (A. & S.) Kummer		1	2		4
I. dulcamara var. *homomorpha* Kühner					4
I. eutheles (B. & Br.) Quél. (*I. tomentosa*) auct.)		1	2	3	4
I. fastigiata (Schaeff.) Quél.		1	2	3	4
I. (*C.*) *fibrosoides* Kühner				3	
I. flocculosa (Berk.) Sacc. (*I. gausapata* Kühner)		1	2	3	4
I. fuscidula Bres.		1			
I. geophylla (Sow.) Kummer		1	2	3	4
I. geophylla var. *lilacina* Gill.		1	2	3	4
I. godeyi Gill.		1	2	3	4
I. (*C.*) *grammata* Quél.		1	2	3	4
I. griseolilacina Lange		1	2	3	4
I. griseovelata Kühner					4
I. haemacta (Berk. & Cooke) Sacc.		1			4
I. hirtella Bres.		1	2	3	4

	1	2	3	4
I. hirtella var. *bispora* Kuyper (*I. amygdalispora* Metrod)	1			
I. hystrix (Fr.) Karst.	1		3	
I. inodora Vel. (*I. albodisca* Kühner)	1			
I. jurana (Pat.) Sacc.	1	2	3	4
I. lacera (Fr.) Kummer	1	2	3	4
I. langei Heim	1	2		4
I. (C.) lanuginosa (Bull.) Kummer (*I. ovatocystis* Boursier & Kuhner)	1	2	3	4
I. leptocystis Atk.	1	2		
I. longicystis Atk.	1	2	3	4
I. lucifuga (Fr.) Kummer		2	3	
I. maculata Boud.	1	2	3	4
I. maculata ssp. *fastigiella* (Atk.) Kühner & Romagn.				4
I. (C.) margaritispora (Berk.) Sacc.	1	2		4
I. microspora Lange	1	2		
I. (C.) mixtilis (Britz.) Sacc.	1	2	3	4
I. mutica (Fr.) Karst.	1	2		4
I. (C.) napipes Lange	1	2	3	4
I. (C.) oblectabilis (Britz.) Sacc.	1	2		4
I. (C.) oblectabilis f. *decemgibbosa* Kühner & Boursier	1			
I. obscura (Pers.) Gill.	1	2	3	4
I. ochroalba Bruylants	1			
I. ovalispora Kauff.	1			
I. patouillardii Bres.	1	2	3	4
I. perlata (Cooke) Sacc.	1	2	3	
I. (C.) petiginosa (Fr.) Gill.	1	2	3	4
I. phaeoleuca Kühner	1	2		
I. posterula (Britz.) Sacc. (*I. xanthodisca* Kühner)	1	2	3	4
I. (C.) praetervisa Quél.	1	2	3	4
I. (C.) proximella Karst.	1			
I. (C.) pseudoasterospora Kühner & Boursier				4
I. pudica Kühner			3	
I. pusio Karst.		2	3	4
I. pyriodora (Pers.) Kummer	1	2	3	4
I. pyriodora var. *incarnata* (Bres.) Heim	1	2		4
I. reducta Lange		2		
I. (C.) rennyi (Berk. & Br.) Sacc.	1			
I. squamata Lange	1			4
I. (C.) striatorimosa Orton	1			
I. subnudipes Kühner	1	2		
I. subtigrina Kühner	1	2		
I. terrigena (Fr.) Kühner	1			4
I. tigrina Heim	1			4
I. (C.) trechispora (Berk.) Karst.	1		3	
I. (C.) umbrina Bres.	1	2	3	4
I. vaccina Kühner	1			4
I. virgatula Kühner	1			4
I. xanthocephala Orton (*I. flavella* Karst.)	1			
I. (C.) xanthomelas Boursier & Kühner		2		

Krombholziella Maire

Mycorrhizal especially with Betulaceae, Fagaceae and Salicaceae; popular edible fungi.

	1	2	3	4
Krombholziella aurantiaca (Bull.) Maire	1	2		4
K. holopus (Rost.) Sutara	1	2		4
K. melaena (Smotlacha) Sutara	1	2		4
K. nigrescens (Richon & Roze) Sutara (*Leccinum crocipodium* Watling)	1	2		4
K. oxydabilis (Sing.) Sutara	1			
K. pseudoscabra (Kallenb.) Sutara (*L. carpini* (Schulz.) Moser)	1	2	3	
K. quercina (Pilat) Sutara	1			

	1	2	3	4
K. roseofracta (Watling) Sutara	1	2		
K. rufescens (Konrad) Sutara (*L. testaceoscabrum* (Secr.) Sing.)	1	2	3	4
K. scabra (Bull.) Maire	1	2	3	4
K. variicolor (Watling) Sutara	1			4

Kuehneromyces Sing. & A.H. Smith
Lignicolous, a segregate from *Pholiota*.

Kuehneromyces mutabilis (Schaeff.) Sing. & A.H. Smith	1	2	3	4

Laccaria Berk. & Br.
Facultatively mycorrhizal with many hosts.

Laccaria amethystea (Bull.) Murr.	1	2	3	4
L. bicolor (Maire) Orton	1	2		4
L. laccata (Scop) Cooke	1	2	3	4
L. proxima (Boud.) Pat.	1	2		4
L. purpureobadia Reid	1			4
L. tortilis (Bolt.) Cooke (*L. echinospora* (Speg.) Sing.)	1	2		

Lachnella Fr.
Saprophytes, especially on dead herbaceous stems and small twigs.

Lachnella alboviolascens (A. & S.) Fr.	1	2	3	4
L. cf. orthospora (Bourd. & Galz.) Reid		2		
L. villosa (Pers.) Gill.	1	2	3	4

Lachrymaria Pat.
Saprophytes, segregated from *Psathyrella* on account of their spore ornament.

Lachrymaria pyrotricha (Holmsk.) Konr. & Maubl.	1	2	3	4
L. velutina (Pers.) Konr. & Maubl.	1	2	3	4

Lactarius DC.
Mycorrhizal with trees.

Lactarius acerrimus Britz.	1	2		
L. acris (Bolt.) Gray	1	2		
L. aspideus (Fr.) Fr.	1	2		
L. azonites (Bull.) Fr.	1	2		
L. blennius (Fr.) Fr.	1	2	3	4
L. blennius f. *albidopallens* Lange			3	
L. britannicus Reid (*L. subsericeus* Hora)	1	2	3	4
L. camphoratus (Bull.) Fr.	1	2	3	4
L. chrysorheus Fr.	1	2	3	4
L. cilicioides Fr.	1	2		
L. cimicarius (Batsch) Gill.	1	2		4
L. circellatus Fr.	1	2	3	4
L. controversus (Fr.) Fr.	1	2		4
L. cyathula (Fr.) Fr.	1		3	4
L. decipiens Quél.	1		3	4
L. deliciosus (L.) Gray Including *L. deterrimus* Groger	1	2	3	4
L. flavidus Boud.	1	2		
L. flexuosus (Pers.) Gray	1	2		4
L. fluens Boud.		2		4
L. fuliginosus (Fr.) Fr.	1	2	3	4
L. fulvissimus Romagn.		2		4
L. glaucescens Crossland		2		4
L. glyciosmus (Fr.) Fr.	1	2	3	4
L. helvus (Fr.) Fr.	1	2	3	4

	1	2	3	4
L. hepaticus Plowr. (*L. theiogalus* auct.)	1	2	3	4
L. hysginus (Fr.) Fr.	1		3	4
L. insulsus (Fr.) Fr.	1	2		
L. lacunarum Romagn.	1	2	3	
L. lignyotus Fr. Confirmation required			3	
L. mairei Malençon Including *L. pearsonii* Z. Schaeffer	1	2		4
L. mammosus (Fr.) Fr.	1			
L. mitissimus (Fr.) Fr.	1	2	3	4
L. mollis Reid				4
L. obscuratus (Lasch) Fr. (*L. obnubilis* (Lasch) Fr.)	1	2	3	
L. pallidus (Pers.) Fr.	1	2		4
L. picinus Fr.	1	2		
L. piperatus (Scop.) Gray	1	2	3	4
L. pterosporus Romagn.	1	2		4
L. pubescens	1	2		4
L. pyrogalus (Bull.) Fr.	1	2	3	4
L. quietus (Fr.) Fr.	1	2	3	4
L. rubrocinctus Fr.		2		
L. rufus (Scop.) Fr.	1	2	3	4
L. ruginosus Romagn.	1	2		
L. salmonicolor Heim & Leclair (if truly distinct from *L. deliciosus*)				4
L. scrobiculatus (Scop.) Fr.	1	2		
L. serifluus (Dc.) Fr.	1	2	3	4
L. spinosulus Quél.	1	2	3	4
L. subdulcis (Pers.) Gray	1	2	3	4
L. tabidus Fr.	1	2	3	4
L. torminosus (Schaeff.) Gray	1	2	3	4
L. trivialis (Fr.) Fr.	1	2		
L. turpis (Weinm.) Fr.	1	2	3	4
L. uvidus (Fr.) Fr.	1	2		4
L. vellereus (Fr.) Fr.	1	2	3	4
L. vellereus var. *velutinus* Bertill.	1			
L. vietus (Fr.) Fr.	1	2	3	4
L. volemus (Fr.) Fr.	1	2		4
L. volemus var. *subrugatus* Neuhoff				4
L. zonarius (Bull.) Fr.	1	2	3	

Lactocollybia Sing.
Saprophytes, tropical or subtropical.

	1	2	3	4
Lactocollybia angiospermarum Sing. Glasshouse alien	1			

Lentinellus Karst.
Lignicolous saprophytes.

	1	2	3	4
Lentinellus cochleatus (Pers.) Karst	1	2	3	4
L. cochleatus var. *inolens* (Konr. & Maubl.)			3	4
L. flabelliformis (Bolt.) Orton (*L. omphalodes* (Fr.) Fr.)			3	4
L. vulpinus (Sow.) Kühner & Maire	1			4
L. vulpinus f. *ursinus* (Fr.) (*L. ursinus* (Fr.) Kühner)				4

Lentinus Fr.
Lignicolous saprophytes.

	1	2	3	4
Lentinus lepideus (Fr.) Fr.	1	2		4
L. tigrinus (Bull.) Fr.	1	2		4

Lepiota (Pers.) Gray
Saprophytes, including *Cystolepiota* Sing. *Leucoagaricus* Sing. & *Macrolepiota* Sing.

Species	1	2	3	4
Lepiota acerina Peck				4
L. (Cyst.) adulterina Moll.	1			
L. alba (Bres.) Sacc.		2		4
L. asper (Pers.) Quél. (*L. friesii* (Lasch) Quél., *L. acutaesquamosa* auct.)	1	2	3	4
L. (Leuc.) badhamii (B. & Br.) Quél.	1	2		4
L. bresadolae Schulz.			3	
L. brunneoincarnata Chodat & Martin	1	2		4
L. (C.) bucknallii B. & Br.	1	2	3	4
L. calcicola Knudsen		2		
L. (Leuc.) carneifolia Gill.	1			
L. castanea Quél.	1	2	3	4
L. (Leuc.) cinerascens Quél.	1			
L. clypeolaria (Bull.) Kummer	1	2	3	4
L. clypeolarioides Rea		2		4
L. cortinarius Lange	1	2		
L. cristata (Fr.) Kummer	1	2	3	4
L. echinacea Lange	1	2		4
L. echinella Quél. & Bernard	1			4
L. efibulis Knudsen	1			4
L. emplastrum (Cooke & Massee) Sacc.	1			
L. erminea (Fr.) Gill.	1			
L. (M.) excoriata (Schaeff.) Kummer		2		4
L. felina (Pers.) Karst.	1		3	4
L. felinoides (Bon) Orton	1			
L. flagellata (B. & Br.) Sacc. Glasshouse alien	1			
L. forquignoni Quél.		2		
L. fulvella Rea	1	2		4
L. fuscovinacea Møll. & Lange	1			4
L. (M.) gracilenta Fr.	1	2		4
L. grangei (Eyre) Lange	1	2		4
L. griseovirens Maire				4
L. heimii Locquin	1			
L. (C.) hetieri Boud.	1	2		4
L. (Leuc.) holosericea (Fr.) Gill.	1			4
L. hymenoderma Reid				4
L. hystrix Lange Needs confirmation	1	2		
L. ignipes Locquin				4
L. ignivolvata Bousset & Joss.	1	2		
L. josserandii Bon & Boiffard	1			
L. (M.) konradii Huijsman	1	2		4
L. langei Knudsen (*L. eriophora* auct. non Peck)	1			
L. latispora (Kühner) Bon		2		
L. (Leuc.) leucothites (Vitt.) Orton (*L. naucina* (Fr.) Kummer)	1	2	3	4
L. lilacea Bres.	1			
L. luteicystidiata Reid Glasshouse alien	1			
L. (Leuc.) marriageae Reid	1			4
L. (M.) mastoidea (Fr.) Kummer	1	2	3	4
L. nympharum (Kalchbr.) Karst.	1			
L. ochraceofulva Orton				4
L. oreadiformis Vel. (*L. laevigata* Lange)	1			4
L. pallida Locquin	1			
L. parvannulata (Lasch) Gill.	1			
L. (M.) permixta Barla				4
L. perplexa Knudsen (*L. acutaesquamosa* var. *typica* Kühner)	1	2		4
L. (Leuc.) pilatianus Desmoulins	1			4
L. (Leuc.) pinguipes Pearson			3	
L. (M.) procera (Scop.) Gray	1	2	3	4
L. pseudoasperula (Knudsen) Knudsen	1			
L. pseudohelveola Kühner	1	2		4
L. (M.) puellaris (Fr.) Rea	1			

	1	2	3	4
L. pulverulenta Huijsman	1			
L. (*Leuc.*) *purpureorimosa* Bon & Boiffard				4
L. (*M.*) *rhacodes*	1	2	3	4
L. rhodorrhiza Romagn. & Locquin	1			
L. (*C.*) *rosea* Rea (*Cystolepiota moelleri* Knudsen)	1			
L. scobinella (Fr.) Gill. (*L. helveola* auct.)				4
L. (*Leuc.*) *serena* (Fr.) Sacc.	1	2		4
L. (*Leuc.*) *sericatellus* Malençon	1			
L. (*Leuc.*) *sericifera* (Locq.) Locquin	1			4
L. setulosa Lange	1	2		4
L. (*C.*) *sistrata* (Fr.) Quél. (*L. seminuda* (Lasch) Kummer)	1	2	3	4
L. subalba Kühner	1	2		4
L. subgracilis Kühner	1	2		4
L. subincarnata Lange	1			4
L. tomentella Lange	1	2		4
L. ventriosospora Reid (*L. metulaespora* auct. non (B. & Br.) Sacc.)	1	2	3	4
L. (*Leuc.*) *wychanski* Pilat (*L. sublittoralis* Kühner)	1			
L. xanthophylla Orton		2		

Lepista (Fr.) W.G. Smith
Terrestrial saprophytes, popular edible fungi.

	1	2	3	4
Lepista caespitosa (Bres.) Sing.				4
L. glaucocana (Bres.) Sing.	1			
L. irina (Fr.) Bigelow	1	2		
L. luscina (Fr.) Sing. (*Tricholoma panaeolum* (Fr.) Quél.)	1	2	3	4
L. nuda (Bull.) Cooke	1	2	3	4
L. saeva (Fr.) Orton (*Tricholoma personatum* auct.)	1	2	3	4
L. sordida (Fr.) Sing. Doubtfully distinct from *L. nuda*	1	2	3	4

Leptoglossum Karst.
Associated with mosses, see also *Phaeotellus* Kühner & Lamoure and *Mniopetalum* Donk.

	1	2	3	4
Leptoglossum lobatum (Pers.) Ricken		2		4
L. muscigenum (Bull.) Karst.	1			4
L. retirugum (Bull.) Ricken	1	2		

Leptonia (Fr.) Kummer
Saprophytes, better treated as a subgenus of *Entoloma*.

	1	2	3	4
Leptonia acuta Rea Doubtful species		2		
L. aethiops (Scop.) Gill.	1			
L. anatina (Lasch) Kummer Needs confirmation	1			
L. andrianae Bres.				4
L. asprella (Fr.) Sensu Lange				4
L. cephalotricha Orton		2		4
L. chalybaea (Pers.) Kummer	1	2	3	
L. corvina (Kühner) Orton	1	2		
L. euchroa (Pers.) Kummer	1	2		4
L. formosa (Fr.) Gill. (*L. fulva* Orton)				4
L. griseocyanea (Fr.) Orton	1	2	3	4
L. griseorubida (Kühner) (*L. griseorubella* auct.)	1	2	3	4
L. incana (Fr.) Gill. (*L. euchlora* (Lasch) Kummer)	1	2		4
L. incana var. *citrina* Reid		2		
L. indutus (Boud.) Orton		2		
L. lampropus (Fr.) Quél.	1	2	3	4
L. lazulina (Fr.) Quél.	1	2		4
L. leptonipes (Kühner & Romagn.) Orton		2		
L. mougeotii (Fr.) Orton	1	2		4
L. pernitrosa Orton	1	2		

	1	2	3	4
L. polita (Pers.) Konr. & Maubl.	1	2		
L. rosea Longyear				4
L. sarcitula Kühner & Romagn.		2	3	
L. sericella (Fr.) Barbier	1	2	3	4
L. serrulata (Pers.) Kummer	1	2	3	4
L. strigosissima (Rea) Orton	1			
L. tjallingiorum (Noordeloos) Orton		2		
L. turci Bres.				4

Leucocoprinus Pat.

Lignicolous or terrestrial saprophytes, mainly tropical.

		1	2	3	4
Leucocoprinus birnbaumii (Corda) Sing. (*L. luteus* (Bolt.) Locquin)	Alien	1			
L. brebissonii (Godey) Locquin		1	2	3	4
L. cepaestipes (Sow.) Pat.		1			4
L. croceovelutinus Bon & Boiffard		1			
L. denudatus (Rab.) Sing.	Glasshouse alien	1			
L. georginae (W.G. Smith) Vasser		1	2		
L. ianthinus (Cooke) Locquin (*L. lilacinogranulosus* (Heim) Locquin)	Alien	1			4
L. tenellus (Boud.) Locquin	Alien	1			
L. zeylanicus (Berk.) Boedijn	Alien	1			

Leucocortinarius (Lange) Sing.

Mycorrhizal. There is an unverified record of *Leucocortinarius bulbiger* (A. & S.) Sing. from Buckinghamshire in the last century. Nothing to suggest its presence in our area.

Leucopaxillus Boursier

Saprophytes, some perhaps facultatively mycorrhizal.

	1	2	3	4
L. amarus (A. & S.) Kühner	1			4
L. giganteus (Sow.) Sing.	1	2		4
L. paradoxus (Cost. & Duf.) Boursier	1	2		4
L. rhodoleucus (Romell) Kühner				4

Limacella Earle

Saprophytes, mainly terrestrial.

	1	2	3	4
Limacella delicata (Fr.) Konr. & Maubl.	1	2		
L. glioderma (Fr.) Maire	1			4
L. guttata (Pers.) Konr. & Maubl. (*L. lenticularis* (Lasch) Maire)	1	2		4
L. roseofloccosa Hora	1			

Lyophyllum Karst.

Saprophytes, mainly terrestrial, see also *Hypsizygus* Sing. & *Tephrocybe* Donk.

	1	2	3	4
Lyophyllum connatum (Schum.) Sing.	1		3	4
L. decastes (Fr.) Sing. (*Tricholoma aggregatum* auct.)	1	2	3	4
L. decastes var. *ovisporum* (Lange) Kühner & Romagn.				4
L. favrei Haller	1			
L. fumosum (Pers.) Orton (*L. cinerascens* (Bull.) Konr. & Maubl.)	1	2	3	4
L. gangraenosum (Fr.) Gulden (*L. leucophaeatum* (Karst.) Karst)	1	2	3	4
L. immundum (Berk.) Kühner	1	2		4
L. infumatum (Bres.) Kühner (*L. deliberatum* Britz.) Kreisel)	1			
L. loricatum (Fr.) Kühner — Doubtfully distinct from *L. decastes*	1	2	3	
L. semitale (Fr.) Kühner	1			

Macrocystidia Heim
Terrestrial saprophyte.

	1	2	3	4
Macrocystidia cucumis (Pers.) Heim (*Nolanea pisciodora* (Ces.) Gill.)	1	2	3	4

Marasmius Fr.
Saprophytes, lignicolous and terrestrial, including *Marasmiellus* Murr.

	1	2	3	4
Marasmius acicola Romagn.	1			
M. alliaceus (Jacq.) Fr.	1	2	3	
M. androsaceus (L.) Fr.	1	2	3	4
M. bulliardii Quél.	1	2		
M. buxi Fr.	1			
M. (Marasmiellus) calopus (Pers.) Fr.	1	2		4
M. cohaerens (A. & S.) Cooke & Quél.	1	2		4
M. collinus (Scop.) Sing.		2	3	
M. epiphylloides (Rea) Sacc. & Trott.	1			4
M. epiphyllus (Pers.) Fr.	1	2		4
M. graminum (Lib.) Berk.	1		3	4
M. hudsonii (Pers.) Fr.	1			
M. lupuletorum (Weinm.) Bres. (*M. torquescens* Quél.)	1	2		4
M. oreades (Bolt.) Fr.	1	2	3	4
M. porreus (Pers.) Fr. Needs confirmation	1	2		4
M. prasiosmus (Fr.) Fr. Needs confirmation	1	2		
M. (Marasmiellus) ramealis (Bull.) Fr. (*M. amadelphus* (Bull.) Fr.)	1	2	3	4
M. recubans Quél.	1	2		
M. rotula (Scop.) Fr.	1	2	3	4
M. saccharinus (Batsch) Fr.	1	2		4
M. scorodonius (Fr.) Fr.		2		
M. siccus Schw.		2		
M. splachnoides (Fr.) Fr. (*M. quercophilus* Pouz.)				4
M. (Marasmiellus) tricolor (A.& S.) Kühner (*M. pruinatus* Rea)	1			
M. undatus (Berk.) Fr. (*Collybia vertirugis* (Cooke) Sacc.)	1	2	3	4
M. vaillantii (Pers.) Fr. (*M. candidus* (Bolt.) Fr.)				4
M. wynnei B. & Br. (*M. globularis* Fr.)	1	2		4

Melanoleuca Pat.
Terrestrial saprophytes.

	1	2	3	4
Melanoleuca adstringens (Pers.) Metrod	1	2		4
M. albifolia Boekhout	1			
M. arcuatum (Fr.) Sing.	1			4
M. atripes Boekhout	1			
M. brevipes (Bull.) Pat.	1	2	3	4
M. cinerascens Reid	1	2		4
M. cognatum (Fr.) Konr. & Maubl.	1			4
M. curtipes (Fr.) Bon	1			
M. excissa (Fr.) Sing.	1		3	
M. grammopodia (Bull.) Pat.	1	2	3	4
M. humilis (Pers.) Pat.	1	2	3	
M. iris Kühner	1	2		4
M. megaphylla (Boud.)		2		
M. melaleuca (Pers.) Murr.	1	2		4
M. paedida (Fr.) Kühner				4
M. polioleuca (Fr.) Kühner & Maire	1			
M. reai Sing.				4
M. schumacheri (Fr.) Sing.	1	2		4
M. strictipes (Karst.) J. Schaeff. (*M. evenosa* (Sacc.) Konr.)	1			4
M. subpulverulenta (Pers.) Sing.	1			4

Melanophyllum Vel.
Terrestrial saprophytes.

	1	2	3	4
Melanophyllum echinatum (Roth) Sing.	1	2		4
M. eyrei (Massee) Sing.	1	2		

Melanotus Pat.
Saprophytes on dead plant tissue.

	1	2	3	4
Melanotus phillipsii (B. & Br.) Sing.	1	2		
M. proteus (B. & Br.) Sing.	1	2		

Micromphale Nees
Lignicolous and terrestrial saprophytes, also *M. cauvetii* Maire & Kühner in Berkshire.

	1	2	3	4
Micromphale brassicolens (Romagn.) Orton	1	2		
M. foetidum (Sow.) Sing.	1	2		4
M. impudicum (Fr.) Orton	1	2		4
M. inodorum (Pat.) Svrček non Dennis nom. illegit.	1			4
M. perforans (Hoffm.) Gray Old record only	1			

Mniopetalum Donk & Sing.
Saprophytes on dead tissue or parasites of bryophyta.

	1	2	3	4
Mniopetalum globisporum Donk	1			

Mycena Pers.
Lignicolous or terrestrial saprophytes, including *Hemimycena* Sing. *Delicatula* Fayod, *Hydropus* (Kühner) Sing. and *Rickenella* Raitelhuber. *Mycena seynei* Quél. is in Dorset and *M. epichloe* Kühner in Middlesex.

	1	2	3	4
Mycena abramsii (Murr.) Murr. (*M. praecox* Vel.)	1	2	3	4
M. (Hemi.) acicula (Schaeff.) Kummer	1	2	3	4
M. (Hemi.) adonis (Bull.) Gray		1	2	3
M. aetites (Fr.) Quèl.	1	2	3	4
M. (Hemi.) alba (Bres.) Kühner	1	2		
M. alcalina (Fr.) Kummer	1	2	3	4
M. amicta (Fr.) Quél. (*M. iris* (Berk.) Quél.)	1	2	3	4
M. (Hemi.) angustispora (Joss.) Orton	1			4
M. atrocyanea (Batsch) Gill. (*M. fusconigra* Orton)	1			
M. aurantiomarginata (Fr.) Quél. (*M. elegans* auct.)	1	2		4
M. bulbosa (Cejp) Kühner	1		3	4
M. (Hemi.) candida (Bres.) Kühner	1			
M. capillaripes Peck	1	2	3	4
M. capillaris (Schum.) Kummer	1	2	3	4
M. carnicolor Orton (*Marasmiellus rosellus* (Lange) Kuiper & Noordel)	1			
M. (Hemi.) cephalotricha Joss.		2		
M. chlorantha (Fr.) Kummer	1	2		4
M. cinerella Karst. (*Omphalia grisea* auct.)	1	2	3	4
M. citrinomarginata Gill.	1	2	3	4
M. clavicularis (Fr.) Gill.	1	2		
M. clavularis (Batsch) Sacc.	1	2		4
M. coccinea (Sow.) Quél.	1	2	3	
M. corticola (Pers.) Gray (*M. hiemalis* (Osb.) Quél.)	1	2	3	4
M. corynephora Maas Geesteranus		2		4
M. (Delic.) crispata Kühner	1			
M. crocata (Schrad.) Kummer	1	2		
M. (Hemi.) cucullata (Pers.) (*M. gypsea* auct.)	1	2		4
M. (Hemi.) cyphelloides Orton (*Omphalia gibba* Pat.)		2		
M. dasypus Maas Geesteranus & Laessoe	1			

	1	2	3	4
M. (Hemi.) delictabilis (Peck) Sacc.	1	2		4
M. epipterygia (Scop.) Gray	1	2	3	4
M. epipterygioides Pearson	1	2	3	
M. erubescens v. Höhn.	1	2		
M. fagetorum (Fr.) Gill.				3
M. (Hemi., Rick.) fibula (Bull.) Kühner (*Gerronema fibula* Sing.)	1	2	3	4
M. filopes (Bull.) Kummer	1	2	3	4
M. flavescens Vel.	1	2		
M. flavoalba (Fr.) Quél.	1	2	3	4
M. (Hydropus) floccipes (Fr.) Kühner				4
M. galericulata (Scop.) Gray (*M. rugosa* (Fr.) Quél., *M. rugulosiceps* (K.) Sm.)	1	2	3	4
M. galopus (Pers.) Kummer	1	2	3	4
M. galopus var. *candida* Lange	1	2	3	4
M. haematopus (Pers.) Kummer	1	2		4
M. inclinata (Fr.) Quél.	1	2	3	4
M. (Delic.) integrella (Pers.) Gray	1	2	3	
M. (Hemi.) lactea (Pers.) Kummer	1	2	3	4
M. lactella Orton				4
M. latifolia (Peck) Sacc.	1			4
M. leptocephala (Pers.) Gill.	1	2	3	4
M. leucogala (Cooke) Sacc. (*M. galopus* var. *nigra* Rea)	1	2		4
M. lineata (Bull.) Kummer	1	2	3	4
M. lineata f. *pumila* Lange		2		
M. longiseta v. Höhn.				4
M. maculata Karst.	1			4
M. (Hemi.) mairei (Gilbert) Kühner	1			
M. (Hemi.) mauretanica (Maire) Kûhner	1	2	3	
M. meliigena (Berk. & Cooke) Sacc. (*M. corticola* auct. non Sing.)	1			4
M. megaspora Kauffm. (*M. uracea* Pearson)	1	2	3	4
M. metata (Fr.) Kummer	1		3	4
M. mirata (Peck) Sacc.	1			4
M. mucor (Batsch) Gill.	1	2		4
M. olida Bres.	1	2		4
M. olivaceomarginata (Massee) Massee (*M. avenacea* auct.)	1	2	3	4
M. oortiana Hora (*M. arcangeliana* Bres. var.)	1	2		4
M. osmundicola Lange Alien, predominantly in glasshouses	1			
M. osmundicola subsp. *imleriana* Kühner Glasshouse alien	1			
M. pearsoniana Dennis (*M. pseudopura* auct.)	1			
M. pelianthina (Fr.) Quél.	1	2	3	
M. pelliculosa (Fr.) Quél.	1	2		
M. picta (Fr.) Needs confirmation	1			
M. polyadelpha (Lasch) Kühner	1			
M. polygramma (Bull.) Gray	1	2	3	4
M. polygramma f. *candida* Lange	1			
M. polygramma f. *pumila* Lange	1			
M. pterigena (Fr.) Kummer Recorded in error from Kent but not unlikely to occur				
M. pura (Pers.) Kummer	1	2	3	4
M. pura var. *alba* Gill.	1			
M. pura var. *carnea* Rea	1			
M. pura var. *multicolor* Bres.		2		
M. pura f. *violacea* (Gill.) Maas Geesteranus	1			
M. renati Quél. (*M. flavipes* Quél.)		2		
M. rorida (Scop.) Quél.	1	2	3	4
M. rosella (Fr.) Kummer	1			4
M. rubromarginata (Fr.) Kummer Most records refer to *M. capillaripes*				4
M. (Resinomycena) saccharifera (B. & Br.) Sacc. (*Del. quisquiliaris* Joss.)	1	2	3	
M. sanguinolenta (A. & S.) Kummer	1	2	3	4
M. (Hydropus) scabripes (Murr.) A.H. Smith	1			
M. sepia Lange	1			

	1	2	3	4
M. smithiana Kühner	1			
M. speirea (Fr.) Gill. (*M. camptophylla* (Berk.) Sing.)	1	2		4
M. stylobates (Pers.) Kummer	1	2	3	4
M. swartzii (Fr.) A.H. Smith (*Gerronema* (*Rickenella*) *setipes* (Fr.) Sing.)	1	2	3	4
M. tenerrima (Berk.) Sacc. (*M. adscendens* (Lasch) Maas Geesteranus)	1	2		4
M. tortuosa Orton	1	2	3	4
M. (Hydropus) trichoderma Joss.	1			
M. viscosa Maire	1		3	4
M. vitilis (Fr.) Quél.	1	2	3	4
M. vulgaris (Pers.) Kummer	1	2	3	4
M. zephyrus (Fr.) Kummer		2		

Mycenella (Lange) Sing.
Lignicolous or terrestrial saprophytes.

	1	2	3	4
Mycenella bryophila (Vogl.)		2		4
M.salicina (Vel.) Sing.	1			

Myxomphalia (Kühner) Hora
Terrestrial, scarcely separable from *Fayodia*.

	1	2	3	4
Myxomphalia maura (Fr.) Hora	1	2		4

Naucoria (Fr.) Kummer
Mycorrhizal, especially with *Alnus* and *Salix*, *Alnicola* Kühner is a synonym. There are several more species in southern England, including *N. silvaenovae* Reid in Hampshire.

	1	2	3	4
Naucoria alnetorum (Maire) Kühner & Romagn.	1		3	
N. amarescens Quél.				4
N. bohemica Vel.	1	2	3	4
N. escharoides (Fr.) Kummer	1	2	3	4
N. luteolofibrillosa (Kühner) Kühner & Romagn.		2		
N. pseudoamarescens (Kühner & Romagn.) Kühner & Romagn.	1			
N. pseudoscolecina Reid	1			
N. salicis Orton (*N. macrospora* Lange non Pat.)	1	2		4
N. scolecina (Fr.) Quél. Nearly all British records are to be rejected	1			
N. striatula Orton	1	2	3	4
N. subconspersa Kühner	1	2	3	4

Nolanea (Fr.) Kummer
Terrestrial saprophytes, the legitimate name is *Latzinaea* O.K., not adopted as the genus is better united with *Entoloma*.

	1	2	3	4
Nolanea araneosa Quél. (*Leptonia fulvostrigosa* (B. & Br.) Orton)	1	2		4
N. cetrata (Fr.) Kummer	1	2	3	4
N. conferenda (Britz.) Sacc. (*N. staurospora* Bres., *N. proletaria* auct.)	1	2	3	4
N. cuneata Bres.	1			
N. cuspidifer (Kühner & Romagn.) Orton			3	
N. dysthales (Peck) Murr. (*Leptonia babingtonii* auct.)	1			4
N. farinolens Orton	1	2		4
N. hebes (Romagn.) Orton	1			
N. hirtipes (Schum.) Kummer	1	2	3	
N. icterina (Fr.) Kummer	1		3	4
N. infula (Fr.) Gill.	1		3	4
N. juncinus (Kühner & Romagnesi) Orton	1	2	3	4
N. lucida Orton	1	2		
N. mammosa (L.) Quél.	1	2		
N. minuta Karst.	1	2		
N. papillata Bres.	1	2	3	4

	1	2	3	4
N. radiata (Lange) Orton	1			
N. rhombispora (Kühner & Boursier)	1	2		
N. sericea (Bull.) Orton	1	2	3	4
N. solsticialis (Fr.) Orton	1	2		
N. tenuipes Orton	1		3	4
N. versatilis (Fr.) Gill.				4
N. xylophila (Lange) Orton		2		

Omphaliaster Lamoure
  Saprophytes.

	1	2	3	4
Omphaliaster asterosporus (Lange) Lamoure	1			
O. borealis (Lange & Skifte) Lamoure	1			

Omphalina Quél.
  Terrestrial saprophytes or algal symbionts.

	1	2	3	4
Omphalina chlorocyanea (Pat.) Sing. (*Omphalia viridis*(Hornemann) Lange)	1			
O. cupulata (Fr.) Orton		2	3	
O. demissa (Fr.) Quél. Needs confirmation	1			
O. ericetorum (Fr.) M. Lange (*O. umbellifera* (L.) Kummer)	1	2		4
O. griseopallida (Desm.) Quél.	1	2	3	4
O. grossula (Pers.) Sing. (*O. wynniae* (B. & Br.) Orton)			3	4
O. hepatica (Fr.) Orton	1	2		4
O. mutila (Fr.) Orton Needs confirmation				
O. obscurata Reid			3	
O. oniscus (Fr.) Quél.	1		3	
O. philonotis (Basch) Quél.	1	2	3	
O. postii (Fr.) Sing.	1			4
O. pyxidata (Bull.) Quél. (*O. muralis* auct.)	1	2	3	4
O. rickenii Sing.	1			
O. rustica (Fr.) Quél.	1	2		
O. sphagnicola (Berk.) Moser	1			
O. umbratilis (Fr.) Quél.	1	2		

Omphalotus Fayod
  Lignicolous saprophyte.

	1	2	3	4
Omphalotus olearius (Fr.) Sing.	1	2		

Oudemansiella Speg.
  Lignicolous saprophytes or weak facultative parasites of *Fagus*; includes *Xerula* Maire.

	1	2	3	4
Oudemansiella (Xerula) badia (Quél.) Moser				4
O. (X.) longipes (Bull.) Moser	1	2		4
O. mucida (Schrad.) v. Höhn.	1	2	3	4
O. radicata (Relhan) Sing.	1	2	3	4

Panaeolina Maire
  Terrestrial saprophyte.

	1	2	3	4
Panaeolina foenisecii (Pers.) Maire	1	2	3	4

Panaeolus (Fr.) Quél.
  Coprophilous or terrestrial saprophytes, annellate species were formerly segregated in *Anellaria* Karst.

	1	2	3	4
Panaeolus acuminatus (Schaeff.) Quél.	1	2	3	
P. ater (Lange) Kühner & Romagn.	1	2		4
P. campanulatus (Bull.) Quél.	1	2	3	4

P. fimicola (Fr.) Quél.	1		3 4
P. olivaceus Moll.	1 2		
P. papilionaceus (Bull.) Quél.	1 2	3 4	
P. phalaenarum (Fr.) Quél.	1 2		
P. retirugis (Fr.) Gill.	1 2		
P. rickenii Hora	1 2	3 4	
P. semiovatus (Sow.) Lundell (*A. separata* (L.) Karst.)	1 2	3 4	
P. sphinctrinus (Fr.) Quél.	1 2	3 4	
P. subbalteatus (B. & Br.) Sacc.	1		4

Panellus Karst.
Lignicolous saprophytes.

Panellus mitis (Pers.) Sing.	1		4
P. serotinus (Schrad.) Kühner	1 2	3 4	
P. stipticus (Bull.) Karst.	1 2	3 4	

Panus Fr.
Lignicolous saprophytes.

Panus torulosus (Pers.) Fr. (*P. conchatus* (Bull.) Fr.)	1 2 3 4

Paxillus Fr.
Facultatively mycorrhizal (*P. involutus)* or lignicolous saprophytes.

Paxillus atrotomentosus (Batsch) Fr.	1 2 3 4
P. involutus (Batsch) Fr.	1 2 3 4
P. panuoides (Fr.) Fr.	1 2 3 4

Pellidiscus Donk
Lignicolous saprophyte.

Pellidiscus pallidus (B. & Br.) Donk (*Cyphella bloxami* Berk. & Phill.)	1 2

Phaeocollybia Heim.
Saprophytes; also *P. lugubris* (Fr.) Heim in N. Wales and *P. cidaris* (Fr.) Romagn. dubious in 1.

Phaeocollybia festiva (Fr.) Heim	Confirmation desirable	1
P. jennyae (Karst.) Heim		4

Phaeogalera Kühner
Saprophytes: a few sphagnicolous species in Scotland.

Phaeogalera oedipus (Cooke) Romagn.	But Singer retains this in *Pholiota*	1 2

Phaeolepiota Maire
Terrestrial, allied to *Cystoderma*.

Phaeolepiota aurea (Mattuschka) Maire	1 . 3 4

Phaeomarasmius Scherffel
Lignicolous saprophytes.

Phaeomarasmius erinaceus (Fr.) Kühner	1 2 4
P. horizontalis (Bull.) Kühner	1

Phaeotellus Kühner & Lamoure
Terrestrial saprophytes, variously referred but perhaps akin to *Leptoglossum*.

	1	2	3	4
Phaeotellus acerosus (Fr.) Kühner & Lamoure	1	2	3	4
P. acerosus var. *latisporus* (Favre) Gulden (*Pleurotellus tremulus* (Schaeff.) K. & M.)	1			4

Pholiota (Fr.) Kummer
Saprophytes, mainly lignicolous, or weakly facultatively parasitic on trees.

	1	2	3	4
Pholiota abstrusa (Fr.) Sing.	1			
P. adiposa (Fr.) Kummer	1	2	3	4
P. albocrenulata (Peck) Sacc.	1			
P. alnicola (Fr.) Sing.	1	2	3	4
P. apicrea (Fr.) Moser	1	2		
P. aromatica Orton	1			
P. aurivella (Batsch) Kummer	1	2	3	4
P. carbonaria (Fr.) Sing. (*P. highlandensis* (Peck) Smith & Hesler)	1	2	3	4
P. comosa (Fr.) Quél.	1			
P. conissans (Fr.) Moser	1	2		
P. curvipes (Fr.) Quél.	1	2		4
P. decussata (Fr.) Moser				4
P. destruens (Brond.) Gill.	1	2		4
P. flammans (Fr.) Kummer	1			4
P. graminis (Quél.) Sing.	1	2		
P. gummosa (Lasch) Sing.	1	2		4
P. inaurata (W.G. Smith) Moser	1			
P. lenta (Pers.) Sing. (*Hebeloma glutinosum* (Lindgren) Sacc.)	1	2	3	4
P. limonella (Peck) Sacc.				4
P. lubrica (Pers.) Sing.				4
P. muelleri (Fr.) Karst.	1			4
P. myosotis (Fr.) Sing.	1	2		4
P. ochrochlora (Fr.) Orton	1	2	3	4
P. persicina Orton	1			
P. spumosa (Fr.) Sing.	1	2		4
P. squarrosa (Müll.) Kummer	1	2	3	4
P. tuberculosa (Schaeff.) Kummer	1	2	3	4

Phylloporus Quél.
Mycorrhizal.

	1	2	3	4
Phylloporus rhodoxanthus (Schw.) Bres.	1	2		4

Phyllotopsis (Gilbert & Donk) Sing.
Lignicolous saprophyte.

	1	2	3	4
Phyllotopsis nidulans (Pers.) Sing.	1			4

Phyllotus Karst.
Lignicolous saprophyte on coniferous timber; *P. porrigens* (Pers.) Karst. in Scotland only.

Pleuroflammula Sing.
Lignicolous saprophytes, mainly tropical but *P. ragazziana* (Bres.) Horak should be looked for in southwestern England.

Pleurotellus Fayod
Lignicolous saprophytes.

	1	2	3	4
Pleurotellus dictyorhizus (DC.) Kühner				4
P. hypnophyllus (Pers.) Fayod	1			4

Pleurotus (Fr.) Kummer

Lignicolous saprophytes; *P. opuntiae* (Dur. & Lev.) Sacc. occurs occasionally on imported plants of Yucca spp.

	1	2	3	4
Pleurotus cornucopiae (Paulet) Rolland (*P. macropus* Bagl.)	1	2	3	4
P. dryinus (Pers.) Kummer (*P. corticatus* (Fr.) Kummer)	1	2	3	4
P. (Pleurocybella) lignatilis (Pers.) Kummer	1		3	4
P. ostreatus (Jacq.) Kummer	1	2	3	4
P. ostreatus var. *euosmus* (Berk.) Massee	1	2		4
P. ostreatus f. *pulmonarius* (Fr.) Pilat	1	2		
P. ostreatus f. *salignus* (Pers.) Pilat	1			

Pluteolus (Fr.) Gill.

Lignicolous or terrestrial saprophytes, united with *Bolbitius* by some authors.

	1	2	3	4
Pluteolus aleuriatus (Fr.) Karst. (*Bolbitius reticulatus* (Pers.) Ricken)	1	2	3	4

Pluteus Fr.

Saprophytes, mainly lignicolous.

	1	2	3	4
Pluteus aurantiorugosus (Trog.) Sacc.	1	2	3	4
P. cervinus (Schaeff.) Kummer (*P. atricapillus* (Batsch) Fayod; *P. curtisii* (B.) Sacc.)	1	2	3	4
P. cinereofuscus Lange	1	2		4
P. depauperatus Romagn.	1	2		4
P. dryophiloides Orton	1			
P. galeroides Orton	1			
P. godeyi Gill.	1	2		4
P. granulatus Bres.	1	2		
P. griseoluridus Orton	1			
P. griseopus Orton	1	2		4
P. hispidulus (Fr.) Gill.	1	2	3	4
P. leoninus (Schaeff.) Kummer	1	2		4
P. luctuosus Boud.	1	2		
P. luteovirens Rea	1	2		4
P. minutissimus Maire	1	2		4
P. murinus Bres.	1	2	3	
P. nanus (Pers.) Kummer	1	2		4
P. olivaceus Orton	1			
P. pallescens Orton				4
P. patriceus (Schulz.) Boud.	1	2		
P. pearsonii Orton	1	2	3	4
P. pellitus (Pers.) Kummer	1	2		4
P. petasatus (Fr.) Gill.	1	2		4
P. phlebophorus (Ditmar) Kummer	1	2	3	4
P. plautus (Weinm.) Gill.	1	2		
P. podospileus Sacc. & Cub.	1	2	3	4
P. punctipes Orton	1	2		
P. rimulosus Kühner		2		4
P. romellii (Britz.) Sacc. (*P. lutescens* (Fr.) Bres.)	1	2		4
P. salicinus (Pers.) Kühner	1	2	3	4
P. semibulbosus (Lasch) Gill. (*P. boudieri* Orton; *P. gracilis* (Bres.) Lange)	1	2	3	4
P. splendidus Pearson	1			
P. thomsonii (B.& Br.) Dennis	1	2		4
P. tricuspidatus Vel. (*P. atromarginatus* (Sing.) Kühner & Romagn.)	1		3	
P. umbrosus (Pers.) Kummer	1	2	3	4
P. villosus (Bull.) Romagn. (*P. ephebius* (Fr.) Gill.)	1			
P. xanthophaeus Orton (*P. chrysophaeus* (Schaeff.) Quél.)	1	2		4

Polyporus Micheli

Stipitate, poroid, lignicolous carpophores, saprophytic or at most feebly parasitic on trees.

	1	2	3	4
Polyporus badius (Pers.) Schw. (*P. picipes* Fr.)	1	2		4
P. brumalis (Pers.) Fr. (*P. ciliatus* Fr.)	1	2	3	4
P. floccipes Rost.) (*P. lentus* Berk.)	1	2	3	4
P. melanopus (Swartz) Fr.	1	2		4
P. squamosus (Huds.) Fr.	1	2	3	4
P. tuberaster Jacq.	1			
P. varius (Pers.) Fr.	1	2	3	4
P. varius var. *nummularius* (Bull.) Fr.	1	2		

Porphyrellus Gilbert
Mycorrhizal.

	1	2	3	4
Porphyrellus porphyrosporus (Fr.) Gilb.	1	2		4

Psathyrella (Fr.) Quél.
Saprophytes, lignicolous or terrestrial.

	1	2	3	4
Psathyrella albidula (Romagn.) Moser	1			
P. ammophila (Dur. & Lév.) Orton		2	3	4
P. atomata (Fr.) Quél.	1	2	3	4
P. bipellis (Quél.) A.H. Smith	1	2		
P. candolleana (Fr.) Maire (*Hypholoma appendiculatum* (Bull.) auct.)	1	2	3	4
P. caudata (Fr.) Quél.	1		3	
P. cernua (Vahl) Moser	1		3	
P. clivensis (B. & Br.) Orton	1			
P. conopilus (Fr.) Pearson & Dennis (*P. subatrata* (Batsch) Gill.)	1	2		4
P. corrugis (Pers.) Konr. & Maubl.	1	2	3	4
P. cotonea (Quél.) Konr. & Maubl.	1		3	4
P. fatua (Fr.) Konr. & Maubl.	1	2		
P. friesii Kits v. Wav. (*P. fibrillosa* (Pers.) Sing.)	1	2		4
P. frustulenta (Fr.) A.H. Smith	1			
P. fusca (Schum.) Moser	1	2		
P. gossypina (Bull.) Pearson & Dennis	1	2		4
P. gracilis (Fr.) Quél.	1	2		4
P. hydrophila (Bull.) Maire (*P. piluliforme* (Bull.) Orton)	1	2	3	4
P. laevissima (Romagn.) Moser	1			
P. leucotephra (B. & Br.) Orton	1	2		
P. lutensis (Romagn.) Moser	1	2		
P. marcescibilis (Britz.) Sing.	1	2		
P. microrhiza (Lasch) Konr. & Maubl.	1	2		
P. multipedata (Peck) A.H. Smith	1			4
P. nolitangere (Fr.) Pearson & Dennis	1		3	4
P. obtusata (Pers.) A.H. Smith	1	2	3	4
P. olympiana A.H. Smith	1			
P. orbitarum (Romagn.) Moser		2		
P. pannucioides (Lange) Moser	1			
P. pennata (Fr.) Pearson & Dennis	1	2	3	
P. polycystis (Romagn.) Moser	1			
P. prona (Fr.) Gill.	1	2		
P. pseudocasca (Romagn.) Moser (*P. casca* ss Lange)				4
P. pseudogordonii Kits v. Wav.	1			
P. pseudogracilis (Romagn.) Moser	1	2		4
P. pygmaea (Bull.) Siong.	1			4
P. sarcocephala (Fr.) Sing.	1		3	
P. scobinacea (Fr.) Sing.		2		4
P. semivestita (B. & Br.) A.H. Smith	1			
P. spadicea (Schaeff.) Sing.	1	2	3	4
P. spadiceogrisea (Fr.) Maire	1	2		4
P. spintrigeroides Orton	1			
P. squamosa (Karst.) Moser	1	2	3	4

P. subnuda (Karst.) A.H. Smith	1		
P. sylvestris (Gill.) Moser			4
P. suavissima Ayer	1		
P. tephrophylla (Romagn.) Moser	1	2	
P. typhae (Kalchbr.) Pearson & Dennis	1		
P. vernalis (Lange) Moser	1	2	

Psilocybe (Fr.) Kummer
Saprophytes, coprophilous or terrestrial, including *Deconica* (W.G. Smith) Karst.

Psilocybe bullacea (Bull.) Kummer	1	2		4
P. callosa (Fr.) Quél.				4
P. coprophila (Bull.) Kummer	1			4
P. crobula (Fr.) Lange (*Tubaria crobula* (Fr.) Karst.)	1	2	3	4
P. cyanescens Wakefield	1	2		
P. fimetaria (Orton) Watling	1	2		
P. graminicola (Orton) Orton	1			
P. inquilina (Fr.) Bres.	1	2	3	4
P. merdaria (Fr.) Ricken	1	2		
P. montana (Pers.) Kummer (*P. atrorufa* (Schw.) Quél.)	1	2	3	4
P. muscorum (Orton) Moser	1			4
P. percevalii (B. & Br.) Reid	1			
P. physaloides (Bull.) Quél.	1	2	3	4
P. pratensis Orton	1			
P. rhombispora (Britz.) Sacc.	1	2		
P. semilanceata (Fr.) Kummer	1	2	3	4
P. squamosa (Fr.) Orton	1	2		4
P. subcoprophila (Britz.) Sacc.	1			
P. thrausta (Schulz.) Bon	1			

Resupinatus Nees
Saprophytes on wood and debris. *R. silvanus* (Sacc.) Sing. has perhaps been overlooked.

Resupinatus applicatus (Batsch) Gray	1	2	3	4
R. cyphellaeformis (Berk.) Sing.	1			4
R. kavinii (Pilat) Moser	1			
R. trichotis (Pers.) Sing. (*R. rhacodium* (B. & C.) Sing.)	1			

Rhodocybe Maire
Terrestrial. *Rhodocybe caelata* (Fr.) Maire is in the east midlands.

Rhodocybe fallax (Quél.) Sing.				4
R. hirneola (Fr.) Orton	1		3	
R. melleopallens Orton	1			
R. mundula (Lasch) Sing.	1			4
R. nitellina (Fr.) Sing.	1	2		
R. popinalis (Fr.) Sing.	1	2		4
R. truncata (Schaeff.) Sing.	1	2		4

Rhodotus Maire
Lignicolous saprophyte.

Rhodotus palmatus (Bull.) Maire	1	2	3	4

Rimbachia Pat.
Saprophytes.

Rimbachia arachnoidea (Peck) Redhead, a *Mniopetalum* for Singer	1

Ripartites Karst.
Saprophytic, terrestrial or on debris; *R. metrodii* Huijsman has been reported from Oxfordshire.

	1	2	3	4
Ripartites helomorpha (Fr.) Karst.	1			
R. tricholoma (A. & S.) Karst.	1	2		4

Rozites Karst.
Mycorrhizal, common in highland coniferous forests, rare in England.

		1	2	3	4
Rozites caperatus (Pers.) Karst.	An old Kentish record is better rejected			3	

Russula Pers.
Mycorrhizal. In addition *R. cremeoavellanea* Sing. has been reported from Hampshire.

		1	2	3	4
Russula acrifolia Romagn. (*R. densifolia* auct. angl. pp.)			2		4
R. adusta (Pers.) Fr.		1	2	3	4
R. aeruginea Lindblad		1	2	3	4
R. albonigra (Kromb.) Fr.		1		3	4
R. alutacea (Pers.) Fr.		1	2	3	4
R. amethystina Quél.				3	
R. amoena Quél. (*R. punctata* Gill.)		1	2		4
R. amoenolens Romagn.		1			
R. aquosa Leclair		1	2	3	
R. atrorubens Quél.	Including *R. laccata* Huijsman	1			4
R. aurata (With.) Fr.		1	2		4
R. azurea Bres.		1	2	3	4
R. badia Quél.					4
R. betularum Hora (*R. fragilis* var. *nivea* auct.)		1	2	3	4
R. borealis Kauffm.			2		
R. brunneoviolacea Crawshay		1	2	3	4
R. caerulea (Pers.) Fr.		1	2	3	4
R. cessans Pearson		1			4
R. claroflava Grove			2	3	4
R. consobrina (Fr.) Fr.			2	3	
R. cuprea Krombh.		1	2	3	
R. curtipes Møll. & Schaeff.			2	3	
R. cutefracta Cooke		1	2		
R. cyanoxantha (Schaeff.) Fr.		1	2	3	4
R. cyanoxantha f. *peltereaui* Sing.		1	2		4
R. decipiens (Sing.) Kühner & Romagn.		1	2		4
R. decolorans (Fr.) Fr.	Confirmation desirable	1			
R. delica Fr. (*R. chloroides* Krombh s.str. 2,4: *Lactarius exsuccus* (Otto) Sm.)		1	2	3	4
R. densifolia Gill.		1	2	3	4
R. emetica Schaeff.) Gray	Old records probably referred to *R. mairei*	1	2	3	4
R. emeticella (Sing.) Hora			2	3	4
R. emiticolor J. Schaeff.				3	
R. farinipes Romell (*R. subfoetens* auct.)		1	2	3	4
R. fellea (Fr.) Fr.		1	2	3	4
R. firmula J. Schaeffer		1	2	3	
R. foetens (Pers.) Fr.		1	2	3	4
R. fragilis (Pers.) Fr. (*R. fallax* (Fr.) Britz.)		1	2	3	4
R. gigasperma Romagn.			2	3	
R. gracillima J. Schaeffer		1	2	3	4
R. grisea (Pers.) Fr.	Old records may be *R. ionochlora* or *R. parazurea*	1	2	3	4
R. heterophylla (Fr.) Fr.	Includes old records as *R. furcata* (Lam.) Fr.	1	2	3	4
R. illota Romagn.		1	2	3	4
R. integra (L.) Fr.		1		3	4

Species	Note	1	2	3	4
R. ionochlora Romagn.		1	2		4
R. laurocerasi Melzer		1	2	3	
R. lilacea Quél. (*R. carnicolor* (Bres.) Rea)		1	2		
R. lutea (Huds.) Gray (*R. armeniaca* Cooke)		1	2	3	4
R. luteotacta Rea		1	2	3	4
R. luteoviridans Martin			2		
R. maculata Quél.		1	2	3	4
R. mairei Sing.		1	2	3	4
R. melliolens Quél.			2		
R. melitodes Romagn.		1			
R. mustelina Fr.	Better rejected until confirmed	1			4
R. nauceosa (Pers.) Fr.		1	2		4
R. nigricans (Bull.) Fr.		1	2	3	4
R. nitida (Pers.) Fr. (*R. venosa* Vel.)		1	2	3	4
R. ochroleuca (Pers.) Fr. (*R. citrina* Gill.)		1	2	3	4
R. odorata Romagn.		1			
R. olivacea (Schaeff.) Fr.		1	2	3	4
R. paludosa Britz.		1			
R. parazurea J. Schaeff.		1	2	3	4
R. pectinata (Bull.) Fr.		1	2		4
R. pectinatoides Peck		1		3	
R. pelargonia Nioll		1		3	
R. persicina Krombh. (*R. intactior* Schaeff.)		1	2		
R. pseudodelica Lange		1			
R. pseudointegra Arn. & Goris		1	2	3	4
R. puellaris Fr.		1	2	3	4
R. puellula Moll. & Schaeff.			2		
R. pulchella Borszczow		1	2	3	4
R. pumila Rouzeau & Massart		1			4
R. queletii Fr.		1		3	
R. queletii var. *flavovirens* (Bomm. & Rouss.) Maire					4
R. querceti Haas & Schaeff.		1			
R. raoultii (Quél.) Sing.		1	2		
R. romellii Maire		1	2		
R. rosea Pers. non Quél. (*R. lepida* Fr.)		1	2	3	4
R. rosea f. lactea (Pers.) Knudsen & Stordal (*R. lactea* Pers.)		1	2		
R. rosea var. *salmonea* (Melzer & Zvara)			2		
R. sanguinea (Bull.) Fr.		1	2	3	4
R. sardonia Fr. (*R. drimeia* Cooke)		1	2	3	4
R. seperina Dupain				3	
R. sericatula Romagn.		1	2		
R. solaris Ferdinandsen & Winge		1	2	3	
R. sororia (Fr.) Romell		1	2	3	4
R. subfoetens W.G. Smith		1			
R. turci Bres.		1		3	4
R. undulata Vel. (*R. atropurpurea* (Krombh.) Britz.; *R. krombholzii* Shaffer)		1	2	3	4
R. velenovskyi Melzer & Zvara		1	2	3	4
R. velutipes Vel. (*R. rosea* (Schaeff.) Quél. non Pers.).		1	2	3	4
R. versicolor J. Schaeff.		1	2	3	4
R. vesca Fr.		1	2	3	4
R. veternosa Fr. (*R. schiffneri* Sing.)			2	3	
R. vinosobrunnea (Bres.) Romagn.					4
R. vinosopurpurea J. Schaeff.				3	4
R. violacea Quél. ss Romagn.	Old records may have referred to *R. fragilis*		2		
R. violeipes Quél.		1	2	3	4
R. violeipes f. *citrina* Quél.			2		4
R. virescens (Schaeff.) Fr.		1	2	3	4
R. viscida Kudrna					4
R. xerampelina (Schaeff.) Fr. (*R. erythropus* Pelt.)		1	2	3	4

R. xerampelina var. *barlae* (Quél.) Melzer & Zvara	1		4
R. xerampelina var. *fusca* (Quél.) Melzer & Zvara		2 3	4
R. xerampelina var. *graveolens* (Rom.) Melzer & Zvara	1		4
R. zonatula Ebbesen & J. Schaeffer	1		
R. zvarae Vel.		3	

Simocybe Karst.
Saprophytes on wood and plant debris; *S. laevigata* (Favre) Orton is in East Anglia.

Simocybe centunculus (Fr.) Sing.	1	2 3	4
S. obscura (Romagn.) Reid	1		
S. rubi(Berk.) Sing. (*Naucoria effugiens* Quél.)	1		4
S. sumptuosa (Orton) Sing.	1	2 3	4

Squamanita Imbach
Terrestrial saprophytes; *S. pearsonii* Bas in Scotland.

Squamanita (*Dissoderma*) *paradoxa* (A.H. Smith & Sing.) Bas	4

Stigmatolemma Kalchbr.
Lignicolous saprophyte, minute, cupulate, sessile on a subiculum.

Stigmatolemma poriaeformis (DC.) W.B. Cook	1 2

Strobilomyces Berk.
Mycorrhizal.

S. strobilaceus (Scop.) Berk.	1 2 3 4

Strobilurus Sing.
Saprophytic on fallen cones of conifera.

Strobilurus esculentus (Wulf.) Sing.	1	2	4
S. stephanocystis (Hora) Sing.	1	2	4
S. tenacellus (Pers.) Sing.	1	2 3	4

Stropharia (Fr.) Quél.
Coprophilous or terrestrial saprophytes, see also *Psilocybe*. *S. albonitens* (Fr.) Quél. in Essex.

Stropharia aeruginosa (Curt.) Quél.	1	2 3	4
S. aurantiaca (Cooke) Imai	1		4
S. coronilla (Bull.) Quél.	1	2 3	4
S. cyanea (Bolt.) Tuomikoski	1		4
S. ferrii Bres. (*S. rugosoannulata* Farlow)	1		
S. inuncta (Fr.) Quél.	1	2	4
S. melanosperma (Bull.) Gill.	1		
S. pseudocyanea (Desm.) Orton (*S. albocyanea* (Fr.) Quél.)	1	3	4
S. semiglobata (Batsch) Quél. (*S. stercoraria* (Schum.) Quél.)	1	2 3	4

Tephrocybe Donk
Terrestrial saprophytes, some associated with burnt soil.

Tephrocybe albofloccosa Orton	1		4
T. anthracophila (Lasch) Orton	1	2 3	4
T. atrata (Fr.) Donk	1	2 3	4
T. baeospermum (Romagn.) Moser	1		
T. boudieri (Kühner & Romagn.) Moser	1		
T. confusa (Orton) Orton	1		4

T. coracina (Fr.) Moser	1		4
T. ferruginella (Pearson) Orton	1		4
T. fusispora (Hora) Moser	1		
T. gibberosa (J. Schaeff.) Orton (*T. ambusta* (Fr.) Donk)	1	2	4
T. impexa (Karst.) Moser	1		
T. inolens (Fr.) Moser	1	2	
T. mephitica (Fr.) Moser	1		
T. misera (Fr.) Moser	1		
T. mutabile (Favre) Moser	1		
T. ozes (Fr.) Moser	1		
T. palustris (Peck) Donk (*Collybia leucomyosotis* (Cke. & Smith) Sacc.)	1	2	3	4
T. putidella (Ort.) Ort. (*T. putida* (Fr.) Moser)	1		
T. rancida (Fr.) Donk	1	2	3	4
T. striaepilea (Fr.) Donk	1	2	
T. tesquorum (Fr.) Moser (*T. tylicolor* (Fr.))	1	2	3	4

Tricholoma (Fr.) Kummer

Terrestrial saprophytes or mycorrhizal, several apparently confined to highland pine woods.

Tricholoma acerbum (Bull.) Quél.	1	2	3	4
T. albobrunneum (Pers.) Kummer　　　Doubtful, confused with *T. ustale*	1	2		4
T. album (Schaeff.) Kummer	1	2	3	4
T. argyraceum (Bull.) Gill. (*T. scalpturatum* (Fr.) Quél.)	1	2	3	4
T. atrosquamosum (Chev.) Sacc.	1	2	3	4
T. cingulatum (Almf.) Jacobasch	1	2		4
T. columbetta (Fr.) Kummer	1		3	4
T. distinguendum Lundell	1		
T. flavovirens (Pers.) Lundell (*T. equestre (L.) Kummer*)	1		
T. fulvum (DC.) Sacc. (*T. flavobrunneum* (Fr.) Kummer)	1	2	3	4
T. gausapatum (Fr.) Quél.	1		
T. imbricatum (Fr.) Kummer	1		3	4
T. inamoenum (Fr.) Gill.		2	
T. inocybeoides Pearson	1	2	
T. lascivum (Fr.) Gill.	1	2		4
T. orirubens Quél.	1	2		4
T. populinum Lange	1		
T. portentosum (Fr.) Quél.	1	2	3	
T. psammopus (Kalchbr.) Quél.	1	2	
T. resplendens (Fr.) Karst.	1	2	
T. saponaceum (Fr.) Kummer	1	2	3	4
T. saponaceum var. *squamosum* (Cooke) Rea	1	2		4
T. sciodes (Pers.) Martin	1	2		4
T. sejunctum (Sow.) Quél.	1	2		4
T. spinulosum Kühner & Romagn.	1	2		4
T. squarrulosum Bres.	1	2		4
T. sulphureum (Bull.) Kummer	1	2	3	4
T. sulphureum var. *bufonium* (Pers.) Quél.	1	2	
T. terreum (Schaeff.) Kummer	1	2	3	4
T. ustale (Fr.) Kummer	1	2	
T. ustaloides Romagn.	1	2	
T. vaccinum (Pers.) Kummer	1		3	4
T. virgatum (Fr.) Kummer	1	2	3	4

Tricholomopsis Sing.

Lignicolous saprophytes, including *Megacollybia* Kotl. & Pouz. *T. decora* (Fr.) Sing. in Scotland.

Tricholomopsis (Megacollybia) platyphylla (Pers.) Sing.	1	2	3	4
T. rutilans (Schaeff.) Sing.	1	2	3	4

Tubaria (W.G. Smith) Gill.
Saprophytes, terrestrial or on plant remains; *T. trigonophylla* (Lasch) Fayod is in Essex.

	1	2	3	4
Tubaria autochthona (B. & Br.) Sacc.	1	2	3	
T. conspersa (Pers.) Fayod	1	2		4
T. furfuracea (Pers.) Gill (*T. pellucida* (Bull.) Gill., *T. hiemalis* Bon)	1	2	3	4
T. minutalis (Romagn.) Kühner & Romagn.	1			
T. pallidispora Lange	1			4

Tylopilus Karst.
Mycorrhizal.

	1	2	3	4
Tylopilus felleus (Bull.) Karst.	1	2	3	4

Uloporus Quél.
Mycorrhizal: *Gyrodon* Opat. is a synonym.

	1	2
Uloporus lividus (Bull.) Quél.	1	2

Volvariella Speg.
Saprophytes, lignicolous or terrestrial, or hyperparasites.

	1	2	3	4
Volvariella bombycina (Schaeff.) Sing.	1	2		4
V. caesiotincta Orton	1			
V. hypopithys (Fr.) Moser (*V. pubescentipes* (Peck) Sing.)	1	2		4
V. media (Schum.) Sing.	1	2		
V. murinella (Quél.) Moser	1	2	3	4
V. parvula (Weinm.) Orton (*V. pusilla* (Pers.) Quél.)	1	2		
V. reidii Heinemann (*V. parvispora* Reid non Heinemann)	1			4
V. speciosa (Fr.) Sing. (*V. gloiocephala* (DC.) Gill.)	1	2		4
V. surrecta (Knapp) Sing. (*V. loveiana* (Berk.) Gill.)	1			
V. taylori (Berk.) Sing.	1	2	3	4
V. volvacea (Bull.) Sing.	1			4

Xeromphalina Kühner & Maire
Lignicolous saprophytes in highland coniferous forests. Records of *X. campanella* (Batsch) Maire from Surrey and *X. cauticinalis* (Fr.) Kühner & Maire from Kent must be rejected until voucher material is forthcoming.

APHYLLOPHORALES including Tulasnellaceae

See also *Microstoma*

Abortiporus Murr.
Saprophytes, including *Heteroporus* Lazaro.

	1	2	3	4
Abortiporus biennis (Bull.) Sing. (*Polyporus rufescens* Fr.)	1	2	3	4

Acanthophysium (Pilat) Cunningham
Saprophytes, segregated from *Aleurodiscus*.

	1	2	3	4
Acanthophysium acerinum (Pers.) Cunningham	1	2	3	4
A. aurantium (Pers.) Cunningham	1			4

Aleurodiscus Rabh.
Saprophytes.

	4
Aleurodiscus amorphus (Fr.) Schroet.	4

Amphinema Karst.
  Saprophytes.

Amphinema byssoides (Fr.) J. Erikss.   1   2   3   4

Amyloporia Bond. & Sing.
  Lignicolous saprophytes.

Amyloporia lenis (Karst.) Bond. & Sing.   2   4
A. xantha (Fr.) Bond. & Sing.   1   2   3   4

Amylostereum Boidin
  Lignicolous saprophytes, especially on coniferae.

Amylostereum chailletii (Pers.) Boidin   1
A. laevigatum (Fr.) Boidin   1   2   3   4

Antrodia Karst.
  Lignicolous saprophytes, including *Coriolellus* Murr.

Antrodia albida (Fr.) Donk   1   2
A. ramentacea (B. & Br.) Donk   1
A. serialis (Fr.) Donk   1   2
A. serpens (Fr.) Karst.   1   2   3

Antrodiella Ryvarden & Johansen
  Saprophytes.

Antrodiella onychoides (Egel.) Niemala   4
A. romellii (Donk) Niemalä (*Poria romellii* Donk)   1
A. semisupina (B. & C) Ryvarden   1   2   4

Aphanobasidium Julich
  Lignicolous saprophytes, see *Xenasmatella*.

Aphelaria Corner
  Clavarioid saprophytes, *A. tuberosa* (Grev.) Corner in East Anglia.

Asterostroma Massee
  Saprophytes; *A. medium* Bres. in Buckinghamshire.

Asterostroma laxum Bres.   1   2

Athelia Pers.
  Saprophytes, see also *Piloderma* and *Ceraceomyces*.

Athelia acrospora Julich   4
A. arachnoidea (Berk.) Juel. (*A. bispora* (Schroet.) Donk)   1   2
A. coprophila (Wakef.) Jul.   1
A. decipiens (v. Höhn. & Litsch.) Erikss.   1
A. epiphylla Pers. (*Corticium centrifugum* auct.)   1   2   3   4
A. fuciforme (Wakef.) Burds.   1
A. hypochnoidea Jul.   2
A. macrospora (Bourd. & Galz.) Christ.   2
A. olivaceoalba (Bourd. & Galz.) Donk (*Leptosporomyces fuscostratus* (Burt) Hjort.)   2
A. nivea Jul.   1   2   3

Athelopsis Oberw.
  Saprophytes, see *Luellia*.

Aurantioporus Murr.
Lignicolous saprophytes.

Aurantioporus fissilis (Berk. & Curt.) Jahn (*Polyporus albosordescens* Rom.)	1	2	3	4

Auriculariopsis Maire
Lignicolous saprophytes.

Auriculariopsis ampla (Lév.) Maire (*Cytidia flocculenta* auct.)	1		3

Auriscalpium Gray
Saprophyte on fallen cones of coniferae.

Auriscalpium vulgare Gray (*Hydnum auriscalpium* (L.) Fr.)	1	2	3	4

Bankera Coker & Beers
Saprophytes in coniferous forests, *B. fuligineoalba* (Schmidt) Pouzar in Berkshire.

Basidioradulum Nobles
Lignicolous saprophyte.

Basidioradulum radula (Fr.) Nobles (*Radulum orbiculare* Fr.; *R. quercinum* Fr.)	1	2	3	4

Bensingtonia Ingold
Saprophyte, basidioid yeast.

Bensingtonia ciliata Ingold	On *Hirneola* carpophores	4

Bjerkandera Karst.
Lignicolous saprophytes or weak facultative parasites of angiosperm trees.

Bjerkandera adusta (Willd.) Karst.	1	2	3	4
B. fumosa (Pers.) Karst.	1		3	4

Boidinia Stolpers & Hjortstam
Saprophyte.

Boidinia furfuracea (Bres.) Stolpers & Hjortstam	3

Boletopsis Fayod
Terrestrial saprophyte; *B. subsquamosa* (Fr.) Kotlaba & Pouzar, rare in Highlands only.

Botryobasidium Donk
Corticioid saprophytes with anamorphs in *Acladium* Link.

Botryobasidium aureum Parmasto (*Oidium aureum* Fr.)	1	2		
B. botryosum (Bres.) J. Erikss.	1	2	3	4
B. candicans J. Erikss. (*Acladium candicans* Sacc.)	1		3	4
B. conspersum J. Erikss. (*A. conspersum* Link)	1	2	3	
B. danicum Erikss. & Hjortstam	1			
B. laeve (J. Erikss.) Parm.		2		
B. pruinatum (Bres.) J. Erikss.	1	2		
B. subcoronatum (v. Höhn. & Lits.) Donk	1	2	3	4

Botryohypochnus Donk
Tomentelloid saprophyte.

Botryohypochnus isabellinus (Fr.) Erikss.	1	2

Brevicellicium Larsson & Hjortstam
Corticioid saprophytes.

Brevicellicium olivascens (Bres.) Larsson & Hjortstam (*Grandinia mutabilis* auct.)   1   2

Buglossoporus Kotlaba & Pouzar
Saprophytic on angiosperm trees.

Buglossoporus quercina (Schrad.) Kotl. & Pouz. (*B. pulvinus* (Pers.) Donk)   1   2

Bulbillomyces Jul.
Ubiquitous saprophyte with *Aegerita* anamorph.

Bulbillomyces farinosus (Bres.) Jul. (*Aegerita candida* Pers.)   1   2   3   4

Bullera Derx.
Basidioid yeast, *B. alba* (Hanna) Derx widespread amongst Erysiphales and Uredinales.

Byssocorticium Bond. & Sing.
Lignicolous corticioid saprophytes.

Byssocorticium atrovirens (Fr.) Bond. & Sing.   1   2   3   4
B. pulchrum (Lundell) Christ.   1
B. terrestre (DC.) Bond. & Sing.   1   2

Byssomerulius Parmasto
Lignicolous saprophytes including *Ceraceomerulius* Erikss. & Ryvard.

Byssomerulius corium (Fr.) Parm.   1   2   3   4
B. serpens (Fr.) Parm. (*Merulius serpens* Fr.)   1   4

Caldesiella Sacc.
Tomentelloid lignicolous saprophytes.

Caldesiella crinalis (Fr.) Rea (*C. ferruginosa* (Fr.) Sacc.)   1   2   4
C. italica Sacc.   1   2

Cartilosoma Kotlaba & Pouzar
Poroid saprophyte, segregate from *Antrodia*; *C. subsinuosa* (Bres.) Kotl. & Pouz. Highlands only.

Cejpomyces Svrček & Pouzar
Corticioid saprophyte.

Cejpomyces terrigenus (Bres.) Svrček & Pouzar   1

Ceraceomyces Jul.
Corticioid saprophytes.

Ceraceomyces sublaevis (Bres.) Jul. (*Corticium microsporum*(Karst.) B. & G.)   1   2   3   4
C. tessulatus (Cooke) Jul.   2   3

Ceratellopsis Konr. & Maubl.
Clavarioid saprophytes.

Ceratellopsis aculeata (Pat.) Corner   1

Ceratobasidium Rogers
Tulaselloid saprophytes.

	1	2	3	4
Ceratobasidium atratum (Bres.) Rogers (*Hydrabasidium subolivaceum* (Peck.) Eriks. & Ryv.)			3	
C. cornigerum (Bourd.) Rogers	1	2	3	

Ceriporia Donk
Lignicolous saprophytes, see also *Meruliopsis*.

	1	2	3	4
Ceriporia excelsa (Lundell) Parm. (*Poria rhodella* auct.)	1			
C. mellita (Bourd.) Bond. & Sing. (*Poria viridans* v. *aurantiocarnescens*)	1	2		
C. reticulata (Hoffm.) Domanski	1	2	3	4
C. viridans (B. & Br.) Donk	1	2	3	4

Ceriporiopsis Domanski
Lignicolous saprophytes, *C. incarnata* (A. & S.) Domanski in Berkshire.

	1	2	3	4
Ceriporiopsis aneirinus (Somm.) Dom. (*Poria metamorphosa* (Fuck.) Cooke)	1			4
C. gilvescens (Bres.) Dom.	1	2	3	4
C. placenta (Fr.) Dom. Needs confirmation	1			

Cerrena Gray
Lignicolous saprophyte.

	1	2	3	4
Cerrena unicolor (Bull.) Murr. (*Daedalea unicolor*) (Bull.) Fr.	1	2		4

Chaetoporellus Bond. & Sing.
Lignicolous saprophyte, *C. latitans* (Bourd. & Galz.) Sing. in Scotland.

Chaetoporus Karst. see *Oxyporus*

Chondrostereum Pouzar
Ubiquitous lignicolous saprophyte and facultative parasite especially of Rosaceous trees.

	1	2	3	4
Chondrostereum purpureum (Pers.) Pouzar	1	2	3	4

Christiansenia Hauerslev see *Syzygospora*

Clavaria Fr.
Saprophytes, mainly terrestrial.

	1	2	3	4
Clavaria acuta Fr.	1	2	3	4
C. affinis Pat. & Doass.	1			
C. argillacea Fr.	1	2		4
C. falcata Fr. var. *citronipes* Quél.				4
C. fumosa Fr.	1			4
C. guilleminii Bourd. & Galz.	1			
C. incarnata Weinm.	1	2		4
C. purpurea Fr.				4
C. rosea Fr. var. *subglobosa* Corner	1			
C. straminea Cotton	1		3	4
C. tenuipes B. & Br.	1	2	3	4
C. vermicularis Fr.	1	2	3	4
C. zollingeri Lév.		2		4

Clavariadelphus Donk
Lignicolous or terrestrial saprophytes.

Clavariadelphus fistulosus (Fr.) Corner	1	2	3	4
C. fistulosus var. *contortus* (Fr.) Corner	1		3	4
C. junceus (Fr.) Corner	1	2		4
C. ligula (Schaeff.) Donk	1			
C. pistillaris (L.) Donk	1	2		4

Clavariella Karst. see Ramaria

Clavicorona Schroet.
Saprophytes.

Clavicorona taxophila (Thom) Doty	1			4

Clavulina Schroet.
Saprophytes.

Clavulina cinerea (Bull.) Schroet.	1	2	3	4
C. cinerea f. *subcristata* Bourd. & Galz.				4
C. cristata (Holmsk.) Schroet.	1	2	3	4
C. cristata var. *subcinerea* Donk	1			
C. cristata subsp. *coralloides* (L.) Corner				4
C. rugosa (Bull.) Schroet.	1	2	3	4

Clavulinopsis van Overeem
Terrestrial saprophytes; *C. rufipes* (Atk.) Corner & *C. subtilis* (Fr.) Corner in Berkshire.

Clavulinopsis asterospora (Pat.) Corner	1			
C. cinereoides (Atk.) Corner	1	2		
C. corniculata (Schaeff.) Corner	1	2	3	4
C. fusiformis (Sow.) Corner	1	2	3	4
C. helvola (Pers.) Corner (*Clavaria inaequalis* auct., angl.)	1	2	3	4
C. luteoalba (Rea) Corner	1	2	3	4
C. luteoalba var. *latispora* Corner				4
C. luteo-ochracea (Cav.) Corner	1			4
C. pulchra (Peck) Corner (*Clavaria persimilis* Cotton)	1	2	3	4
C. tenerrima (Massee) Corner	1			
C. umbrinella (Sacc.) Corner	1	2	3	4

Coltricia Gray
Stipitate poroid saprophytes sometimes associated with burnt ground.

Coltricia montagnei (Fr.) Bond.	Very doubtful record	1			
C. perennis (L.) Murr.		1	2	3	4

Coniophora DC.
Lignicolous saprophytes.

Coniophora arida (Fr.) Karst.	1	2	3	4
C. arida var. *flavobrunnea* Bres.				4
C. betulae Karst.	1	2		
C. bourdotii Bres.		2		
C. laxa Quél.		2		
C. mustialaensis (Karst.) Massee		2		
C. puteana (Schum.) Karst.	1	2	3	4
C. suffocata (Peck) Massee				4
C. sulfurea (Fr.) Massee	1			

Coniophorella Karst.
Lignicolous saprophyte.

Coniophorella olivacea (Fr.) Karst.	1	2	3	

Coriolus Quél.
Poroid lignicolous saprophytes or feebly facultatively parasitic, often reunited with *Trametes*. *C. pubescens* (Schum.) Quél. has been reported from Hampshire.

Coriolus hirsutus (Wulf.) Quél.	1	2	3	4
C. velutinus (Pers.) Quél.	1			
C. versicolor (L.) Quél.	1	2	3	4
C. zonatus (Nees) Quél.	1	2		

Cotylidia Karst.
Stereoid, saprophytic.

Cotylidia pannosa (Sow.) Reid		2	4

Cristella Pat.
Corticioid saprophytes, largely lignicolous, equals *Trechispora* auct. See also *Brevicellicium*.

Cristella alnicola (Bourd. & Galz.) Donk	1			
C. amianthina (Bourd. & Galz.) Donk	1			
C. byssinella (Bourd.) Donk			3	4
C. candidissima (Schw.) Donk	1	2	3	4
C. cohaerens (Schw.) Julich & Stalpers	1			
C. confinis (Bourd. & Galz.) Donk	1	2	3	
C. farinacea (Pers.) Donk	1	2	3	4
C. fastidiosa (Pers.) Brinkmann	1			4
C. filia (Bres.) Donk			3	
C. nivea (Pers.) Christ. Needs confirmation	1			
C. praefocata (Bourd. & Galz.) Christ.	1			
C. sphaerospora (Maire) Donk	1	2	3	4
C. submutabilis (v. Höhn. & Donk	1			
C. sulphurea (Pers.) Donk (*Phlebiella vaga* (Fr.) Karst.)	1	2	3	4

Cristinia Parmasto
Corticioid saprophytes.

Cristinia helvetica (Pers. Parm.	1		3	4
C. mucida (Bourd. & Galz.) Erikss. & Ryvarden				4

Crustoderma Parmasto
Corticioid saprophyte.

Crustoderma dryinum (Berk. & Curt) Parm.	4

Crustomyces Julich
Corticioid saprophyte.

Crustomyces expallens (Bres.) Hjortstam	1

Cylindrobasidium Julich
Corticioid lignicolous saprophyte.

Cylindrobasidium evolvens (Fr.) Julich (*Corticium laeve* auct. angl.)	1	2	3	4

Cyphellostereum Reid
  Saprophyte.

Cyphellostereum laeve (Fr.) Reid   1

Dacryobolus Fr.
  Lignicolous odontioid saprophyte.

Dacryobolus sudans (Fr.) Fr. (*Porothelium confusum* B. & Br.)   1

Daedalea Fr.
  Lignicolous saprophyte.

Daedalea quercina (L.) Fr.   1 2 3 4

Daedaleopsis Schroet.
  Lignicolous saprophyte, especially on *Alnus* and *Salix*.

Daedaleopsis confragosa (Bolt.) Schroet. (*Trametes rubescens* (A. & S.) Fr.)   1 2 3 4

Datronia Donk
  Irregularly poroid, coriaceous, resupinate saprophyte.

Datronia mollis (Sommerf.) Donk (*Trametes mollis* (Sommerf.) Fr.)   1 2 3 4

Dichomitus Reid
  Poroid lignicolous saprophyte; *D. campestris* (Quél.) Dom. & Orl.

Dichostereum Pilat
  Lignicolous saprophyte, segregate from *Vararia*.

Dichostereum durum (Bourd. & Galz.) Pilat   2

Donkiopora Kotlaba & Pouzar
  Lignicolous saprophyte, *D. expansa* (Desm.) Kotlaba & Pouzar on household timber in
E. Anglia.

Echinotrema Parker Rhodes
  Echinotrema clanculare Parker Rhodes in nesting burrows of seabirds in Wales.

Epithele (Pat.) Pat.
  Corticioid saprophyte on *Carex* and *Typha*; *E. typhae* (Pers.) Pat. in Hampshire.

Exobasidiellum Donk
  Parasite; *E. culmigenum* Webster & Reid on *Dactylis glomerata* in the Midlands.

Exobasidium Woronin
  Obligate parasites; *E. vaccinii* (Fuck.) Wor. on *Vaccinium vitis-idaea* in highland areas.

Exobasidium camelliae Shirai	Casual alien on flowers of *Camellia* spp.			3	
E. japonicum Shirai	Alien on *Rhododendron*, usually imported pot plants	1	2	3	4
E. myrtilli Siegmund on *Vaccinium myrtillus*		1			
E. rhododendri (Fuck.) Cramer	Alien on *Rhododendron*	1		3	

Fibrodontia Parmasto
  Lignicolous saprophyte.

Fibrodontia gossypina Parm. (*Hyphodontia stipata* (Fr.) Gilb.)   1

Fibroporia Parmasto
Poroid resupinate saprophyte.

Fibroporia vaillantii (DC.) Parm. (*Fibuloporia vaillantii* (DC.) Bond. & Sing.)   1 2

Fibulomyces Julich
Corticioid saprophyte.

Fibulomyces mutabilis (Bres.) Jul. (*Corticium mutabile* Bres.)   2

Fibuloporia Bond. & Sing.
Saprophyte.

Fibuloporia wynnei (B. & Br.) Bond. & Sing.   1 2   4

Fistulina Bull.
Saprophyte and facultative parasite of *Castanea* & *Quercus*.

Fistulina hepatica (Schaeff.) Fr.   1 2 3 4

Flaviporus Murr.
Lignicolous saprophyte probably alien. *F. brownei* (Humb.) Donk. Recorded from Wiltshire and mine timbers elsewhere.

Fomes (Fr.) Fr.
Saprophyte and facultative parasite of *Fagus*.

Fomes fomentarius (L.) Kickx   All old records apply to *Ganoderma* spp.   1   4

Fomitopsis Karst.
Lignicolous saprophytes, see also *Perenniporia*.

Fomitopsis rosea (A. & S.) Karst.   Alien on glasshouse or domestic timber   1   4
F. pinicola (Sw.) Karst.   1

Funalia Pat.
Poroid lignicolous saprophyte.

Funalia gallica (Fr.) Bond. & Sing.   1 2   4

Ganoderma Karst.
Woody perennial facultative parasites on trees.

Ganoderma adspersum (Schultz.) Donk (*G. europaeum* Stey.; *G. australe* auct.)   1 2   4
G. applanatum (Pers.) Pat.   Commonly confused with the above   1 2   4
G. atkinsonii Jahn, Kotlaba & Pouzar (*G. valesiacum* auct. angl.)   1 2   4
G. lucidum (Leyss) Karst.   1 2   4
G. pfeifferi Bres.   1 2   4
G. resinaceum Boud.   1 2   4

Gloeocystidiellum Donk
Corticioid lignicolous saprophytes see also *Boidinia, Gloiothele, Vesiculomyces*.

Gloeocystidiellum leucoxanthum (Bres.) Boidin   1 2
G. luridum (Bres.) Boidin   1
G. porosum (Berk. & Curt.) Donk   1 2 3 4

Gloeophyllum Karst.
Lamello-poroid woody lignicolous saprophytes on coniferous timbers.

Gloeophyllum abietinum (Bull.) Karst. (*Lenzites abietina* (Bull.) Fr.)   1   2
G. saepiarium (Wulf.) Karst. (*L. saepiaria* (Wulf.) Fr.)   1   2   4

Gloeoporus Mont.
    Poroid, lignicolous saprophyte, *G. dichrous* (Fr.) Bres. in the Highlands.

Gloiothele Bres.
    Segregate from *Gloeocystidiellum*.

Gloiothele lactescens (Berk.) Hjortstam   1   2   4

Grifola Gray
    Stipitate poroid saprophytes.

Grifola frondosa (Schrank) Gray (*Polyporus intybaceus* Fr.)   1   2   4
G. umbellata (Pers.) Pilat   1   4

Hapalopilus Karst.
    Imbricate, poroid, lignicolous saprophyte.

Hapalopilus nidulans (Fr.) Karst. (*Polyporus rutilans* (Pers.) Fr.)   1   2   3   4

Henningsomyces I. Kuntze
    Solenioid lignicolous saprophytes.

Henningsomyces candidus (Pers.) O. Kuntze (*Solenia candida* Fr.)   1   2   3   4

Hericium Pers.
    Hydnoid lignicolous saprophytes or facultative parasites, including *Creolophus* Karst.

Hericium (Creolophus) cirrhatum (Pers.) Nikol.   1   2   4
H. coralloides (Scop.) Gray   1
H. (Cr.) diversidens (Fr.) Nicol.   1   3
H. erinaceus (Bull.) Pers.   1   2   4
H. ramosum (Bull.) Let.   Perhaps only a form of *H. coralloides*   1

Heterobasidion Bref.
    Active poroid root parasite of woody plants.

Heterobasidion annosum (Fr.) Bref. (*Trametes radiciperda* Hart.; *Fomes annosus* Cke.)   1   2   3   4

Hydnellum Karst.
    Stipitate hydnoid saprophytes, additional species in the Highlands.

Hydnellum ferrugineum (Fr.) Karst.   1   3
H. spongiosipes (Peck) Pouzar   1   4
H. velutinum (Fr.) Karst.   1
H. velutinum var. *scrobiculatum* (Fr.) Maas Geesteranus   1   4
H. velutinum var. *zonatum* (Fr.) Maas Geesteranus   1   3

Hydnum L.
    Mycorrhizal.

Hydnum repandum L.   1   2   3   4
H. repandum var. *rufescens* (Fr.) Barla   1   2   3   4

Hymenochaete Lév.
    Resupinate to effusoreflexed corticioid saprophytes with finely setose hymenium.

Hymenochaete cinnamomea (Pers.) Bres.   1   2

	1	2	3	4
H. corrugata (Fr.) Lév. (*H. fuliginosa* (Pers.) Lév.)	1	2	3	4
H. cruenta (Pers.) Donk (*H. mougeotii* (Fr.) Massee)	1			
H. rubiginosa (Schrad.) Lév.	1	2	3	4
H. tabacina (Sow.) Lév.	1	2	3	4

Hyphoderma Wallr.
Corticioid lignicolous saprophytes.

	1	2	3	4
Hyphoderma argillaceum (Bres.) Donk	1	2	3	
H. definitum (Jacks.) Donk	1			
H. guttuliferum (Karst.) Donk	1			
H. pallidum (Bres.) Donk	1			
H. poloniense (Bres.) Donk		2		
H. praetermissum (Karst.) Erikss. & Ryvard. (*H. tenue* (Pat.) Donk)	1	2	3	4
H. puberum (Fr.) Wallr.	1	2		4
H. reticulata (Wakef.) Donk	1	2	3	4
H. roseocremeum (Bres.) Donk	1	2	3	
H. salicicola (Christ.) Christ.		2		
H. sambuci (Pers.) Jul. (*Hyphodontia sambuci* (Pers.) Erikss.)	1	2	3	4
H. setigerum (Fr.) Donk	1	2	3	4

Hyphodermella Erikss. & Ryvard.
Odontioid lignicolous saprophyte.

	1	2	3	4
Hyphodermella corrugata (Fr.) Erikss. & Ryvard.		2		4

Hyphodontia J. Erikss.
Corticioid-odontioid lignicolous saprophytes, see also *Fibrodontia, Lagarobasidium.*

	1	2	3	4
Hyphodontia alutaria (Burt) Erikss.	1	2		
H. arguta (Fr.) Erikss.	1	2		4
H. barbajovis (Fr.) Erikss.	1	2	3	4
H. breviseta (Karst.) Erikss.	1			
H. bugellensis (Ces.) Erikss.		2		
H. crustosa (Fr.) Erikss.	1	2	3	
H. nespori (Bres.) Erikss. & Hjortstam (*H. papillosa* auct. angl.)	1	2	3	4
H. pallidula (Bres.) Erikss.	1	2	3	4
H. quercina (Fr.) Erikss.	1	2	3	4
H. subalutacea (Karst.) Erikss.	1	2	3	

Hypochnella Schroet.
Lignicolous saprophyte.

	1	2	3	4
Hypochnella violacea (Auersw.) Schroet.	1	2		4

Hypochniciellum Hjortstam & Ryvarden
Lignicolous saprophytes.

	1	2	3	4
Hypochnieciellum molle (Fr.) Hjortstam	1	2	3	4
H. ovoideum (Julich) Hjortstam & Ryvard. (*Leptosporomyces ovoideus* (Jul.))			3	
H. subillaqueatum (Litsch.) Hjortstam (*Corticium subillaqueatum* Lits.)			3	

Hypochnicium Erikss.
Corticioid lignicolous saprophytes; *H. (Gloeohypochnicium) analogum* (Bourd. & Galz.) in Essex.

	1	2	3	4
Hypochnicium albostramineum (Bres.) Hallenb.	1			
H. bombycinum (Sommerf.) Erikss.	1	2		
H. eichleri (Bres. Erikss. & Ryvard.	1			
H. geogenium (Bres.) Erikss.		2	3	4

H. punctulatum (Cooke) Erikss. (*Corticium wakefieldiae* Bres.)	1	2	3	
H. erikssonii Hallenb. & Hjortstam (*H. sphaerosporum* (v. Höhn. & LIts. ss Erikss.)	1	2		4
H. (Granulobasidium) vellereum (Ell. & Crag.) Parmasto	1	2		4

Hypochnopsis Karst.
See *Coniophora mustialaensis*.

Incrustoporia Domanski
Resupinate to effusoreflexed poroid lignicolous saprophytes; *I. tschulymica* (Pilat) Dom. in Hampshire, *I. alutacea* (Lowe) Reid in Scotland.

Incrustoporia nivea (Jungh.) Ryvarden (*Leptotrimitus semipileatus* (Peck) Pouz.)	1	2	3	4

Inonotus Karst.
Polyporoid to fomoid facultative parasites of woody Angiosperms.

Inonotis cuticularis (Bull.) Karst.	1	2		4
I. dryadeus (Pers.) Murr.	1	2		4
I. hispidus (Bull.) Karst.	1	2	3	4
I. nodulosus (Fr.) Karst.		2		4
I. obliquus (Pers.) Pilat	1			
I. radiatus (Sow.) Karst.	1	2	3	4

Irpex Fr.
Lignicolous saprophyte.

Irpex lacteus (Fr.) Fr.	Needs confirmation	2

Irpicodon Pouzar
Lignicolous saprophyte on coniferae, *I. pendulus* (A. & S.) Pouz. reported from Yorkshire.

Ischnoderma Karst.
Polyporoid lignicolous saprophyte on coniferae.

Ischnoderma resinosum (Schrad.) Karst. (*I. benzoinum* (Wahl.) Karst.)	1	2	3	4

Itersonilia Derx
Basidiomycetoid yeasts. Reputed cause of widespread "canker" in *Pastinaca*.

Itersonilia pastinacae Cannon

I. perplexans Derx	1	2	4

Jaapia Bres.
Corticioid lignicolous saprophyte.

Jaapia argillacea Bres.	1

Junghunia Corda
Poroid lignicolous saprophyte.

Junghunia nitida (Pers.) Ryvard. (*Poria (Chaetoporus) eupora* (Karst.) Cke.)	1	3

Knieffiella Karst.
Tomentelloid saprophyte *K. bombycina* Karst.

Laeticorticium Donk
Corticioid lignicolous saprophyte.

Laeticorticium roseum (Pers.) Donk   1

Laetiporus Murr.
    Polyporoid facultative parasite on *Taxus* and angiosperm trees.

Laetiporus sulphureus (Bull.) Bond. & Sing.   1   2   3   4

Laetisaria Burdsall
    Facultative parasite of Graminiae.

Laetisaria fuciformis (Mc. Alp.) Burdsall   1   3   4

Lagarobasidium Julich
    Peniophoroid lignicolous saprophytes.

Lagarobasidium detriticum (Bourd. & Galz.) Jul.   1   2   3

Laxitextum Lentz
    Sterioid lignicolous saprophyte.

Laxitextum bicolor (Pers.) Lentz   1   2   4

Lazulinospora Burdsall & Larsen see *Tomentella*

Lentaria Corner
    Lignicolous clavarioid saprophytes, *L. delicata* (Fr.) Corner in Buckinghamshire.

Lenzites Fr.
    Lamellate, coriaceous, dimidiate saprophyte on woody angiosperms.

Lenzites betulina (L.) Fr.   1   2   3   4
L. betulina f. *flaccida* (Fr.) Bond.   1   3   4
L. betulina f. *variegata* (Fr.) Bond.   1   4

Leptosporomyces Julich
    Corticioid lignicolous saprophytes, compare *Athelia*.

Leptosporomyces galzinii (Bourd.) Julich   1   2   3

Leucogyrophana Pouzar
    Merulioid lignicolous saprophytes.

Leucogyrophana mollusca (Fr.) Pouzar (*Merulius molluscus* Fr.)   1   2   3   4
L. pinastri (Fr.) Ginns & Weresub (*M. pinastri* (Fr.) Burt)   1   4
L. pulverulenta (Fr.) Ginns (*M. pulverulentus* Fr. *Serpula tignicola* (Harms.) Christ.)   3

Lindtneria Pilat
    Poroid saprophyte.

Lindtneria trachyspora (Bourd. & Galz.) Pilat   1

Litschauerella Oberw.
    Peniophoroid lignicolous saprophytes, *L. clematidis* (Bourd. & Galz.) Erikss. & Ryvarden.

Litschauerella clematidis (Bourd. & Galz.) Erikss. & Ryvard.   2

Luellia Larsson & Hjortstam
Corticioid lignicolous saprophytes; *L. cystidiata* Hauerslev in Wales.

	1	2	3	4
Luellia (Athelopsis) lembospora (Bourd.) Julich		2		
L. recondita (Jackson) Larsen & Hjortstam			3	

Membranomyces Julich
Corticioid saprophyte.

	1	2	3	4
Membranomyces spurius (Bourd.) Julich (*Corticium spurium* Bourd.)				4

Meripilus Karst.
Polyporoid, lignicolous, facultative parasite of angiospermous trees, especally *Fagus*.

	1	2	3	4
Meripilus giganteus (Pers.) Karst.	1	2	3	4

Meruliopsis Bondartzev
Merulioid lignicolous saprophytes.

	1	2	3	4
Meruliopsis (Merulioporia) purpurea (Fr.) ? Not distinct from the next Bond.	1	2		
M. taxicola (Pers.) Bond.	1	2	3	4

Merulius Fr.
Lignicolous saprophyte.

	1	2	3	4
Merulius tremellosus (Schrad.) Fr.	1	2	3	4

Mucronella Fr.
Lignicolous clavarioid saprophytes.

	1	2	3	4
Mucronella aggregata (Fr.) Fr.	1	2	3	
M. calva (A. & S.) Fr.	1		3	

Multiclavula Petersen
Terrestrial clavarioid associate of Cyanophyceae; *M. vernalis* (Schw.) Petersen in Shetland.

Mycoacia Donk
Lignicolous, resupinate, hydnoid saprophytes.

	1	2	3	4
Mycoacia aurea (Fr.) Erikss. & Ryvarden (*M. stenodon* (Pers.) Donk)	1	2	3	4
M. fuscoatra (Fr.) Donk	1	2	3	
M. nothofagi (Cunningham) Erikss. & Ryvarden (*Odontia nothofagi* Cunn.)	1	2		
M. uda (Fr.) Donk	1	2	3	4

Oxyporus (Bourd. & Galz.) Donk
Resupinate to dimidiate, poroid, lignicolous saprophytes or facultative parasites, includes *Chaetoporus* Karst. in part.

	1	2	3	4
Oxyporus (Chaetoporus) corticola (Fr.) Ryvarden	1			4
O. latemarginatus (Dur. & Mont.) Donk (*Chaetoporus ambiguus* (Bres.) Bond. & Sing.)	1	2		4
O. obducens (Pers.) Donk	1	2	3	4
O. pearsonii (Pilat) E. Komar.		2		
O. populinus (Schum.) Donk	1	2	3	

Parvobasidium Julich
Corticioid lignicolous saprophyte.

	1	2	3	4
Parvobasidium cretatum (Bourd. & Galz.) Julich		2		

Paullicorticium Erikss.
 Sistotremoid lignicolous saprophyte.

	1			
Paullicorticium pearsonii (Bourd.) Erikss.	1			

Peniophora Cooke
 Resupinate lignicolous saprophytes.

	1	2	3	4
Peniophora boidinii Reid	1			
P. cinerea (Fr.) Cooke	1	2	3	4
P. eichleriana (Bres.) Bourd. & Galz.		2		
P. erikssonii Boidin (*P. aurantiaca* (Bres.) v. Höhn. & Lits. ss auct. angl.)	1			
P. incarnata (Pers.) Karst.	1	2	3	4
P. laeta (Fr.) Boidin	1			
P. limitata (Fr.) Cooke (*P. fraxinea* (Pers.) Lundell)	1	2	3	4
P. lycii (Pers.) v. Höhn. & Litsch.	1	2	3	4
P. nuda (Fr.) Bres.	1	2		4
P. pini (Schleich.) Boidin	1	2		
P. polygonia (Pers.) Bourd. & Galz.	1	2		
P. proxima Bres.	1			
P. quercina (Pers.) Cooke	1	2	3	4
P. reidii Boidin & Lanquetin		2		
P. rufomarginata (Pers.) Litsch.	1	2		
P. tamaricicola Boidin & Malençon		2		
P. versicolor (Bres.) Sacc. & Syd. (auct. angl. = *P. boidinii* pr. max. part.)	1			
P. violaceolivida (Somm.) Massee	1	2		

Perenniporia Murr.
 Fomoid lignicolous saprophytes or facultative parasites.

	1	2		4
Perenniporia fraxinea (Fr.) Ryvard.	1	2		4
P. medullapanis (Fr.) Donk	1			

Phaeocoriolellus Kotlaba & Pouzar
 Effusoreflexed poroid saprophyte on coniferae.

	1		3	
Phaeocoriolellus trabeus (Pers.) Kotl. & Pouz. (*Trametes trabeus* (Pers.) Bres.)	1		3	

Phaeolus Pat.
 Active facultative parasite of coniferous trees.

	1	2	3	4
Phaeolus schweinitzii (Fr.) Pat. (*Polyporus spongia* Fr.)	1	2	3	4

Phanerochaete Karst.
 Peniophoroid lignicolous saprophytes.

	1	2	3	4
Phanerochaete filamentosa (Berk. & Curt.) Burdsal	1			
P. laevis (Fr.) Erikss. & Ryvard. (*P. affinis* (Burt) Parmasto)	1			
P. leprosa (Bourd. & Galz.) Parmasto	1			
P. magnoliae (Berk. & Curt.) Burdsal				4
P. sanguinea (Fr.) Parmasto	1	2	3	
P. sordida (Karst.) Erikss. & Ryvard. (*P. cremea* (Bres.) Parmasto)	1	2	3	4
P. tuberculata (Karst.) Parmasto (*Corticium tuberculatum* Karst.)		2		
P. velutina (DC.) Parmasto	1	2	3	4

Phellinus Quél.
 Polyporoid, porioid or fomoid lignicolous saprophytes or facultative parasites of trees.
 Phellinus robustus (Karst.) Bourd. & Galz. is on *Quercus* in Berkshire & Hampshire.

	1	2		4
Phellinus conchatus (Pers.) Quél.	1	2		4

P. contiguus (Pers.) Pat. — 1
P. ferreus (Pers.) Bourd. & Galz. — 1 2 3 4
P. ferruginosus (Schrad.) Bourd. & Galz.   Old records are confused with P. ferreus — 1 2 3 4
P. hippophaeicola Jahn (P. robustus f. hippophaes Donk) — 4
P. igniarius (L.) Quél. — 1 2 3 4
P. laevigatus (Fr.) Bourd. & Galz. — 2
P. pini (Brot.) A. Ames — 1 ... 4
P. pomaceus (Pers.) Maire (P. tuberculosus (Baumgarten) Niemala) — 1 2 3 4
P. ribis (Schum.) Bond. & Sing.   Usually as f. euonymi (Kalchbr.) Pilat — 1 2 3 4
P. torulosus (Pers.) Bourd. & Galz. — 1

Phellodon Karst.
Stipitate hydnoid saprophytes.

Phellodon confluens (Pers.) Pouzar — 1
P. melaleucum (Fr.) Karst. (Hydnum graveolens (Delast.) Fr. — 1 2 ... 4
P. niger (Fr.) Karst. — 1 ... 4
P. tomentosus (L.) Banker (Hydnum cyathiforme (Schaeff.) Fr. — 1 2 ... 4

Phlebia Fr.
Corticioid lignicolous saprophytes, including *Scopuloides* (Massee) v. Höhn. & Litsch. pp.

Phlebia (Scopuloides) hydnoides (Cooke & Massee) Christ. — 1 2 3
P. livida (Pers.) Bres. — 1 2 3 4
P. radiata Fr. (P. merismoides Fr.) — 1 2 3 4
P. rufa (Pers.) Christ. (Merulius rufus (Pers.) Fr.) — 1 2 3 4

Phlebiopsis Julich
Corticioid lignicolous saprophytes.

Phlebiopsis gigantea (Fr.) Julich — 1 ... 3 4
P. ravenelii (Cooke) Hjortstam (P. roumeguerii (Bres.) Julich) — 1 2

Physisporinus Karst. see *Podoporia*

Piloderma Julich
Corticioid saprophytes on forest debris.

Piloderma byssinum (Karst.) Julich (Corticium byssinum Karst.) — 1 ... 3 4
P. croceum Erikss. & Hjortstam — 1

Piptoporus Karst.
Actively parasitic polypore confined to *Betula*.

Piptoporus betulinas (Bull.) Karst. — 1 2 3 4

Pistillaria Fr.
Small clavarioid saprophytes on plant debris.

Pistillaria culmigena Mont. & Fr. — 1
P. micans (Pers.) Fr. — 1 2
P. pusilla (Pers.) Fr. — 1
P. setipes Grev. (Typhula grevillei Fr., T. candida Fr.) — 1 ... 4
P. uncialis (Grev.) Cost. & Duf. — 1

Pistillina Quél.
Saprophytes akin to *Pistillaria*, *P. patouillardii* Quél. in E. Anglia.

Plicaturopsis Reid
Small narrowly lamellate carpophores saprophytic on timber in the north, *P. crispa* (P.) Reid.

Podoporia Karst.
Resupinate poroid saprophytes on soil or plant debris.

Podoporia confluens Karst. (*Physisporinus sanguinolentus* (A. & S) Pilat)	1	2		4
P. vitrea (Pers.) Donk (*Polyporus adiposus* B. & Br.; *Phys. vitreus* (P.) Karst.)	1	2	3	4

Podoscypha Pat.
Terrestrial, sessile, infundibuliform stereoid saprophytes.

Podoscypha multizonata (B. & Br.) Pat. (*Stereum multizonatum* (B. & Br.) Massee)	1	2	3	4

Pseudomerulius Julich
Lignicolous merulioid saprophyte.

Pseudomerulius aureus (Fr.) Julich	1

Pterula Fr.
Slender furcate clavarioid saprophytes.

Pterula gracilis (Berk. & Desm.) Corner	1	2		4
P. multifida Fr.	1			

Ptychogaster Corda
Anamorphs of polypores so better listed here than amongst the other imperfecti.

Ptychogaster albus Corda	Assigned to *Tyromyces ptychogaster* (Ludw.) Donk	1	2		4
P. cubensis Pat	Assigned to *Inonotus rickei* (Pat.) Reid	1			
P. rubescens Boud.		1			
P. sp.		1	2		4

Pulcherricium Parmasto
Corticioid lignicolous saprophyte.

Pulcherricium caeruleum (Schrad.) Parm.	1	2	3	4

Pycnoporus Karst.
Bright red poroid saprophytes, there is an old record of *P. cinnabarinus* (Jacq.) Karst. from Essex.

Radulomyces Christ.
Corticioid lignicolous saprophytes formerly referred to *Radulum* Fr., the type species of which is an ascomycete. See also *Basidioradulum*.

Radulomyces confluens (Fr.) Christ. (*Cerocorticium confluens* (Fr.) Julich & Stalp.)	1	2		4
R. molaris (Chaill.) Christ. (*Cerocorticium molare* (Chaill.) Julich & Stalpers)	1	2		4
R. rickii (Bres.) Christ.			3	4

Ramaria (Fr.) Bon.
Branching clavarioid saprophytes, lignicolous or terrestrial.

Ramaria abietina (Pers.) Quél. (*R. ochraceovirens* (Jungh.) Donk)		1		3	4
R. aurea (Fr.) Quél.	Requires confirmation, may have been *R. flava*				4
R. botrytis (Fr.) Ricken		1	2	3	4

R. botrytis var. *alba* (Pearson) Corner		2	4
R. broomei (Cotton & Wakef.) (*R. nigrescens* Brinkm.) Donk)	1		
R. (*Clavariella*) *condensata* (Fr.) Quél.	1		4
R. crispula (Fr.) Quél. (*R. decurrens* (Pers.) Petersen)	1		
R. flaccida (Fr.) Ricken	1	2	4
R. flava (Fr.) Quél.	1		
R. formosa (Pers.) Quél.	1	2	
R. fumigata (Peck) Corner (*R. versatilis* Quél.)		2	
R. gracilis (Fr.) Quél.	1		
R. invalii (Cotton & Wakefield) Donk (*R. eumorpha* (Karst.) Corner)		2	4
R. mairei Donk (*R. pallida* (Bres.) Ricken)	1		
R. spinulosa (Pers.) Quél.	1		
R. stricta (Pers.) Quél.	1	2	4

Ramariopsis (Donk) Corner

Slender, branched, clavarioid, terrestrial saprophytes, differing from *Ramaria* in having colourless spores.

Ramariopsis biformis (Atk.) Petersen	1	2	
R. crocea (Fr.) Corner	1		
R. kunzei (Fr.) Donk	1	2	4
R. minutula (Bourd. & Galz.) Petersen	1		
R. pulchella (Boud.) Corner (*Clavaria bizozzeriana* Sacc.)			4
R. tenuiramosa Corner		3	4

Repetobasidium Eriksson & Hjortstam

Corticioid saprophytes, *R. vile* (Bourd.) Galz.) Erikss. reported from Devon.

Repetobasidium cf. *americanum* Erikss. & Hjort.	1

Resinaceum Parmasto

Odontioid lignicolous saprophyte.

Resinaceum bicolor (A. & S.) Parm.	1	2	3

Rigidoporus Murr.

Fomoid lignicolous saprophytes. See also *Podoporia*.

Rigidoporus lineatus (Pers.) Ryvarden Glasshouse alien	1		
R. ulmarius (Fr.) Imazeki	1	2	4

Sarcodon Karst.

Stipitate hydnoid terrestrial saprophytes; *S. regalis* Maas Geesteranus in Berkshire with *S. scabrosum* (Fr.) Karst.

Sarcodon imbricatum (L.) Karst.	1	2	4

Sarcodontia Schulzer

Hydnoid lignicolous saprophyte.

Sarcodontia crocea (Fr.) Kotlaba	1	4

Schizophyllum Fr.

Lamellate flabellate, coriaceous saprophyte, lignicolous, with lamellae revolute at margin.

Schizophyllum commune Fr.	1	2	3	4

Schizopora Vel.

Ubiquitous saprophyte with lacerate pores.

Schizopora paradoxus (Schrad.) Donk   1   2   3   4

Scopuloides (Massee) v. Höhn. & Litsch. see *Phlebia*

Scytinostroma Donk
     Coriaceous, resupinate, lignicolous saprophytes.

	1	2	3	4
Scytinostroma ochroleucum (Bres. & Torrend) Donk		2		
S. portentosum (Berk. & Curt) Donk				4

Serpula Pers.
     Merulioid lignicolous saprophytes with brown spores.

	1	2	3	4
Serpula himantioides (Fr.) Bond.	1	2	3	4
S. lachrymans (Wulf.) Schroet.	1	2		4

Sistotrema Fr.
     Stipitate to resupinate, poroid, irpicoid or corticioid, lignicolous to terrestrial saprophytes with urniform basidia.

	1	2	3	4
Sistotrema brinkmannii (Bres.) Bourd. & Galz.	1	2	3	
S. commune J. Erikss. (*S. octosporum* (Schroet.) Hallenberg)		2		
S. confluens (Pers.) Fr.	1			4
S. coroniferum (v. Höhn. & Lits.) Donk	1	2	3	
S. diademiferum (Bourd. & Galz.) Donk	1	2		
S. hispanicum Dueñas et al.	1			
S. oblongisporum Christ. & Hauerskev	1			
S. octosporum (Schroet.) Hallenberg	1			
S. sernanderi (Litsch.) Donk		2		
S. variecolor (Bourd. & Galz.) Wakef. & Pearson			3	

Sistotremastrum J. Erikss.
     Corticioid saprophyte with multisporous nonurniform basidia.

	1	2	3	4
Sistotremastrum niveocremeum (v. Höhn. & Lits.) Erikss.	1	2	3	

Sistotremella Hjortstam
     Saprophytes, *S. perpusilla* Hjort. in Wales.

Skeletocutis Kotlaba & Pouzar
     White, effusoreflexed to dimidiate lignicolous saprophyte; if not on conifer see *Incrustoporia*.

	1	2	3	4
Skeletocutis amorphus (Fr.) Kotlaba & Pouzar	1	2		
S. carneogrisea David	1			

Sparassis Fr.
     Fleshy stipitate stereoid terrestrial saprophytes with lamellate more or less imbricate branches.

	1	2	3	4
Sparassis crispa (Wulf.) Fr.	1	2	3	4
S. laminosa Fr.	1			4
S. simplex Reid Doubtfully distinct from the preceding	1			

Spongipellis Pat.
     Whitish, fleshy, dimidiate, poroid to irpicoid lignicolous saprophytes.

	1	2	3	4
Spongipellis delectans (Peck) Murr. (*Leptoporus bredecelensis* Pilat)	1	2		4
S. (Irpex) pachyodon (Pers.) Kotlaba & Pouzar (*Irpiciporus pachyodon* (P.) K. & P.)		2		

S. spumeus (Sow.) Pat.   1 2   4

Sporobolomyces Kluyver & v. Niel
Basidioid yeasts, saprophytic or merely epiphyllous.

Sporobolomyces roseus Kluyver & v. Niel   Presumably ubiquitous   1

Steccherinum S.F. Gray
Coriaceous, effused to dimidiate, hydnoid to odontioid lignicolous saprophytes.

Steccherinum fimbriatum (Pers.) Erikss. (*Mycoleptodon fimbriatum* (P.)   1 2   4
Bourd. & Galz.)
S. ochraceum (Pers.) S.F. Gray (*Mycoleptodon ochraceum* (Pers.) Pat.)   1 2   4

Stereopsis Reid
Stipitate stereoid saprophytes, *S. vitellina* (Plowr.) Reid in the Highlands.

Stereum Pers.
Thin coriaceous, resupinate to effusoreflexed with smooth hymenium, sometimes
bleeding when scratched (*Haematostereum* Pouzar), lignicolous saprophytes or weak
facultative parasites.

Stereum (*Haematostereum*) *gausapatum* Fr.   1 2 3 4
S. hirsutum (Willd.) S.F. Gray   1 2 3 4
S. rameale (Pers.) Fr.   1   4
S. (*H.*) *rugosum* (Pers.) Fr.   1 2 3 4
S. (*H.*) *sanguinolentum* (A. & S.) Fr.   1 2   4
S. subtomentosum Pouzar   1 2
S. sulphuratum Berk. & Rav.   1 2 3 4

Strangulidium Pouzar
Poroid lignicolous saprophytes, segregated from *Tyromyces*.

Strangulidium rennyi (Berk. & Br.) Pouzar   1
S. sericeomolle (Romell) Pouzar   1   4

Stromatoscypha Donk
Minute cupulate receptacles aggregated on a membranous, lignicolous, white stroma.
Stromatoscypha fimbriata (Pers.) Donk reported a few times from the north and west.

Subulicystidium Parmasto
Corticioid lignicolous saprophytes.

Subulicystidium longisporum (Pat.) Parm.   1 2 3

Subulicium Hjorstam & Ryvarden
Segregate from *Peniophora*

Subulicium lautum (Jacks.) Hjort. & Ryv.   1

Syzygospora Martin
Tulasnelloid hyperparasites of Agaricales (*Christiansenia* Hauerslev).

Syzygospora pallida (Hauerslev) Ginns   2
S. tumefaciens (Ginns & Sunhede) Ginns   4

Thanatephorus Donk
Hypochnoid facultative parasites of angiosperms.

Thanatephorus cucumeris (Frank) Donk (*Corticium solani* (Prill. & Del.) Bourd. & Galz.)   1 2 3 4

Thelephora Ehrhart
Stipitate and lacerate, dimidiate or effusoreflexed, lignicolous or terrestrial, saprophytic or possible very feebly parasitic on seedlings.

	1	2	3	4
Thelephora anthocephala (Bull.) Fr.	1	2	3	4
T. caryophyllea Schaeff.	1			4
T. mollissima (Pers.) Fr.	1		3	
T. palmata Scop.	1	2		
T. spiculosa (Fr.) Fr.	1	2		4
T. terrestris Ehrh. (*T. laciniata* (Pers.) Fr.)	1	2	3	4

Tilletiopsis Derx
Basidioid yeasts, saprophytes.

	1
Tilletiopsis minor Nyland	1
T. washingtonensis Nyland	1

Tomentella Pat.
Hypochnoid lignicolous or terrestrial saprophytes, usually strongly coloured.

	1	2	3	4
Tomentella avellanea (Burt) Bourd. & Galz.		2		
T. botryoides (Schw.) Bourd. & Galz.	1		3	
T. bresadolae (Brinkm.) v. Höhn. & Litsch.			3	
T. bryophila (Pers.) Larsen (*T. pallidofulva* (Peck) Litsch.)	1	2		
T. coerulea (Bres.) v. Höhn & Litsch. (*T. jaapii* (Bres.) B. & G.; *T. sordida* Wakef.)			3	
T. (Lazulinospora) cyanea (Wakef.) Bourd. & Galz.	1			
T. (Tomentellopsis) echinospora (Ellis) Bourd. & Galz.	1	2	3	4
T. ellisii (Sacc.) Julich & Stalpers (*T. microspora* (Karst.) v. Höhn. & Litsch.)	1	2	3	
T. ferruginea Bres. (*T. coriaria* (Peck) Bourd. & Galz.; *T. fusca* (Pers.) Schroet.)	1	2	3	4
T. (Tomentellastrum) fuscocinerea (Pers.) Donk		2		
T. griseoviolacea Litsch.	1			
T. hoehnelii Skovsted		2		
T. hydrophila (Bourd. & Galz.) Litsch.	1			
T. (Pseudotomentella) mucidula (Karst.) v. Höhn. & Litsch.	1		3	
T. neobourdotii Larsen (*T. bourdotii* Svrček)	1	2	3	
T. ochracea (Sacc.) v. Höhn. & Litsch.				4
T. pilosa (Burt) Bourd. & Galz.	1	2		
T. punicea (A. & S.) Schroet. (*T. epiphylla* (Schw.) Litsch.)	1		3	
T. ramosissima (Berk. & Curt.) Wakef.				4
T. rubiginosa (Bres.) Maire (*T. atrovirens* (Bres.) v. Höhn. & Litsch.)	1	2	3	
T. sublilacina (Ell. & Holw.) Wakefield	1			
T. (Tomentellopsis) submollis (Svrček) Wakefield	1			4
T. testaceogilva Bourd. & Galz.	1			
T. (Pseudotomentella) tristis (Karst.) v. Höhn. & Litsch.	1	2	3	
T. viridula (Bourd. & Galz.) Svrček		2		
T. (Tomentellopsis) zygodesmoides (Ell.) v. Höhn. & Litsch.	1	2	3	4

Trametes Fr.
Dimidiate, coarsely poroid, lignicolous saprophytes, including *Pseudotrametes* Bond. & Sing.

	1	2	3	4
Trametes (Ps.) gibbosa Fr.	1	2	3	4
T. suaveolens (Fr.) Fr.	1			

Trechispora See *Cristella*

Trichaptum Murr.
Resupinate, effusoreflexed to dimidiate poroid lignicolous saprophytes with pore surface at first violaceous (*Hirschioporus* Donk).

Trichaptum abietinum (Dicks.) Ryvarden	1	2	3	4
T. fuscoviolaceus (Ehrenb.) Ryvarden (*Irpex fuscoviolaceus* Fr.)	1			

Tubulicrinis Donk
Corticioid lignicolous saprophytes.

Tubulicrinis accedens (Bourd. & Galz.) Donk	1	2	3	
T. glebulosa (Bres.) Donk	1	2	3	4
T. karstenii (Bres.) Donk	1	2		
T. subulata (Bourd. & Galz.) Donk	1	2	3	

Tulasnella Schroet.
Membranous, subgelatinous, corticioid, mainly lignicolous saprophytes with distinctive enlarged sterigmata bearing basidiospores germinating by conidia, hence regarded by many as heterobasidial. Additional species reported from the west of England.

Tulasnella albida Bourd. & Galz.	1			
T. allantospora Wakef. & Pearson	1	2	3	
T. calospora (Boud.) Juel		2	3	4
T. cystidiophora v. Höhn. & Litsch.		2	3	4
T. eichleriana Bres.	1			
T. incarnata Bres. (*T. violea* (Quél.) Bourd. & Galz.) (*T. microspora* W. & P.)	1	2	3	4
T. (*Gloeotulasnella*) *inclusa* (Christ.) Donk	1			
T. pruinosa Bourd. & Galz.		2		
T. rubropallens Bourd. & Galz.	1			
T. sordida Bourd. & Galz.			3	
T. tomaculum Roberts	1			
T. tremelloides Wakef. & Pearson	1			
T. violacea (Olsen) Juel	1			

Tylospora Donk
Corticioid resupinate lignicolous saprophytes with distinctive basidiospore shape.

Tylospora asterophora (Bon) Donk	3	4

Typhula Fr.
Clavarioid carpophores arising from sclerotia, saprophytes or parasites. *T. graminum* Karst. in Berkshire.

Typhula erythropus (Pers.) Fr.	1		3	4
T. gyrans Fr.	1		3	
T. phacorrhiza (Reich.) Fr. (*Sclerotium scutellatum* A. & S.)	1		3	4
T. quisquiliaris (Fr.) Corner	1	2	3	4
T. sclerotioides Fr.	1	2		
T. variabilis Riess				4

Tyromyces Karst.
Resupinate to dimidiate poroid lignicolous saprophytes, white or light coloured, see also *Aurantioporus* and *Strangulidium*, corresponds to *Leptoporus* Quél. pro max. part. *T. destructor* (Schrad.) Bond. & Sing. is a doubtful species and records from 1 under that name have been rejected. For records of *T. ptychogaster* (Ludwig) Donk see *Ptychogaster*.

Tyromyces alborubescens (Bourd. & Galz.) Bond.				4
T. balsameus (Peck.) Murr. (*T. kymatodes* auct. angl.)	1	2		4
T. caesius (Schrad.) Murr.	1	2	3	4

75

T. chioneus (Fr.) Karst. (*T. albellus* (Peck) Bond. & Sing.)	1	2	4
T. cinerascens (Bres.) Bond. & Sing.	1	2	4
T. floriformis (Quél.) Bond. & Sing. Glasshouse alien	1		
T. fragilis (Fr.) Donk	1	2 3	4
T. guttulatus (Peck) Murr.			4
T. lacteus (Fr.) Murr.	1	2 3	4
T. leucomallus Murr. (*T. gloeocystidiatus* Kotl. & Pouzar)	1		
T. mollis (Pers.) Karst.	1		
T. revolutus (Bres.) Bond. & Sing.	1	2	4
T. stipticus (Pers.) Kotlaba & Pouzar	1	2 3	4
T. subcaesius David	1		
T. tephroleucus (Fr.) Donk	1	2 3	4
T. wakefieldae Kotlaba & Pouzar, *T. wynnei* var. *ellipsospora* Pilat	1		4

Uthatobasidium Donk
Corticioid saprophytes.

Uthatobasidium ochraceum (Massee) Donk	1	2	3
U. fusisporum (Schroet.) Donk	1		3

Vararia Karst.
Stereoid lignicolous saprophytes, see also *Dichostereum*.

Vararia ochroleuca (Bourd. & Galz.) Donk	1

Veluticeps (Cooke) Pat.
Stereoid lignicolous saprophytes, *V. abietina* (Pers.) Hjortstam & Terreria an alien on imported timber.

Vesiculomyces Hagström
Corticioid lignicolous saprophytes.

Vesiculomyces citrinus (Pers.) Hagstrom (*Gloeocystidiellum citrinum* (Pers.) Donk)	1	2

Vuilleminia Maire
Corticioid lignicolous saprophyte.

Vuilleminia comedens (Fr.) Maire	1	2 3	4
V. coryli Boidin et al.	1		

Xenasma Donk
Corticioid saprophytes.

Xenasma pruinosum (Pat.) Donk (*Phanerochaete chordalis* (v. Höhn. & Lits.) Rhodes)	1	2	4
X. pseudotsugae (Burt.) J. Erikss.	1		4
X. pulverulentum (Litschauer) Donk	1	2	
T. tulasnelloideum (v. Höhn. & Lits.) Donk	1	2 3	

Xenasmatella Oberwinkler
Perhaps inseparable from *Phlebiella*.

Xenasmatella albida Hauerslev On *Taxus*	1	
X. allantospora Oberw.	1	

Xenosperma Oberw.
Corticioid lignicolous saprophytes; *X. ludibundum* (Rogers & Liberta) Oberw. & Julich reported from Devon.

Xylobolus Karst.
  Stereoid lignicolous saprophytes.

Xylobolus frustulatus (Pers.) Boidin   Casual alien not seen for over a century   1

PHALLALES

Aseroe Labill.
  Saprophytes.

A. rubra Labill.   Casual alien reported once in the last   1
                                          century and again in 1993

Clathrus Pers.
  Terrestrial saprophytes, associated with woody debris.

Clathrus archeri (Berk.) Dring   Alien, introduced from Australasia   3 4
C. ruber Pers.   Probable alien, introduced from   1 2
                                    southern Europe

Ileodictyon Tul
  Segregate from *Clathrus*.

Ileodictyon cibarius Tul.   Alien, introduced from New Zealand   1

Lysurus Fr.
  Saprophytes, tropical to subtropical.

Lysurus cruciatus (Lepr. & Mont.) Lloyd   Casual glasshouse alien   2

Mutinus Fr.
  Terrestrial but associated with woody debris, saprophytes.

Mutinus caninus (Huds.) Fr.   1 2 3 4

Phallus L.
  Saprophytes, occasionally suspected of parasitising roots.

Phallus impudicus L.   1 2 3 4
P. impudicus var. *togatus* (Kalchbr.) Cost. & Duf.   1

HYSTERANGIALES

Hysterangium Vitt.
  Hypogeous saprophytes akin to Phallales; *H. nephriticum* Berk. and *H. thwaitesii* B. & Br.
  in the west of England.

GAUTIERIALES

Gautieria Vitt.
  Hypogeous, *G. morchelliformis* Vitt. in Gloucestershire.

HYMENOGASTRALES

Gastrosporium Mattir.
Hypogeous, *G. simplex* Mattir. probably alien, reported from Hertfordshire.

Gymnomyces Massee & Rodway
Hypogeous; *G. xanthosporus* (Hawker) A.H. Smith in Wales.

Hydnangium Wallr.
Hypogeous, probably mycorrhizal especially with *Eucalyptus* spp.

Hydnangium carneum Wallr.	Alien in gardens	1	3

Hymenogaster Vitt.
Hypogeous, probably mycorrhizal; at least four more species in Gloucestershire.

Hymenogaster arenarius Tul					4
H. citrinus Vitt.					4
H. klotzschii Tul.	Doubtful record				4
H. luteus Vitt.		1	2		4
H. olivaceus Vitt.		1			
H. tener B. & Br.		1	2	3	4
H. vulgaris Tul.			2		

Octavianina (Vitt.) Kuntze
Hypogeous, *O.asterosperma* Vitt. in Hampshire.

Rhizopogon Fr.
Hypogeous, at least in origin, mycorrhizal with Gymnosperms; *R. reticulatus* Hawker in Somerset, *R. vulgaris* (Vitt.) M. Lange in Hampshire.

Rhizopogon luteolus Fr.	1			
R. roseolus (Corda) Th. Fr.	1	2		4

Sclerogaster Hesse
Hypogeous, mycorrhizal; *S. broomeianus* Zeller & Dodge in Gloucestershire.

Sclerogaster lanatus Hesse	4

Stephanospora Lat.
Hypogeous, mycorrhizal.

Stephenospora caroticolor (Berk.) Pat.	1	3

Wakefieldia Corner & Hawker
Segregate from *Sclerogaster*; *W. macrospora* (Hawker) Hawker in Gloucestershire.

Zelleromyces Singer & Smith
Segregate from *Octavianina*; *Z. stephensii* (Berk.) A.H. Smith in Somerset (Avon)

LYCOPERDALES

Bovista Pers.
Saprophytes, "puffballs" lacking a sterile base, cf. also *Lycoperdon ericetorum. B. limosa* Rostr. reported from Dyved & Lancashire.

Bovista nigrescens Pers.	1	2	3	4
B. plumbea Pers.	1	2	3	4

Bovistella Morgan
   Saprophyte.

Bovistella radicata (Dur. & Mont.) Pat.	1

Calvatia Fr.
   Saprophytes, puffballs lacking an apical pore.

Calvatia excipulaeformis (Schaeff.) Perdeck (*C. saccata* (Vahl) Morgan)	1	2	3	4
C. utriformis (Bull.) Jaap (*C. caelata* (Bull.) Morg.)	1	2	3	4

Geastrum Pers.
   Saprophytes, "earth-stars".

Geastrum badium Pers. (*G. elegans* Vitt.)	1			
G. campestre Morgan (*G. berkeleyi* Massee)	1			4
G. corollinum (Batsch) Hollos (*G. recolligens*)				4
G. nanum Pers.	1			
G. pectinatum Pers.	1		3	4
G. sessile (*G. fimbriatum* Fr.)	1	2	3	4
G. striatum DC. (*G. bryantii* Berk.)	1		3	4
G. triplex Jungh.	1	2		4
G. vulgatum Vitt.	1			4

Langermannia Rostk.
   Saprophyte, giant "puffballs".

Langermannia gigantea (Batsch) Rostk.	1	2	3	4

Lycoperdon Tournef.
   Saprophytes, lignicolous or terrestrial; "puffballs" with single apical pore and a sterile base, *L. caudatum* Schroet. in Scotland.

Lycoperdon atropurpureum Vitt.	1			
L. decipiens Dur. & Mont.	1			
L. echinatum Pers.	1	2	3	4
L. ericetorum Pers. (*Bovista pusilla* (Batsch) Pers.)				4
L. foetidum Bon.	1	2	3	4
L. lividum Pers. (*L. spadiceum* Pers.)	1	2	3	4
L. mammaeforme Pers. (*L. velatum* Vitt.)	1	2		4
L. molle Pers.	1	2		4
L. perlatum Pers.	1	2	3	4
L. polymorphum Vitt. (*Bovista aestivalis* (Bon.) Demoulin)	1			
L. pyriforme Schaeff.	1	2	3	4

Myriostoma Desv.
   Saprophyte, terrestrial; simulating *Geastrum* but endoperidium opening by numerous pores.

Myriostoma coliforme (Dicks.) Corda	Thought to be extinct in Britain	4

Vascellum Smarda
   Terrestrial "puffballs", saprophyte or parasite of turf grasses in "fairy rings".

Vascellum pratense (Pers.) Kreisel	1	2	3	4

MELANOGASTRALES

Melanogaster Corda
Saprophytes? hypogeous or mixed with forest litter.

Melanogaster ambiguus (Vitt.) Tul.	1	2	3	4
M. broomeianus Berk.	1		3	4
M. variegatus (Vitt.) Tul.	1			

Nia R.T. Moore & Meyers
Saprophyte on timber in sea water *N. vibrissa* Moore & Meyers, reported from Man.

SCLERODERMATALES

Astraeus Morgan
Terrestrial saprophyte, simulating *Geastrum* but with 3-layered hygroscopic exoperidium.

Astraeus hygrometricus (Pers.) Morgan	1	2	3	4

Battarraea Pers.
Terrestrial saprophyte with basal volvoid exoperidium supporting a stipitate endoperidium.

Battarraea phalloides (Dicks.) Pers.	1	4

Pisolithus A. & S.
Mycorrhizal, massive terrestrial "puffball" with gleba containing numerous peridioles.

Pisolithus arhizus (Pers.) Rauschert (*P. arenarius* A. & S.)	3

Queletia Fr.
Saprophyte.

Queletia mirabilis Fr.	Apparently a casual alien	1

Scleroderma Pers.
Terrestrial saprophytes, probably facultatively mycorrhizal with trees; "puffballs" with thick irregularly dehiscent peridium.

Scleroderma areolatum Ehrenb.	1	2		4
S. bovista Fr.	1	2	3	4
S. cepa Pers.	1	2	3	4
S. citrinum Pers. (*S. aurantium* (L.) Pers.; *S. vulgare* (Hornem.) Fr.)	1	2	3	4
S. polyrhizum Pers.	1			
S. verrucosum (Bull.) Fr.	1	2	3	4

Tulostoma Pers.
Saprophytes, small longstalked "puffballs" on arid soils.

Tulostoma brumale Pers. (*T. mammosum* Fr.)	1	2	3	4

NIDULARIALES

Crucibulum Tul.
  Lignicolous saprophyte.

Crucibulum laeve (Huds.) Kambly (*C. vulgare* Tul.)	1	2	3	4

Cyathus Haller
  Saprophytes, lignicolous or terrestrial; "bird's nest fungi".

Cyathus olla Batsch	1	2	3	4
C. olla var. *agrestis* (Pers.) Tul.	1			4
C. striatus (Huds.) Pers.	1	2	3	4

Mycocalia Palmer
  Minute terrestrial or lignicolous saprophytes with few peridioles, several species elsewhere.

Mycocalia denudata (Fr.) Palmer	1
M. sphagneti Palmer	1

Nidularia Fr.
  Lignicolous saprophyte with numerous peridioles.

Nidularia farcta (Roth) Fr.	1

Sphaerobolus Tode
  Coprophilous or lignicolous saprophyte with single peridiole violently ejected.

Sphaerobolus stellatus Tode	1	2	3	4

DACRYMYCETALES

Calocera (Fr.) Fr.
  Clavarioid lignicolous saprophytes, including *Dacryomitra* Tul.

Calocera cornea (Batsch) Fr.	1	2	3	4
C. furcata (Fr.) Fr.	1			4
C. (Dacryomitra) glossoides (Pers.) Fr. (*D. pusilla* Tul.)	1		3	4
C. pallidospathulata Reid	1	2		4
C. viscosa (Pers.) Fr.	1	2	3	4

Dacrymyces Nees
  Pulvinate gelatinous lignicolous saprophytes; *D. palmatus* (Schw.) Bres. in the Midlands, *D. variisporus* McNabb in Gloucestershire, *D. macnabbii* Reid, *D. minor* Peck, *D. ovispora* Bref. in Scotland. As the species are superficially alike the "missing" ones are likely to have been overlooked in the south.

Dacrymyces capitatus Schw.	1		3	4
D. enatus (Berk. & Curt.) Massee	1	2		
D. estonicus Raitviir				4
D. punctiformis Neuhoff	1			
D. stillatus Nees (*D. deliquescens* (Bull.) Duby)	1	2	3	4

Dicellomyces Olive
  Parasite (?obligate) of *Scirpus sylvaticus*.

Dicellomyces scirpi Raitviir	1

Ditiola Fr.
Saprophytes, lignicolous: including *Femsjonia* Fr.

Ditiola pezizaeformisu (Lév.) Reid (*Femsjonia luteoalba* Fr.)	1	2	3	4
D. radicata (A. & S.) Fr.	Needs confirmation	1		

Guepiniopsis Pat.
Gelatinous pezizoid lignicolous saprophytes, *G. alpina* (Tracy & Earle) Brasf. in Scotland.

Guepiniopsis buccina (Pers.) Kennedy		1	
G. chrysocoma (Bull.) Brasf.	Very doubtful records	1	4

AURICULARIALES

Achroomyces Bon.
Resupinate lignicolous saprophytes (*Platygloea* Schroet.), also *A. sebaceus* (B. & Br.) Wojewoda in the West of England.

Achroomyces effusus (Schroet.) Mig.	1	2	3	4
A. peniophorae (Bourd. & Galz.) Wojewoda	1	2		
A. vestita (Bourd. & Galz.) Wojewoda	1		3	4

Auricularia Bull
Effusoreflexed to imbricate gelatinous saprophytes on dead wood.

Auricula mesenterica (Dicks.) Pers.	1	2	3	4

Eocronartium Atk.
Clavarioid parasite of mosses: *E. muscicola* (Pers.) Fitzp. in the West of England.

Helicobasidium Pat.
Resupinate saprophytes and facultative parasites of fleshy angiosperm organs.

Helicobasidium brebissonii (Desm.) Donk (*H. purpureum* (Tul.) Pat.), mycelial state *Rhizoctonia crocorum* Fr.)	1	2	4
H. compactum Boedijn		2	

Helicogloeum Pat.
Resupinate lignicolous saprophyte.

Helicogloeum lagerheimii Pat. (*Saccoblastia sebacea* Bourd. & Galz.)	1	3

Herpobasidium Lind
Parasites (?obligate) on pteridophyta; *F. filicinum* (Rostr.) Lind in the West of England.

Hirneola Fr.
Saprophytes, segregate from *Auricularia*.

Hirneola auriculajudae (Bull.) Berk.	1	2	3	4
H. auriculajudae var. *lactea* (Quél.)	1	2		4

Mycogloea Olive
Hyperparasites, *M. macrospora* (B. & Br.) McNabb on *Diatrype* in the West of England.

Pachnocybe Berk.
Lignicolous saprophyte.

Pachnocybe ferruginea (Sow.) Berk.	4

Phleogena Link
Lignicolous saprophytes (*Pilacre* Fr.)

Phleogena faginea (Fr.) Link   1   2   4

Pilacrella Schroet.
Saprophyte. *P. solani* (Cohn.) Schroet. in Scotland.

Saccoblastia Möller
Probably inseparable from *Helicogloeum.*

Saccoblastia farinacea (v. Höhn.) Donk.   1

Stilbum Tode

Stilbum vulgare Tode   Doubtful records in need of confirmation   1   4

TREMELLALES

Basidiodendron J. Rick
Lignicolous saprophytes.

Basidiodendron caesocinereunm (v. Höhn. & Lits.) LuckAllen (*Bourdotia cinerella*)   1   3   4
B. & G.)
B. eyrei (Wakef.) LuckAllen (*Sebacina eyrei* Wakef.)   1   2   3

Bourdotia (Bres.) Trott.
Lignicolous saprophytes.

Bourdotia galzinii (Bres.) Trott.   2

Ceratosebacina Roberts
Lignicolous saprophyte, *C. longispora* (Hauersl.) Roberts in Devon.

Eichleriella Bres.
Raduloid, effused, lignicolous saprophytes.

Eichleriella deglubens (B. & Br.) Lloyd   1   2

Endoperplexa Roberts
Lignicolous saprophytes, two species in Devon.

Exidia Fr.
Resupinate, pulvinate or pendulous, highly gelatinous, lignicolous saprophytes.

Exidia albida (Huds.) Bref. (*E. thuretiana* (Lév.) Fr.)   1   2   3   4
E. glandulosa (Bull.) Fr.   Including the states called *E. plana* &   1   2   3   4
                     E. truncata Fr.
E. recisa (Ditm.) Fr.   1
E. saccharina (A.&.) Fr.   Doubtful record   1

Heterochaetella (Bourd.) Bourd. & Galz.
Resupinate lignicolous saprophytes; *H. brachyspora* LuckAllen in the West of England.

Heterochaetella dubia (B. & G.) B. & G.   1

Microsebacina Roberts
Lignicolous saprophyte, *M. microbasidia* Roberts in Hampshire.

Myxarium Wallr.
Segregate from *Exidia*; *M. laccatum* (Bourd. & Galz.) Reid in East Anglia.

Myxarium crystallinum Reid	1			
M. hyalinum (Pers.) Donk (*Exidia nucleata* (Schw.) Rea)	1	2	3	4
M. podlachicum (Bres.) Raitviir	1	2	3	4
M. sphaerosporum (Bourd. & Galz.) Reid			3	4
M. subhyalinum (Pearson) Reid	1			

Protodontia V. Höhn.
Effused, odontioid, lignicolous saprophytes.

Protodontia ellipsospora Reid	1	2
P. subgelatinosa (Karst.) Pilat (*P. uda* v. Höhn.)	1	2

Pseudohydnum Karst.
Dimidiate to spathulate, hydnoid, highly gelatinous lignicolous saprophyte. (*Tremellodon P.*)

Pseudohydnum gelatinosum (Scop.) Karst.	1	2	3	4

Pseudostypella Reid & Minter
Pulvinate, highly gelatinous, hyperparasite on *Lophodermium*; *P. translucens* (Gordon) Reid & Minter in East Anglia.

Sebacina Tul.
Effused lignicolous saprophytes, including *Exidiopsis* (Bref.) Moller and see *Myxarium*. *S. microbasidia* is in Hampshire.

Sebacina calcea (Pers.) Bres.	1	2	3	
S. (*Exid.*) *effusa* (Bref.) Maire	1	2	3	
S. epigaea (B. & Br.) Neuhoff	1	2	3	
S. (*Exid.*) *fugacissima* Bourd. & Galz.	1	2		4
S. (*Exid.*) *grisea* (Pers.) Bres.	1			
S. incrustans (Pers.) Tul.	1	2		4

Sebacinella Hauersler
Lignicolous saprophytes.

Sebacinella citrispora Hauersler	1

Serendipita Roberts
Lignicolous saprophyte

Serendipita sigmaspora Roberts	1

Stypella Möller
Effused lignicolous saprophyte.

Stypella vermiformis (B. & Br.) Reid (*Heterochaetella crystallina* Bourd. & Galz.)	1	2

Tremella Pers.
Saprophytes and hyperparasites, mostly lignicolous, including *Naematelia* Fr.

Tremella candida Pers.	1	2		4
T. cerebrina Bull.				4
T. encephala Pers.	1		3	4
T. exigua Desm. (*T. atrovirens* Fr.)	1	2	3	
T. foliacea Pers. (*T. frondosa* Fr., *Phaeotremella pseudofoliacea* Rea)	1	2	3	4
T. globospora (Wheldon) Reid		2		

T. glacialis Bourd. & Galz. (*T. grilletii* Boud.)	1			
T. intumescens Smith	1			
T. mesenterica Retz. (*T. lutescens* Pers.)	1	2	3	4
T. moriformis Sm.	1			
T. obscura (Olive) Christ.	1	2		
T. polyporina Reid	1			
T. steidleri (Bres.) Bourd. & Galz.		2		
T. tubercularia Berk.	Doubtful record	1		

In addition *T. albescens* (Sacc. & Maub.) Sacc. is reported from Warwick & *T. indecorata* Somm. from Yorkshire.

Tremiscus (Pers.) Lév.
Terrestrial saprophyte, *T. helvelloides* (DC.) Donk now widespread in the Severn basin.

Trimorphomyces Bandoni & Oberwinkler
?Hyperparasite; *T. papilionaceus* Band. & Oberw. reported from East Anglia.

Xenolachne Rogers
Hyperparasites.

Xenolachne longicornis Hauerslev   1

USTILAGINALES

All obligate parasites of angiosperms.

Anthracoidea Bref.
About another ten species on Cyperaceae, especially in moorland areas of Britain.

Anthracoidea subinclusa (Körn.) Bref.   3

Doassansia Cornu
Foliar parasites of aquatic monocotyledons and Scrophulariaceae; *D. alismatis* (Nees) Cornu in E. Anglia, *D. martianoffiana* (Thum.) Schroet. in Hampshire.

Doassansia sagittariae (West.) Fisch.   1 2 3

Entorrhiza Weber
Forming root galls on Cyperaceae and Juncaceae; *E. scirpicola* (Correns) Sacc. & Syd. in East Anglia.

Entyloma de Bary
Foliar parasites, some with anamorphs in *Entylomella* v. Höhn.; *E. dactylidis* (Pass.) Cif. common elsewhere on various genera of Gramineae.

Entyloma calendulae (Oud.) de Bary		1	2		4
E. calendulae f. *dahliae* (Syd.) Viegas		1			
E. fergussonii (B. & Br.) Plowr.	On *Myosotis* spp.	1			
E. ficariae (Cornu & Rose) Fisch. v. Wald. (*Entylomella ficariae* (Berk.) v. Höhn.)		1	2	3	4
E. fuscum Schroet.	On *Papaver* spp.	1			
E. helosciadii Magnus	On *Apium* & *Oenanthe* (*E. helosciadii-repentis* v. Höhn.)	1			
E. microsporum (Ung.) Schroet.	On *Ranunculus* spp. (*E. microsporum* (Sacc.) Cif.)	1			4
E. (*Rhamphospora*) *nymphaeae* (Cunn.) Setch.			2		

Farysia Racib.
Sori in spikelets of *Carex riparia*.

Farysia thuemenii (Fisch. v. Wald.) Nannf. (*F. olivacea* (DC.) Syd.)   1   2

Graphiola Poit.
Sori in leaves of *Phoenix*, systematic position disputed.

Graphiola phoenicis Poit.   Glasshouse alien   1

Melanopsichium Beck
Sori forming galls on inflorescence of *Polygonum aviculare*.

Melanopsichium nepalensis (Liro) Zundel   Apparently a casual alien   1

Melanotaenium De Bary
Sori in leaf and stem galls; *M. endogenum* (Ung.) de Bary on *Galium* in E. Anglia, *M. lamii*
Beer on *Lamium album* in Buckinghamshire, others on *Linaria* spp.

Schizonella Schroet.
Foliar sori in Carices; *S. melanogramma* (DC.) Schroet. on *Carex ericetorum* in East Anglia.

Schroeteria Wint.
Sori in capsules of *Veronica arvensis*; *S. delastrina* (Tul.) Wint. in Oxfordsh. & E. Anglia.

Sphacelotheca de Bary
Sori in florets or bulbils of Polygonaceae.

Sphacelotheca hydropiperis (Schum.) de Bary   1   2   3   4

Sporisorium Ehrenb.
Sori in inflorescences of Gramineae.

Sporisorium sorghi Link   Casual alien   1

Thecaphora Fingerhuth
Sori in seeds or tissue of inflorescence; *T. deformans* Dur. & Mont. in *Ulex minor* in
Hampsh.

Thecaphora seminisconvolvuli (Desm.) Liro (*Gloeosporium antherarum* Oud.)   1   2   3   4

Tilletia Tul & Tul.
Sori in ovaries, less often leaves, of Gramineae; *T. menieri* Har. & Pat. on *Phalaris* in E.
Anglia, *T. sphaerococca* (Rab.) Fisch. v. Wald. widespread on *Agrostis* spp.

Tilletia holci (West.) Schroet. (*T. rauwenhoffii* Fisch. v. Wald.)   1
T. olida (Riess) Schroet.   In leaves of *Brachypodium pinnatum*   1
T. tritici (Bjerk.) Wolff. (*T. caries* (DC.)   Perhaps extinct   1   4
    Tul.)

Urocystis Rab.
Spores united in small groups with associated sterile cells; sori usually in vegetative
tissues but in ovaries of Primula; *U. cepulae* Frost. widespread on *Allium* spp. cult.

Urocystis agropyri (Preuss) Schroet.   1
U. anemones (Pers.) Wint. (*U. pompholygodes* (Schlecht.) Rab.)   1   2   3   4
U. eranthidis (Pass.) Ainsw. & Sampson   1
U. ficariae (Liro) Moesz   On *Trollius europaeus*   4

U. gladiolicola Ainsw.		1			
U. occulta Rab.	Casual. On *Secale cereale*	1			
U. primulicola Magnus	On *Primula vulgaris* (anamorph *Paepalopsis irmischiae* Kühn)	1			
U. sorosporioides Körn.	On *Aquilegia* and *Thalictrum*	"London"			
U. violae (Sow.) Fisch. v. Wald.		1	2		

Ustilago (Pers.) Roussel

Sori in floral or vegetative tissue, spores separate without sterile cells. *U. anomala* Kunze on *Polygonum* spp. in E. Anglia and elsewhere.

Ustilago bullata Berk. (*U. bromivora* (Tul.) Fisch. v. Wald.)	On *Bromus* spp.	1			4
U. grandis Fr.	On culms of *Phragmites*				4
U. hypodytes (Schlecht.) Fr.	On culms of *Agropyron, Bromus, Elymus* etc.	1	2		4
U. intermedia Schroet.	In anthers of *Scabiosa columbaria*	1	2		
U. longissima (Sow.) Meyen	On leaves of *Glyceria fluitans* & *G. maxima*	1	2	3	
U. salvei B. & Br. (*U. striiformis* (West.) Niessl)	On *Holcus* etc.	1	2	3	
U. scabiosae (Sow.) Wint.	In anthers of *Knautia arvensis*	1	2		
U. segetum (Bull.) Rouss. (*U. hordei* (Rab.) Lager.)					4
U. segetum var. *avenae* (Pers.) Brun.		1	2	3	4
U. segetum var. *tritici* (Pers.) Brun.		1	2	3	
U. serpens (Karst.) Lindeb. (*U. macrospora* Desm.)	On *Agropyron* & *Bromus*	1			
U. succisae Magn.	In anthers of *Succisa pratensis*	1	2		
U. tragopogonis-pratensis (Pers.) Rouss.		1	2		4
U. utriculosa (Nees) Tul.	In flowers of *Polygonum* spp.	1		3	
U. vaillantii Tul.	In anthers of *Chionodoxa, Muscari* & *Scilla* spp.	1			4
U. violacea (Pers.) Rouss.	In anthers of *Lychnis* & *Silene* spp.	1	2	3	4
U. violacea var. *stellariae* (Sow.) Sav.		1			
U. zeae (Beckm.) Ung.		1			4

UREDINALES

All obligate parasites.

Chrysomyxa Ung

Heteroecious, alternating between *Picea* spp. and *Empetrum, Pirola* or *Rhododendron*. Four species in the north and west.

Coleosporium Lév.

Heteroecious, alternating between *Pinus* spp. and Campanulaceae, Compositae or Scrophulariaceae.

Coleosporium tussilaginis (Pers.) Lév.		1	2	3	4
	0I on *Pinus sylvestris*	1	2	3	4
	II III on *Calanthe reflexa* (Orchidaceae)	1			
	On *Campanula glomerata, C. persicifolia, C. rapunculoides, C. rotundifolia* and *C. trachelium*	1	2		
	On *Euphrasia* spp.		2	3	
	On *Melampyrum pratense*	1			
	On *Notonia grandiflora*	1			
	On *Odontites verna*	1	2		

	1	2	3	4
On *Petasites hybridus*				
On *Rhinanthus*				4
On *Senecio pulcher, S. sylvaticus, S. vulgaris*	1	2	3	4
On *Sonchus arvensis, S. asper, S. oleraceus*	1	2		4
On *Tropaeolum canariensis*	1			
On *Tussilago farfara*	1	2	3	4

Cronartium Fr.

Heteroecious, alternating between *Pinus* spp. and Paeoniaceae or Grossulariaceae.

		1	2	3	4
Cronartium flaccidum (A. & S.) Wint.	0I on twigs of *Pinus sylvestris*				
	II III on *Paeonia mascula*		2	3	
C. ribicola Fisch.	0I on twigs of *Pinus strobus* and other 5-needled species				
	II III on *Ribes nigrum* and other species	1		3	4

Cumminsiella Art.

Autoecious on *Mahonia*.

		1	2	3	4
Cumminsiella mirabilissima (Peck) Nannf.	0I, II, III on *Mahonia aquifolium* since 1932	1	2	3	

Endophyllum Lév.

Microcylic with aecidioid III.

		1	2	3	4
Endophyllum euphorbiae-sylvaticae (DC.) Wint.	0 III on *Euphorbia amygdaloides*	1			4
E. sempervivi de Bary	0 III casual on *Sempervivum tectorum* and allied species	1			

Frommea Art.

Autoecious on Rosaceae.

		1	2	3	4
Frommea obtusa (Str.) Art.	0I II III on *Potentilla reptans*			3	

Gymnosporangium Hedwig f.

Heteroecious between Rosaceae and Cupressineae.

		1	2	3	4
Gymnosporangium clavariiforme (Pers.) DC.	0I on *Crataegus* spp.	1	2		4
	III on *Juniperus communis*	1			
G. confusum Plowr.	0I on *Crataegus*	1			
	On *Cydonia oblonga*				4
	On *Mespilus germanica*		2	3	4
	III on *Juniperus sabina*				
G. cornutum Kern	0I on *Sorbus aucuparia*	1			
	III on *Juniperus communis*				
G. fuscum DC.	0I on *Pyrus communis*	1	2		
	III on *Juniperus sabina*	1			

Hyalopsora Magn.

Some heteroecious between *Abies* and Pteridophyta, others apparently microcyclic *H. aspidiotus* (Magn.) Magn in Scotland, *H. polypodii* (Diet) Magn. Derbyshire northwards.

Kuehneola Magn.

Autoecious on Rosaceae.

		1	2	3	4
Kuehneola uredinis (Link) Art.	0I II III on *Rubus fruticosus*	1	2	3	4

Melampsora Cast.

Heteroecious or Autoecious.

		1	2	3	4
Melampsora allii-populina Kleb.	(0I on *Allium* & *Arum*) II III on *Populus*				4
M. capraearum Thuem.	0I on *Larix*, II III on *Salix caprea* or *S. cinerea*	1	2	3	4
M. epitea Thuem.	0I on *Larix* indistinguishable from the preceding, 0I on *Euonymus europaeus* in Hampshire, Berkshire; II III on *Salix* spp.	1	2	3	4
M. euphorbiae Cast.	0I II III on *Euphorbia exigua*, *E. helioscopiae*, *E. peplus*	1	2	3	4
M. hypericorum Wint.	I III on *Hypericum androsaemum*		2	3	4
	On *H. calycinum* & *H.X inodorum*	1			
	On *H. hirsutum*		2		
M. laricipopulina Kleb.	(0I on *Larix*) II III on *Populus* spp.	1		3	4
M. lini (Ehrenb.) Lév.	0I II III on *Linum catharticum*	1		3	4
M. lini var. *liniperda* Körn.	0I II III on *Linum usitatissimum*	1		3	4
M. populnea (Pers.) Karst.	0I on Mercurialis perennis	1	2	3	4
	II III on *Populus alba*	1			
	II III on *P. tremula*	1	2	3	4
	0I on *Pinus sylvestris*	1			

Melampsorella Schroet.
Heteroecious between *Abies* and Caryophyllaceae or Boraginaceae.

		1	2	3	4
Melampsorella caryophyllacearum Schroet.	0I on *Abies alba*	1			
	II III on *Cerastium tomentosum* etc.	1			4
	II III on *Stellaria holostea*	1			4
M. symphyti Bub.	II on *Symphytum officinale* & *S. X uplandicum*	1	2	3	

Melampsoridium Kleb.
Heteroecious between *Larix* and *Alnus* or *Betula*.

		1	2	3	4
Melampsoridium betulinum (Fr.) Kleb.	0I on *Larix potaninii*				4
	II III on *Betula* spp.	1	2	3	4

Milesina Magn.
Heteroecious between *Abies* and Pteridophyta but 0I seldom seen in Britain.

		1	2	3	4
Milesina blechni Syd.	II III on *Blechnum spicant*		2	3	4
M. dieteliana (Syd.) Magn.	II III on *Polypodium vulgare*	1	2		
M. kriegeriana (Magn.) Magn.	II III on *Dryopteris dilatata*, *D. filixmas* & *D. spinulosa*	1	2	3	
M. scolopendrii (Faull) Henderson	II III on *Phyllitis scolopendrium*	1	2	3	

Miyagia Miyabe
Autoecious on Compositae.

		1	2	3	4
Miyagia pseudosphaeria (Mont.) Jørst.	I II III on *Sonchus arvensis*, *S. asper* & *S. oleraceus*	1			4

Nyssopsora Art.
Apparently microcyclic; *N. echinata* (Lév.) Art. on *Meum athamanticum* in Scotland.

Ochropsora Diet.
Heteroecious between Ranunculaceae and Rosaceae.

		1	2	3	4
Ochropsora ariae (Fuck.) Ramsb.	0I on *Anemone nemorosa*	1		3	
	II III on *Sorbus aucuparia*				

Phragmidium Link
Autoecious on Rosaceae.

		1	2	3	4
Phragmidium bulbosum (Strauss) Schlecht.	0I II III on *Rubus fruticosus* (Corylifolii)	1	2	3	4
P. fragariae (DC.) Karst.	0I II III on *Potentilla sterilis*	1	2		4
P. potentillae (Pers.) Karst.	Needs confirmation, confusion with *Frommea* possible		2		
P. mucronatum (Pers.) Schlecht.	0I II III especially on native *Rosa* spp.	1	2	3	4
P. rosae-pimpinellifoliae Diet.	01 II III on *Rosa pimpinellifolia*			3	
P. rubi-idaei (DC.) Karst.	0I II III on *Rubus idaeus*	1	2		4
P. sanguisorbae (DC.) Schroet.	0I II III on *Poterium sanguisorba*	1	2	3	4
P. tuberculatum J. Müller	0I II III on *Rosa* cultivars	1	2		4
P. violaceum (Schultz) Wint.	0I II III on *Rubus fruticosus* & cultivars	1	2	3	4

Puccinia Pers.

Life cycles various, from Heteroecious to microcyclic but always on Angiosperms.

		1	2	3	4
Puccinia acetosae Koern.	II rarely III on *Rumex acetosa*	1	2	3	4
P. adoxae DC.	III on *Adoxa moschatellina*	1	2		4
P. aegopodii (Strauss.) Röhl.	III on *Aegopodium podagraria*	1			
P. albescens Plowr.	0I II III on *Adoxa moschatellina*	1	2		4
P. allii Rud.	(0I) II III on *Allium porrum* & *A. schoenoprasum*	1	2	3	4
P. annularis (Strauss.) Röhl.	III on *Teucrium scorodonia*	1	2		4
P. antirrhini Diet. & Holway	II & III on *Antirrhinum majus* since 1934	1	2	3	4
P. apii Desm.	0I II III on *Apium graveolens*, probably extinct			3	4
P. arenariae (Schum.) Wint.	III on *Dianthus barbatus*	1		3	4
	On *Moehringia trinervia*	1	2	3	
	On *Silene dioica*	1	2	3	
	On *Stellaria graminea*			3	
	On *S. media*	1	2		4
P. argentata (Schultz.) Wint.	0I on *Adoxa moschatellina*	1			
	II III on *Impatiens capensis*	1			
P. asparagi DC.	0I II III on *Asparagus officinalis*	1			4
P. behenis Otth	II III on *Silene cucubalus*	1	2		
P. betonicae DC.	III on *Stachys officinalis*	1	2	3	
P. brachypodii Otth	II III on *Brachypodium sylvaticum*	1	2		4
P. bupleuri	(0I II) III on *Bupleurum tenuissimum*		2		
P. buxi DC.	III on *Buxus sempervirens*	1	2		
P. calcitrapae DC.	0I II III on *Arctium*				4
	On *Carduus acanthoides, C. crispus, C. nutans*	1	2	3	4
	On *Carduus tenuiflorus*			3	4
	On *Carlina vulgaris*				4
	On *Centaurea nigra, C. scabiosa*	1	2	3	4
	On *Cirsium palustre*	1	2	3	4
P. calthae Link	0I II III on *Caltha palustris*	1			
P. calthicola Schroet.	Needs confirmation	1			
P. campanulae Berk.	III on *Campanula* spp.	1			
P. caricina DC.	0I on *Urtica dioica*	1	2	3	4
	0I on *Ribes uvacrispa*			3	4
	II III on *Carex acutiformis*	1			
	On *C. hirta*	1			
	On *C. laevigata*				4
	On *C. pendula*	1			4
	On *C. pseudocyperus*	1			
	On *C. rostrata*		2		
	On *C. vesicaria*	"Sussex"			
P. caricina var. *paludosa* (Plow.) Hend.	II III on *C. panicea*	1			
P. chrysanthemi Roze	II III on *Chrysanthemum* cultivars only	1			
P. chrysosplenii Grev.	III on *Chrysosplenium oppositifolium*		2		
P. circaeae Pers.	III on *Circaea lutetiana*	1	2	3	4

		1	2	3	4
P. cnici Mart.	0I II III on *Cirsium vulgare*	1		3	
P. cnicioleracei Pers.	III on *Aster tripolium*		2		4
	On *Achillea millefolium*	1		3	
	On *Cnicus palustris*	1	2	3	4
P. conii Lagerh.	(0I) II III on *Conium maculatum*	1			4
P. coronata Corda	0I on *Rhamnus catharticus* (*Frangula alnus* in E. Anglia etc.)	1	2	3	4
	II III on *Agrostis* spp.	1			
	On *Arrhenatherum elatius*		2		4
	On *Avena sativa*	1			
	On *Festuca arundinacea* & *F. gigantea*	1		3	4
	On *Holcus*			3	
	On *Lolium perenne*	1			4
P. crepidicola Syd.	II III on *Crepis capillaris* & *C. taraxacifolia*	1	2		
P. cyani Pass.	0I II III on *Centaurea cyanus*	1	2		
P. deschampsiae Art.	II on *Deschampsia cespitosa*	1		3	
P. difformis Kunze	0I II III on *Galium aparine*	1			4
P. dioicae Magn.	(0I on Compositae) II III on *Carex montana*				4
P. epilobii DC.	III on *Epilobium palustre*	1	2		
P. festucae Plowr.	0I on *Lonicera periclymenum*	1	2		4
	II III on *Festuca rubra*		2		
P. galiicruciatae Duby	0I II III on *Galium cruciata*		2		4
P. galiiverni Ces.	III on *Galium cruciata*		2		4
	On *G. saxatile*	1			
P. gentianae Röhl.	(0I) II III on *Gentiana acaulis* (cult.)	1	2	3	
P. glechomatis DC.	III on *Nepeta hederacea*	1	2	3	4
P. glomerata Grev.	III on *Senecio jacobaea*	1			
P. graminis Pers.	0I on *Berberis vulgaris*				4
	II III on *Agropyron repens*	1			
	On *Agrostis gigantea*	1			
	On *Arrhenatherum elatius*	1			
	On *Deschampsia cespitosa*				4
	On *Phleum pratense*			3	
	On *Triticum aestivum*				4
P. heraclei Grev.	0I II III on *Heracleum sphondylium*	1		3	
P. hieracei Mart.	0I (Uredinioid) II III on *Hieracium* spp.	1			4
	On *Centauria nigra*	1	2		
	On *Cichorium intybus*	1			4
	On *Leontodon autumnalis* & *L. hispidus*	1	2		
	On *Pieris hieracioides*				4
	On *Taraxacum*	1		3	4
P. hieracii var. *hypochaeridis* (Oud.) Jørst.	0I II III on *Hypochaeris radicata*	1	2		4
P. hieracii var. *piloselloidarum* (Probst.) Jørst.	*0I II III on Hieracium pilosella*	1			
P. hordei Otth	(*0I on Ornithogalum pyrenaicum* in Berkshire) II III on *Hordeum murinum*	1			4
	On *H. vulgare*	1	2		4
P. hysterium (Strauss.) Röhl.	0I II on *Tragopogon pratensis*	1	2	3	4
P. iridis Rab.	(0I absent?) II III on *Iris foetidissima* & *I. germanica*	1			4
P. kusanoi Diet.	(I on *Deutzia* in Asia) II III on *Arundinaria fastuosa*			3	
P. lagenophorae Cooke	I III on *Senecio squalidus* & *S. vulgaris* since 1961	1	2	3	4
P. lapsanae Fuck.	0I II III on *Lapsana communis*	1	2	3	4
P. liliacearum Duby	0I III on *Ornithogalum pyrenaicum* in Berks. on *O. umbellatum*	1			
P. longicornis Pat. & Har.	On *Arundinaria fastuosa* & *Sasa veitchii*	1			
P. luzulae Lib.	II III on *Luzula pilosa*		2		

Species	Host/Notes	1	2	3	4
P. maculosa (Strass) Röhl.	(0) I II III on *Mycelis muralis*	1	2		
P. magnusiana Körn.	0I on *Ranunculus* spp. II III on *Phragmites communis*				4
P. malvacearum Bert.	Introduced 1873 III on *Althaea rosea*	1	2	3	4
	on *Malva sylvestris*	1			4
P. menthae Pers.	0I II III on *Clinopodium vulgare*	1	2		
	On *Mentha aquatica, M. arvensis, M. cardiaca, M.* × *gentilis, M.* × *piperata, M.* × *smithiana,M. spicata* & *M.* × *verticillata*	1	2	3	4
	On *Origanum vulgare*	1			
P. obscura Schroet.	0I on *Bellis perennis*	1		3	
	II III on *Luzula campestris, L. forsteri, L. pilosa*	1	2	3	4
	On *L. sylvatica*			3	
P. opizii Bub.	(0I on *Lactuca sativa* & *L. virosa*) II III on *Carex muricata*		2		
P. oxalidis Diet. & Ellis	II III on *Oxalis corymbosa, O. floribunda* since 1973	1	2	3	
P. pazschkei Diet.	III on *Saxifraga longifolia* cult.	1			
P. pelargoniizonalis Doidge	II III on *Pelargonium zonalis* since 1879, 1965	1	2	3	
P. phragmitis (Schum.) Körn.	0I on *Rheum rhaponticum*				4
	On *Rumex crispus* & *R. obtusifolius*		2		
	II III on *Phragmites communis*		2		
P. pimpinellae (Strauss.) Röhl.	0I II III on *Pimpinella saxifraga*		2		
P. poaenemoralis Otth	II III on *Anthoxanthum odoratum*	1			
	On *Arrhenatherum elatius*	1			
	On *Poa angustifolia, P. annua, P. nemoralis (P. compressa, P. pratensis, P. trivialis* in Berk.) on *Puccinellia maritima*		2		
P. poarum Niels	0I on *Tussilago farfara*	1	2	3	4
	II III on *Poa pratensis* & *P. trivialis* necessarily present				
P. polygoniamphibii Pers.	(0I on *Geranium*) II III on *Polygonum amphibium*	1	2		
P. polygoniamphibii var. *convolvuli* Art.	0I on *Geranium dissectum*		2		
	II III on *Polygonum convolvulus*	1	2		4
P. primulae Duby	I II III on *Primula vulgaris*	1	2	3	4
P. prostii Duby	0 III on *Tulipa australis* since 1963	1			
P. pulverulenta Grev.	0I II III on *Epilobium hirsutum*	1	2		4
	On *E. montanum*		2		
P. punctata Link	0I II III on *Asperula odorata*				4
	On *Galium cruciatum, G. mollugo, G. saxatile, G. verum*	1	2	3	4
P. recondita Rob. & Desm.	0I on *Lycopus arvensis*	1			4
	0I on *Aquilegia vulgaris*			3	
	II III on *Agropyron repens*	1	2		4
	On Agrostis spp.	1			
	On *Bromus erectus, B. lepidus, B. mollis, B. ramosus, B. secalinus, B. sterilis*	1	2		4
	On *Holcus lanatus* & *H. mollis*	1	2		4
	On *Secale cereale*	1			
	On *Trisetum flavescens*			3	
	On *Triticum aestivum*	1			
P. rugulosa Tranzsch.	0I II III on *Peucedanum officinale*				4
P. saniculae Grev.	0I II III on *Sanicula europaea*	1	2		4
P. satyrii Syd.	III on *Satyrium aureum* casual alien	1			
P. schroeteri Pass.	III on *Narcissus pseudonarcissus* perhaps extinct	1			

Species	Host	1	2	3	4
P. sessilis Schroet.	0I on *Allium ursinum*		2	3	
	On *Arum maculatum*	1	2		4
	II III on *Phalaris arundinacea*	1	2	3	4
P. smyrnii Biv. Bern.	0I II III on *Smyrnium olusatrum*			3	4
P. striiformis West.	II III on *Triticum aestivum*, ?also *Hordeum*	1	2		4
P. tanaceti DC.	(0I) II III on *Artemisia absinthium*, *A. vulgaris*	1	2		4
	On *Chrysanthemum coccineum*		2		
	On *Tanacetum vulgare*	1			
P. thesii Duby	0I II III on *Thesium humifusum*	1	2	3	
P. thymi (Fuck.) Hend.	III on *Origonum vulgare*	1			4
P. tumida Grev.	II III on *Conopodium majus*	1	2	3	
P. umbilici Duby	III on *Umbilicus rupestris*	1			
P. variabilis Grev.	I II III on *Taraxacum officinale*	1	2		
P. veronicae Schroet.	III on *Veronica montana*	1	2	3	
P. vincae Berk.	0I II III on *Vinca major*	1	2	3	4
P. violae DC.	0I II III on *Viola hirta, V. odorata, V. riviniana*	1	2	3	4
	On Viola cultivars	1	2		
P. virgae-aureae (DC.) Lib.	III on *Solidago virgaurea*	1	2		

Pucciniastrum Otth.

Heteroecious between *Abies* or *Tsuga* and various Dicotyledons; *P. guttatum* (Schroet.) Hyl.et al occurs on *Galium* in Hertfordshire.

Species	Host	1	2	3	4
Pucciniastrum agrimoniae (Diet.) Tranzsch.	II (III) on *Agrimonia eupatoria*	1		3	
P. circaeae (Wint.) de Toni	(0I on *Abies alba*) II III on *Circaea lutetiana*	1	2	3	4
P. epilobii Otth	(0I on *Abies grandis*) II III on *Chamaenerion angustifolium*	1	2		4
	II on *Fuchsia* spp. cult.	1			
P. vaccinii (Wint.) Jørst.	(0I on *Tsuga*) II III on *Vaccinium myrtillus*	1	2		

Trachyspora Fuck.

Autoecious, *T. intrusa* (Grev.) Art. is on *Alchemilla vestita* in Hertfordshire.

Tranzschelia Art.

Heteroecious or microcyclic, on Ranunculaceae or Rosaceae.

Species	Host	1	2	3	4
Tranzschelia anemones (Pers.) Nannf.	III on *Anemone nemorosa*	1	2	3	4
	On *Thalictrum* spp.	1			
T. discolor (Fuck.) Tranz. & Litv.	0I on *Anemone coronaria*	1	2	3	
	II III on *Prunus armeniaca*			3	4
	On *P. domestica* & ssp. *insititia*	1	2	3	4
	On *P. spinosa*		2		

Triphragmium Link

Autoecious on Rosaceae.

Species	Host	1	2	3	4
Triphragmium filipendulae Pass.	I II III on *Filipendula vulgaris*			3	
T. ulmariae (DC) Link	0I II III on *Filipendula ulmaria*	1	2	3	4

Uredinopsis Magn.

Heteroecious between *Abies* and Pteridophyta, *U. filicina* Magn. on *Thelypteris* in Hertfordshire.

Uredo Pers.

Imperfectly known species with III not yet identified.

Species	Host	1
Uredo behnickiana Henn.	Casual alien on *Epidendrum vitellinum*, *Phajus* sp.	1

		1	2	3	4
U. oncidii Henn.	Alien on *Oncidium aureum, O. pulchellum, Ornithidium coccineum* & *Spiranthes* sp.	1		3	
U. quercus Duby	On *Quercus ilex* & *Q. robur*	1		3	

Uromyces Ung.
Separated from *Puccinia* solely by nonseptate teleutospores.

		1	2	3	4
Uromyces acetosae Schroet.	0I II III on *Rumex acetosa.* confirmation desirable	1	2		
U. aecidiiformis (Str.) Rees	0I III alien on *Lilium candidum*	1			
U. airaeflexuosae Ferd. & Winge	II III on *Deschampsia flexuosa*	1			
U. anthyllidis Schroet.	II III on *Anthyllis vulneraria*		2	3	
U. appendiculatus (Pers.) Ung.	0I II III on *Phaseolus coccineus* & *P. vulgaris*				4
U. aritriphylli (Schw.) Seeler	0I only, alien on *Arisaema triphyllum*	1			
U. armeriae Kickx	0I II III on *Armeria maritima*			3	
U. behenis (DC.) Ung.	I III on *Silene maritima* & *S. vulgaris*		2	3	
U. betae (Pers.) Tul.	0I II III on *Beta vulgaris* ssp. *maritima*		2		4
U. chenopodii (Duby) Schroet.	0I II III on *Suaeda maritima*		2		
U. colchici Massee	III alien on *Colchicum autumnale, C. speciosum*	1			
U. dactylidis Otth.	0I on *Ranunculus acris,* (*R. bulbosus* in Berks.)	1			
	On *R. repens*	1	2		4
	On *R. ficaria*	1	2	3	4
	II III on *Dactylis glomerata*	1	2	3	
	II III on *Poa trivialis* (*P. compressa, P. palustris* & *P. pratensis* in Berkshire)	1			
U. dianthi (Pers.) Niessl	II III on *Dianthus barbatus*	1	2		
	on *D. caryophyllus*	1			
U. ervi West.	0I II III on *Vicia hirsuta*	1	2		
U. eugentianae Cumm.	II III on *Gentianella amarella*		2		
U. fallens (Desm.) Kern.	0I II III on *Trifolium pratense*	1		3	4
U. ficariae (A. & S.) Lév.	III on *Ranunculus ficaria*	1	2	3	4
U. geranii (DC.) Lév.	0I II III on *Geranium molle*	1			
	G. pratense			3	
	G. pyrenaicum	1			4
U. junci (Desm.) Tul.	(0I on *Pulicaria dysenterica* in Hants.) II III *Juncus articulatus*	1			
U. limonii (DC.) Lév.	0I II III on *Limonium vulgare*		2		4
	On *L. latifolium* & *L. tataricum* v. *angustifolium*		"Sussex"		4
U. lineolatus (Desm.) Schroet.	0I on *Oenanthe crocata*		2		
	On *Glaux maritima*		2		
	II III on *Scirpus maritimus*		2		
U. minor Schroet.	I III on *Trifolium dubium*	1			
U. muscari (Duby) Lév.	III on *Muscari polyanthum*	1			
	On *Endymion hispanicum* and hybrid	1			
	On *Endymion nonscriptum*	1	2	3	4
U. nerviphilus (Grogn.) Hots.	III on *Trifolium repens*	1	2	3	4
U. pisi (DC.) Otth	0I on *Euphorbia cyparissias*				4
	II III on *Cytisus battandieri*			1	
	On *Laburnum anagyroides*				4
	On *Lathyrus pratensis* (possible confusion with *U. viciaefabae*)	1	2		
	On *Lotus corniculatus*	1			
	On *Medicago arabica*	1	2		4
	On *M. lupulina*				4
	On *M. sativa*			3	
	On *Onobrychis viciifolia*	1			4
	On *Pisum sativum*				4

		1	2	3	4
	On *Sarothamnus scoparius*				4
U. polygoniaviculariae (Pers.) Karst.	0I II III on *Polygonum aviculare*	1	2		4
U. rumicis (Schum.) Wint.	II III on *Rumex conglomeratus, R. crispus, R. hydrolapathum, R. obtusifolius, R. palustris*	1	2	3	4
U. salicorniae de Bary	I II III on *Salicornia europaea & S. ramosissima*		2	3	4
U. trifolii (DC.) Lév.	0I II III on *Trifolium hybridum*	1			
	On *T. repens*	1			
U. scrophulariae Fuck.	0I III on *Scrophularia aquatica & S. nodosa*	1			4
U. scutellatus (Schrank) Lév.	III on *Euphorbia cyparissias*				4
U. valerianae (DC.) Lév.	0I II III on *Valeriana officinalis*	1	2	3	4
U. viciaefabae (Pers.) Schroet.	0I II III on *Lathyrus* see *U. pisi*				
	On *Vicia cracca*	1			4
	On *V. faba*		2	3	4
	On *V. sativa*	1	2		
	On *V. sepium*	1			
U. viciaefabae var. *orobi* (Schum.) Jørst.	II III on *Lathyrus montanus*	1		3	

Xenodochus Schlecht.
Autoecious on *Sanguisorba*.

Xenodochus carbonarius Schlecht.	I III on *Sanguisorba officinalis*	1	2		

Zaghouania Pat.
Autoecious on Oleaceae.

Zaghouania phillyreae Pat.	0I II (III not seen) on *Phillyrea latifolia* casual alien probably extinct.		2	3	

ASCOMYCETES

TAPHRINALES

Taphrina Fr.
Obligate parasites of Pteridophyta and angiosperm foliage and inflorescences. *Taphrina amentorum* (Sadeb.) Rostr. on *Alnus* in E. Anglia, several additional species on various hosts mainly in the north and west.

Taphrina betulina Rostr.	Causing witches' broom on *Betula*	1			4
T. bullata (Berk.) Tul.	Leaf blister on *Pyrus communis* & *Pyronia* × *veitchii*	1			
T. carpini (Rostr.) Johans.	Witches' broom on *Carpinus*	1	2		
T. cerasi (Fuck.) Sadeb.	Brooms and leaf curl on *Prunus avium* & *P. cerasus*	1	2	3	4
T. crataegi Sadeb.	Leaf curl on *Crataegus*	1			
T. deformans (Berk.) Tul.	Leaf curl on *Prunus amygdalus* and *P. persica*	1	2	3	4
T. populina Fr.	Leaf spot of *Populus* × *canadensis* v. *serotina, P. italica, P. nigra, P. tristis*	1	2	3	4
T. potentillae (Farl.) Johans.	On *Potentilla erecta*	1			
T. pruni (Fuck.) Tul.	On leaves & fruit of *Prunus domestica* and *P. spinosa*	"Sussex"			
T. sadebeckii Johans.	Leafspot on *Alnus glutinosa*	1			
T. tosquinetii (West.) Magn.	Leafcurl of *Alnus glutinosa*	1	2	3	
T. ulmi (Fuck.) Johans.	Leafcurl of *Ulmus*		2		4

PROTOMYCETALES

Burenia Reddy & Kramer
Segregate from *Protomyces*.

Burenia inundata (Dang.) Reddy & Kramer	Leaf gall on *Apium nodiflorum*	2

Protomyces Ung.
Obligate parasites causing galls on foliage of angiosperms.

P. macrosporus Ung.	On *Aegopodium podagraria* and *Anthriscus sylvestris* (on *Angelica, Conopodium, Oenanthe, Sium* elswhere)	1	2	3	4
T. pachydermus Thuem.	On *Taraxacum officinale*	1		3	

Protomycopsis Magn.
Parasitic on Compositae, *P. leucanthemi* Magn. In Derbyshire, *P. bellidis* Krieg. in Scotland.

Taphridium Lagerh. & Juel
Obligate parasites of Umbelliferae.

T. umbelliferarum (Rost.) Lagerh. & Juel	On foliage of *Heracleum sphondylium*	1

PEZIZALES

Aleuria Fuck.
  Terrestrial saprophytes.

	1	2	3	4
Aleuria aurantia (Fr.) Fuck.	1	2	3	4
A. cestrica Ell. & Ev.		2		
A. luteonitens (B. & Br.) Gill.				4

Allescheria Sacc. & Syd. See *Pseudallescheria* and *Thielavia*.

Anthracobia Boud.
  Saprophytic on charcoal or burned soil.

		1	2	3	4
Anthracobia macrocystis (Cooke) Boud.		1	2		4
A. maurilabra (Cooke) Boud	Probably only a colour phase of the following	1	2	3	4
A. melaloma (A. & S.) Boud.		1		3	4

Ascobolus Pers.
  Saprophytes, many coprophilous, others on plant debris or bare soil. Amongst others *A. degluptus* v. Brumm. is in the Midlands, *A. hansenii* Paulsen & Dissing in Middlesex, *A. rhytidosporus* v. Brumm. in E. Anglia. *A. asininus* Cke. & Mass. is to be rejected as inadequately described.

		1	2	3	4
Ascobolus albidus Crouan	Coprophilous	1	2	3	
A. boudieri Quél.	Usually coprophilous		2		
A. brassicae Crouan	On rabbit dung or decaying vegetable matter	1			
A. carbonarius Karst.	On burned soil or charcoal	1	2		
A. crenulatus Karst.	Coprophilous	1		3	
A. denudatus Fr.	On rotten wood, leaf mould or soil	1	2	3	
A. epimyces (Cke.) Seaver	On rotten wood or fallen leaves	1			
A. foliicola B. & Br.	On rotting leaves	1			4
A. furfuraceus Pers.	Coprophilous	1	2	3	4
A. immersus Pers.	Coprophilous, in dung of herbivores	1	2	3	4
A. lignatilis A. & S.	On rotten wood or debris	1			
A. roseopurpurascens Rehm	Coprophilous on dung of herbivores	1	2		4
A. stictoidens Speg.	On dung	1			

Ascodesmis v. Tiegh.
  Saprophytes, often coprophilous.

	1	2	3	4
Ascodesmis nigricans v. Tiegh.	1			
A. porcina Seaver			3	
A. volutelloides Mass. & Salm.	1			

Ascozonus (Renny) Boud.
  Coprophilous.

	1	2	3	4
Ascozonus woolhopensis (Renny) Boud.	1			

Boubovia Svrček
  Saprophytes.

	1	2	3	4
Boubovia nicholsonii (*Humaria nicholsonii* Massee, Naturalist 1901, 188) (Massee) Spooner & Yao	1			

Boudiera Cooke
  Terrestrial saprophytes, *B. areolata* Cooke & Phill. in Wales.

Byssonectria Karst.
Terrestrial saprophytes; *Inermisia* Rifai is a synonym.

Byssonectria fusispora (Berk.) Rogerson & Korf	On peaty soil	1		3	4

Caloscypha Boud.
Terrestrial saprophyte; *C. fulgens* (Pers.) Boud. casual record in E. Anglia.

Cheilymenia Boud.
Coprophilous, terrestrial or on leaf mould.

Cheilymenia fibrillosa (Currey) Le Gal	Terrestrial	1		3	4
C. fimicola (de Not. & Bagl.) Dennis	Coprophilous		2		
C. raripila (Phill.) Boud.	Coprophilous	1			
C. rubra (Phill.) Boud.	On leaf mould	1			
C. stercorea (Pers.) Boud.	Coprophilous	1	2	3	4
C. theleboloides (A. & S.) Boud.	On leaf mould	1	2	3	
C. vitellina (Pers.) Dennis	Terrestrial	1			4

Coprobia Boud.
Coprophilous.

Coprobia granulata (Bull.) Boud.	1	2	3	4

Coprotus Korf
Coprophilous.

Coprotus aurorus (Crouan) Kimbr. et al.		2		4
C. cf. *disculus* Kimbr., Luck Allen & Cain			3	
C. granuliformis (Crouan) Kimbr. (*Ascophanus argenteus* (Curr.) Boud.)				4
C. lacteus (Cooke & Phill.) Kimbr. et al.	1	2		
C. ochraceus (Crouan) Larsen		2		
C. rhyparobioides (Heimerl) Kimbr.	1			
C. sexdecemsporus (Crouan) Kimbr. & Korf	1			

Desmazierella Lib.
Saprophyte on fallen gymnosperm needles.

Desmazierella acicola Lib. (Anam. *Verticicladium trifidum* Preuss) on *Pinus*	1

Discina (Fr.) Fr.
Probably mycorrhizal, *D. leucoxantha* Bres. & *D. perlata* (Fr.) Fr. In the north and west.

Disciotis Boud.
Terrestrial saprophyte.

Disciotis venosa (Pers.) Boud.	1	2	3	4

Eleutherascus v. Arx
Terrestrial saprophyte, *E. tuberculatus* Samson & Luiten reported in the midlands.

Fimaria Vel.
Coprophilous.

Fimaria cervaria (Phill.) van Brumm.	1		
F. hepatica (Batsch) van Brumm.			4
F. theioleuca (Roll.) van Brumm.	1	2	

Flavoscypha Harmaja
Terrestrial saprophytes, segregate from *Otidea*; *F. cantharella* (Fr.) Harm. in E. Anglia, *F. phlebophora* (B. & Br.) Harm. in Middlesex.

Geopora Harkness
Terrestrial saprophytes (=*Sepultaria* (Cooke) Boud.); *G. arenicola* (Lév.) Kers. in Essex.

Geopora arenosa (Fuck.) Ahmad		1			4
G. foliosa (Schaeff.) Ahmad		1			
G. sumneriana (Cooke) de la Torre	Possibly mycorrhizal with *Cedrus*	1	2		4
G. tenuis (Fuck.) T. Schumacher		1			

Geopyxis (Pers.) Sacc.
Terrestrial saprophytes, associated with burning.

Geopyxis carbonaria (A. & S.) Sacc.		1	3

Greletia Donadini
Terrestrial saprophyte, *G. planchonis* (Dun.) Donadini in the west midlands.

Gyromitra Fr.
Terrestrial, associated with gymnosperms; *G. infula* (Schaeff.) Quél. in E. Anglia

Gyromitra esculenta (Pers.) Fr.	1	2	3	4

Helvella L.
Terrestrial, including *Cyathipodia* Boud, *Leptopodia* Boud, *Macroscyphus* Gray & *Paxina* O.K.

Helvella (Paxina) acetabulum (L.) Quél.	1	2	3	4
H. (Cyathipodia) corium (Weberb.) Massee	1			4
H. crispa Fr.	1	2	3	4
H. (Leptopodia) elastica Bull.	1	2	3	4
H. (L.) ephippium Lév.	1	2	3	
H. fusca Gill.	1			
H. lacunosa Afz.	1	2	3	4
H. latispora Boud.				4
H. (P.) leucomelaena (Pers.) Nannf.	1		3	4
H. (Macroscyphus) macropus (Pers.) Karst.	1	2	3	4
H. (L.) pezizoides Afz.				4
H. (L.) stevensii (Peck)	1	2	3	
H. (C.) villosa (Hedw.) Dissing & Nannf.	1			

Humaria Fuck.
Terrestrial saprophytes.

Humaria hemisphaerica (Web.) Fuck.	1	2		4

Iodophanus Korf
Saprophytes on soil, plant debris or rotting textiles or even dung.

Iodophanus carneus (Pers.) Korf	1	2	3	4
I. testaceus (Moug.) Korf	1			

Jafneadelphus Rifai
Terrestrial saprophytes, *J. amethystinus* (Phill.) van Brumm. in the midlands.

Kotlabaea Svrček
Terrestrial saprophytes.

Kotlabaea deformis (Karst.) Svrček	On bare soil	1		3

Lasiobolus Sacc.
Coprophilous saprophytes.

Lasiobolus ciliatus (Schmidt) Boud.	1	2	3	4
L. cf. *cuniculi* Vel.	1			

Leucoscypha Boud.
Terrestrial saprophytes, often closely associated with bryophyta; *L. erminea* (Bomm. & Rouss.) in E. Anglia. Here taken to include *Neottiella* (Cooke) Sacc with *N. ricciae* (Crouan) Le Gal in E. Anglia.

Leucoscypha hetieri (Boud.) Rifai	1		
L. leucotricha (A. & S.) Boud.	1	3	
L. rutilans (Fr.) Dennis & Rifai	1		4
L. semi-immersa (Karst.) Svrček		2	
L. vivida (Nyl.) Dennis & Rifai	1		4

Marcelleina van Brumm. Korf & Rifai
Terrestrial saprophytes; *M. persoonii* (Crouan) van Brumm. in E. Anglia.

Marcelleina atroviolacea (Del.) van Brumm.	1
M. rickii Rehm) Graddon	1

Melastiza Boud.
Terrestrial saprophytes; *M. flavovirens* (Rehm) Pfister in E. Midlands, *M. scotica* Graddon in Hampshire.

Melastiza chateri (W.G. Smith) Boud.	1	2	3	4

Microstoma Bernst.
Saprophytes, *M. protracta* (Fr.) Kanouse formerly in the Highlands.

Miladina Svrček
Lignicolous saprophyte.

Miladina lecithina (Cooke) Svrček	(Anamorph *Actinospora megalospora* Ingold)	1	2	4

Monascus v. Tiegh.
Saprophytes on animal and vegetable products, *M. pilosus* Sato in the north.

Monascus purpureus Went	1
M. ruber v. Tiegh.	1

Morchella Pers.
Terrestrial saprophytes.

Morchella esculenta (L.) Pers.	1	2	3	4
M. esculenta var conica (Pers.) Fr.	1	2	3	4
M. (Mitrophora) semilibera DC.	1	2	3	4
M. vaporaria de Brond.	1			

Octospora Hedw.
Terrestrial saprophytes, often intimately associated with individual genera of musci; here includes *Lamprospora* de Not. *O. pseudomuralis* Graddon is in the west country, *O. rustica* (Vel.) Moravec & *O. tetraspora* (Fuck.) Korf in E. Anglia.

Octospora crec'hqueraultii (Crouan) Caillet & Moyne		2	4
O. dictydiola (Boud.) Caillet & Moyne	1	2	4

O. humosa (Fr.) Dennis	1	2	4
O. leucoloma Hedw.			4
O. miniata (Crouan) Caillet & Moyne	1	2	4
O. polytrichi (Schum.) Caillet & Moyne	1		
O. roxheimii Dennis & Itzerott	1		
O. rubens (Boud.) Moser	1		

Octosporella Dobbeler
?Parasites, on Hepaticae, *O. jungermanniarum* (Crouan) Dobbeler in the north.

Orbicula Cooke
Saprophyte, of uncertain systematic position.

Orbicula parietina (Schrad.) Hughes (*Anixia cyclospora* (Cooke) Sacc.)	1	2

Otidea (Pers.) Bon.
Terrestrial saprophytes.

Otidea alutacea (Pers.) Massee	1	2		4
O. bufonia (Pers.) Boud.	1	2	3	4
O. cochleata (Bull.) Fuck.	1	2	3	4
O. leporina (Batsch) Fuck.	1	2	3	4
O. grandis (Pers.) Rehm. Needs confirmation	1			
O. onotica (Pers.) Fuck.	1	2	3	4

Pachyella Boud.
Saprophytes; *P. violaceonigra* (Rehm) Pfister is in E. Anglia.

Pachyella babingtonii (Berk.) Boud.	1	2

Peziza Willd. non L.
Saprophytes; *P. ammophila* Dur. & Mont. is in E. Anglia; *P. brunneoatra* Desm. in Middlesex; *P. epixyla* Rich. in Hertfordshire.

Peziza ampelina Quél. Needs confirmaton		2		
P. ampliata Pers.	1	2	3	4
P. apiculata Cooke	1			
P. arvernensis Boud.	1	2		4
P. asterigma Vuill.	1			
P. aurata (LeGal) Yao & Spooner	1			
P. badia Pers.	1	2	3	4
P. badioconfusa Korf (*Galactinia olivacea* (Boud.) Boud.)	1	2	3	
P. badiofusca (Boud.) Dennis Needs confirmation	1	2		
P. bovina Phill.				4
P. buxea Quél.	1			4
P. cerea Bull.	1		3	4
P. depressa Pers. (*P. castanea* Quél.)	1			
P. domiciliana Cooke (*P. adae* Cooke)	1			
P. echinospora Karst. Anamorph *Oedocephalum elegans* Preuss	1	2	3	4
P. emileia Cooke (*Galactinia howsei* Boud.)	1	2	3	4
P. limnaea Maas Geesteranus (*Galactinia limosa* (Grelet) LeGal)	1			4
P. linteicola Phill. & Plowr.	1			
P. michelii (Boud.) Dennis (*P. plebeia* (LeGal) Nannf.)	1	2		4
P. micropus Pers. (*P. repanda* auct.)	1	2		4
P. obtusiapiculata Moravec	1			
P. ostracoderma Korf Anamorph *Chromelosporium ollare* (Pers.) Henneb.	1	2	3	
P. palustris Boud.	1			
P. petersii Berk. & Curt (*Galactinia sarrazinii* Boud.)	1	2		4
P. proteana (Boud.) Seaver	1			4

		1	2	3	4
P. proteana var *sparassoides* (Boud.) Dur. (*Gyromitra phillipsii* Massee)		1			4
P. pseudoviolacea Donadini		1	2	3	
P. reperta (Boud.)	Doubtful record	1			
P. saccardiana Cooke		1			
P. saniosa Schrad.			2		4
P. sepiatra Cooke		1	2		4
P. sterigmatizans Phill.			2		
P. succosa Berk.		1	2		4
P. succosella (LeGal & Romagn.) Dennis			2		
P. vacinii (Vel.) Svrček		1			
P. varia Hedw.		1	2		4
P. vesiculosa Bull.		1	2	3	4
P. violacea Pers. (*P. praetervisa* Bres.)		1	2		4

Pithya Fuck.
  Saprophytes on plant remains.

		1	2	3	4
Pithya arethusa Vel.	On *Ligustrum*		2		
P. cupressina (Pers.) Fuck.	Anamorph *Molliardomyces cupressina* Paden, on *Juniperus*				4

Plectania Fuck.
  Terrestrial saprophyte.

		1	2	3	4
Plectania melastoma (Sow.) Fuck.			2	3	

Plicaria Fuck.
  Saprophytes on burnt ground, distinguished from *Peziza* by spherical ascospores.

		1	2	3	4
Plicara carbonaria (Fuck.) Fuck. (*P. trachycarpa* v. *muricata* Grelet)		1			
P. endocarpoides (Berk.) Rifai (*P. leiocarpa* (Curr.) Boud.)		1	2	3	4
P. trachycarpa (Curr.) Boud.	Anamorph *Chromelosporium trachycarpum* Henneb.	1	2		4

Pseudombrophila Boud.
  Saprophytes on plant remains, compost or urea-enriched soil.

		1	2	3	4
Pseudombrophila deerata (Karst.) Seaver		1			

Pseudoplectania Fuck.
  Saprophytes, *P. nigrella* (Pers.) Fuck. under conifers and *P. sphagnicola* (Pers.) Kriesel on *Sphagnum* in the north and west.

Pseudotis Boud.
  Saprophytes, doubtfully separable from *Otidea*.

		1	2	3	4
Pseudotis apophysata Cooke & Phillips) Boud.	On wet soil	1			

Pulvinula Boud.
  Saprophytes mainly on burnt soild; *P. cinnabarina* (Fuck.) Boud. in Hampshire.

		1	2	3	4
Pulvinula carbonaria (Fuck.) Boud.		1			
P. convexella (Karst.) Pfister (*P. constellatio* (B.&Br.) Boud.)		1	2		4

Pyronema Carus
  Saprophytes on burnt ground, damp plaster and steam-sterilised soil.

		1	2	3	4
Pyronema domesticum (Sow.) Sacc.		1	2	3	

P. glaucum (Boud.) Sacc. 1
P. omphalodes (Bull.) Fuck. 1 2 4

Rhizina Fr.
Saprophyte on burnt soil and facultative parasite on young conifers.

Rhizina undulata Fr. 1 2 3 4

Saccobolus Boud.
Coprophilous saprophytes.

	"Sussex"		
Saccobolus caesariatus Renny			
S. depauperatus (B.& Br.) Hansen	1	3	4
S. dilutellus Fuck.) Sacc.	1		
S. glaber (Pers.) Lamb. (*S. kerverni* (Crouan) Boud.)	1		
S. globuliferellus Seaver	1		
S. obscurus (Cooke) Phill.	1		
S. quadrisporus Massee & Salmon	1		
S. versicolor (Ksrat.) Karst. (*S. violascens* Boud.)	1 2		4

Sarcoscypha (Fr.) Boud.
Saprophytes on woody plant remains.

Sarcoscypha coccinea (Jacquin) Lamb.   Anamorph *Molliardiomyces* 1 2 3 4

Sarcosphaera Auersw.
Terrestrial saprophyte, perhaps mycorrhizal, with semisubterranean apothecia.

Sarcosphaera crassa (Santi) Pouzar (*S. eximia* (Dur.& Lév.) Maire) 1 2 4

Scutellinia (Cooke) Lamb.
Lignicolous or terrestrial saprophytes; *S. subhirtella* Svrček in Hertfordshire.

Scutellinia asperior (Nyl.) Dennis ~~techispora~~	1 2 3 4	
S. barlae (Boud.) LeGal		4
S. crinita (Bull.) Lamb. (*S. hirtella* (Rehm) O.K.; *S. cervorum* (Vel.) Svrček)	1 2	
S. crucipila (Cooke & Phill.) Moravec	1	4
S. olivascens (Cooke) O.K. (*S. ampullacea* (Limm.) O.K.)	1 2	
S. pseudotrechispora (Schroet.) LeGal		4
S. scutellata (L.) Lamb. Published records probably apply largely to *S. crinita*	1 2 3 4	
S. setosa (Nees) O.K.	1 2	
S. stenosperma LeGal (*S. hirta* auct.)	1 2 3	
S. trechispora (B.& Br.) Lamb.	2 3	
S. umbrarum (Fr.) Lamb.	1 2	

Sowerbyella Nannf.
Terrestrial saprophyte.

Sowerbyella radiculata (Sow.) Nannf. 1 2 3 4

Sphaerosoma Klotzsch
Terrestrial saprophytes; *S. echinulatum* Seaver on wet ground.

Sphaerosporella Svrček & Kubička
Terrestrial saprophytes akin to *Trichophaea*.

Sphaerosporella brunnea (A.& S.) Svrček & Kubička 1 2

Sphaerozone Zobel
Terrestrial saprophytes; *S. ostiolatum* (Tul.) Setchell in the west midlands.

Spooneromyces Schumacher & Moravec
Saprophytes; *S. laeticolor* (Karst.) Schumacher & Moravec in Hampshire.

Tazzetta (Cooke) Lamb.
Terrestrial saprophytes = *Pustularia* Fuck. *T. gaillardiana* (Boud.) Korf & Rogers in the west midlands.

Tazzetta catinus (Holmsk.) Korf & Rogers including *Pustularia ochracea* Boud.	1	2	3	4
T. cupularis (L.) Lamb.	1	2	3	4
T. rosea (Rea) Dennis	1		3	

Thecotheus Boud.
Saprophytes on various substrata; *T. agranulosus* Kimbr. & *T. rivicola* (Vaček) Kimbr. & Pfist. in E. Anglia, *T. pelletieri* (Crouan) Boud. in Yorkshire.

Thecotheus apiculatus Kimbr.	1	
T. cinereus (Crouan) Chenantais	1	2

Thelebolus Tode
Coprophilous saprophytes = *Rhyparobius* in part.

Thelebolus caninus (Auersw.) Jeng & Krug	On dog dung	1	
T. crustaceus (Fuck.) Kimbr.	On dung of rabbit &c.	1	
T. microsporus (B.& Br.) Kimbr.	On dung of rabbit &c.	1	
T. myriosporus (Crouan) van Brumm.	On dung of herbivores	1	2
T. nanus Heimerl	On dung of rabbit &c.	1	
T. stercoreus Tode	Doubtful	1	

Tricharina Eckblad
Saprophytes, mainly terrestrial or associated with burning.

Tricharina cretea (Cooke) Thind & Waraitch	On soil or plaster	1	
T. fimbriata (Quél.)	On soil		2
T. praecox (Karst.) Dennis	On burnt ground in spring	1	2

Trichobolus (Sacc.) Kimbr. & Cain
Coprophilous, *Thelebolus* with setose apothecia.

Trichobolus pilosus (Schroet.) Kimbr.		3
T. zukalii (Heimerl.) Kimbr.	1	

Trichophaea Boud.
Saprophytes, terrestrial or associated with burning.

Trichophaea abundans (Karst.) Boud. (*Anthracobia humillima* Malençon)	Anamorph *Dichobotrys abundans* Henneb., on charred wood or burnt soil	1	2		
T. gregaria (Rehm) Boud.	On soil			3	
T. hemisphaerioides (Mout.) Graddon	On burnt ground	1	2	3	
T. woolhopeia (Cooke & Phill.) Boud.	On burnt ground	1	2		4

Trichophaeopsis Korf & Erb
Segregated by having forked hairs, *T. bicuspis* (Boud.) Korf & Erb in Essex.

Verpa Swartz
Terrestrial saprophyte.

	1	2		4
Verpa digitaliformis Pers.	1	2		4
V. conica (Mull.) Swartz may be the same				

TUBERALES

Subterranean carpophores, perhaps derivatives of Pezizales, probably all mycorrhizal.

Balsamia Vitt.
Also in the west of England *B. platyspora* B.& Br. and *B. vulgaris* Vitt.

	1			
Balsamia fragiformis Tul.	1			

Choiromyces Vitt.

	1		3	4
Choiromyces meandriformis Vitt.	1		3	4

Genea Vitt.
Also *G. sphaerica* Tull. & *G. verrucosa* Vitt. in west of England.

	1	2		4
Genea hispidula B. & Br.		2		4
G. klotzschii B. & Br.	1			

Gyrocratera P. Henn.

	1			
Gyrocratera ploettneriana Henn. (*Hydnotrya ploettneriana* (Henn.) Hawker)	1			

Hydnobolites Tul.

		2		
Hydnobolites cerebriformis Tul.		2		

Hydnotrya B. & Br.

	1			
Hydnotrya tulasnei B. & Br.	1			

Pachyphloeus Tul.

Pachyphloeus citrinus B. & Br., *P. conglomeratus* B. & Br. & *P. melanoxanthus* (Tul) Tul. in the west of England.

Paurocotylis Berk.
Alien, introduced from New Zealand, *P. pila* Berk. established in the midlands.

Stephensia Tul.

Stephensia bombycina (Vitt.) Tul. in the west and north.

Tuber Micheli ex Wigg

		1	2	3	4
Tuber aestivum Vitt.	Edible and formerly exploited e.g. in Hampshire	1	2	3	4
T. brumale Vitt.			2		
T. excavatum Vitt.		1	2		4
T. foetidum Vitt.			2		
T. macrosporum Vitt.			2		
T. maculatum Vitt.			2	3	
T. puberulum B.& Br.		1			4
T. rufum Pico			2	3	4

HELOTIALES

Allophylaria Karst.
Saprophytes on plant remains, akin to *Cyathicula*; *A. clavuliformis* Karst. in Hertfordshire, *A. crystallifera* Graddon & *A. macrospora* (Kirschst.) Nannf. in E. Anglia.

Allophylaria nana (Sacc.) Sacc.   1

Amorphotheca Parbery
Saprophytes.

Amorphotheca resinae Parbery   Anam. *Hormoconis resinae* (Lindau) v. Arx
                                         & de Vries   1

Apostemidium Karst.
Saprophytes on watersoaked wood, akin to *Vibrissea* but sessile apothecia; mainly in the north and west but *A. fiscellum* (Karst.) Karst in Essex, *A. guernisacii* (Crouan) Boud in Hertfordshire and *A. leptospora* (B. & Br.) Boud. in Buckinghamshire.

Arachnopeziza Fuck.
Lignicolous saprophytes; *A. obtusipila* Grelet in the north and west.

Arachnopeziza aurata Fuck.	1	2	3	4
A. aurata var *alba* Grelet (*A. nivea* Lort.)			3	
A. aurelia (Pers.) Fuck.	1	2	3	
A. eriobasis (Berk.) Korf	1			
A. trabinelloides (Rehm) Korf	1			

Arachnoscypha Boud.
Saprophyte.

Arachnoscypha aranea (deNot.) Boud.   On fallen cupules of *Castanea*   1   4

Ascocorticium Bref.
Lignicolous saprophyte.

Ascocorticium anomalum (Ell. & Harkn.) Schroet.   1   3   4

Ascocoryne Groves & Wilson
Lignicolous saprophytes or endophytes in timber.

Ascocoryne cylichnium (Tul.) Korf	1	2		4
A. sarcoides (Jacq.) Groves & Wilson Amamorph *Coryne dubia* (Pers.) Gray	1	2	3	4
A. solitaria (Rehm) Dennis	1			

Ascotremella Seaver
Lignicolous saprophyte.

Ascotremella faginea (Peck) Seaver   1   2   3   4

Belonioscypha Rehm
Saprophyte, especially on dead grass culms, akin to *Cyathicula*.

Belonioscypha culmicola (Desm.) Dennis   1   2

Belonium Sacc.
Saprophytes on Gramineae; *B. hystrix* (de Not.) v. Höhn. on *Molinia* in Hampshire, *B. psammicola* (Rostr.) Nannf. on *Ammophila* in E. Anglia.

Belonopsis Rehm
Saprophytes on Gramineae or marsh plants, perhaps inseparable from *Niptera* Fr.; *B. excelsior* on *Phragmites* in the north and west, *B. graminea* (Karst.) Sacc. & Syd., *B. iridis* (Crouan) Graddon, *B. mediella* (Karst.) Aeli.

Belonopsis filispora (Cooke) Nannf.	On *Brachypodium*	1			
B. knieffii (Wallr.) c.n. (*B. rhenopalaticum* (Rehm) Dennis, *Tapesia retincola* (Rab.) Karst.1					

Betulina Vel.
Foliicolous saprophytes, *B. fuscostipitata* Graddon in the west.

Bisporella Sacc.
Saprophytes, mainly lignicolous, also *B. fuscotincta* (Graddon) Dennis in the north and west.

Bisporella citrina (Batsch) Korf & Carpenter	Especially on *Fagus*	1	2	3	4
B. ochracea (Boud.) Korf		1			
B. pallescens (Pers.) Carpenter & Korf (*Calycella monilifera* (Fuck.) Dennis)		1			
B. scolochloae (de Not.) Spooner	On culms of Gramineae	1			
B. subpallida (Rehm) Dennis		1	2	3	4
B. sulfurina (Quél.) Carpenter	On effete lignicolous perithecia	1	2	3	
Calycella terrestris Boud. perhaps belongs here		1			

Blumeriella v. Arx
Facultative foliar parasite of *Prunus* spp.

Blumeriella jaapii (Rehm) v. Arx	As anamorphs *Microgloeum pruni* Petr. & *Phleosporella padi* (Lib.) v. Arx	1

Bryoscyphus Spooner
On Hepaticae, ?parasitic; *B. conocephali* (Boyd) Spooner on *Conocephalum* in E. Anglia.

Bulgaria Fr.
Lignicolous saprophyte with massive gelatinised apothecia.

Bulgaria inquinans (Pers.) Fr.	Especially on *Betula, Fagus* & *Quercus*	1	2	3	4

Bulgariella Karst.
Lignicolous saprophyte.

Bulgariella pulla (Fr.) Karst.	1

Calloria Fr.
Saprophytes on *Urtica* stems.

Calloria carneoflavida Rehm a *Laetinaevia* teste Hein				3	
C. neglecta (Lib.) Hein (*C. fusarioides* (Berk.) Fr.)	Anam. *Cylindrocolla urticae*	1	2		4

Calycellina v. Höhn.
Saprophytes, mainly foliicolous, *C. chlorinella* (Ces.) Denn., *C. indumenticola* Graddon and *C. ochracea* (Grel. & Croz.) Denn. are in the west and north; *C. caricina* Dennis and *C. microspis* (Karst.) Denn. in E. Anglia and *C. spiraeae* (Rob.) Denn. in Herts.

Calycellina leucella (Karst.) Dennis		1	
C. populina (Fuck.) v. Höhn.		1	
C. punctiformis (Grev.) v. Höhn. (*Urceolella puberula* (Lasch) Boud.)		1	3

Catinella Boud.
Lignicolous saprophyte.

Catinella olivacea (Batsch) Boud.   1   2

Cenangium Fr.
Saprophytes on dead twigs or coniferous needles; *C. acuum* Cooke & Peck on *Pinus* in the north *C. graddonii* Dennis in Hampshire.

Cenangium ferruginosum Fr.   On *Pinus*   1

Chaenothecopsis Vainio
Lignicolous saprophytes; *C. caespitosa* (Phill.) Hawksw. in the west and north.

Chaenothecopsis debilis (Turner & Borrer) Tibell   2

Chlorencoelia Dixon
Lignicolous saprophyte; *C. versiformis* (Pers.) Dixon in Berkshire.

Chloroscypha Seaver
Saprophytes on foliage of gymnosperms; *C. seaveri* Seaver on *Cupressus* and *Thuja* in the north.

Chloroscypha sabinae (Fuck.) Seaver   On *Juniperus*   1   2

Chlorosplenium Fr.
Saprophytes; *C. aeruginellum* (Nyl.) Karst. in E. Anglia.

Chlorosplenium aeruginascens (Nyl.) Karst.   1   2   3   4

Ciboria Fuck.
Saprophytes or facultative parasites of senescent floral parts in angiosperms; *C. acerina* Whetzel & Buchwald on *Myrica gale* in Hampshire, *C. aschersoniana* (Henn. & Ploettn.) Whetzel on *Carex* in E. Anglia, *C. alni* (Maul) Whetzel on *Alnus* in Scotland.

Ciboria amentacea (Balb.) Fuck.	*Alnus* race	1		3		
	Salix race	1	2	3		
C. americana Durand	On *Castanea*	1				
C. batschiana (Zopf) Buchwald	On *Quercus*	1			3	4
C. betulae (Wor.) White	On *Betula*	1	2			
C. conformata (Karst.) Svrček			2			4
C. viridifusca (Fuck.) v. Höhn.		1				

Ciliolarina Svrček
Saprophyte especially on debris of conifers.

Ciliolarina laricina (Raitviir) Svrček   1   4

Cistella Quél.
Saprophytes on wood or vegetable debris, largely segregates from *Lachnum* with *Clavidisculum* Kirschst. as a synonym.

Cistella acuum (A.& S.) Svrček. (*C. kriegeriana* Kirschst. is a synonym)   1   4
C. fugiens (Bucknall) Matheis   1
C. bullii (W.G. Smith) Dennis   1   2
C. cf. geelmuydenii Nannf.   1
C. grevillei (Berk.) Raschle   1

Claussenomyces Kirschst.
Lignicolous saprophytes with small gelatinised apothecia = *Corynella* Boud. non DC.

Claussenomyces atrovirens (Pers.) Korf & Abawi		1	
C. prasinula (Karst.) Korf & Abawi	Anamorph *Dendrostilbella prasinula* v. Höhn.	1	2

Coleosperma Ingold
Saprophyte on submerged Cyperaceae; *C. lacustre* Ingold in the north.

Coronellaria Karst.
Saprophyte on monocotyledons; *C. amaena* Boud. in the north.

Crocicreas Fr.
Saprophytes on Gramineae; *C. gramineum* (Fr.) Fr. & var *incertellum* (Rehm) Carp. in Scotland.

Crumenulopsis Groves
Facultative parasites of *Pinus* spp.; *C. pinicola* (Fr.) Groves in E. Anglia, *C. sororia* (Karst.)Groves in Hampshire.

Cryptodiscus Corda
Lignicolous saprophytes; *C. foveolaris* (Rehm) Rehm in the north and west.

Cryptodiscus rhopaloides Sacc.	On *Buxus, Rubus, Salix, Sambucus* &c.	1

Cudonia Fr.
Terrestrial saprophytes; *C. circinans* (Pers.) Fr. & *C. confusa* Bres. in Scotland.

Cudoniella Sacc.
Saprophytes, segregated from *Hymenoscyphus*.

Cudoniella acicularis (Bull.) Boud.	On wood, especially *Quercus*	1	2	3	
C. clavus (A.& S.) Dennis	On sodden debris	1	2	3	4
C. rubicunda (Rehm) Dennis	On cones of *Pinus*	1			

Cyathicula de Not.
Saprophytes on plant remains; *C. strobilina* (Fr.) Korf & Dixon, *C. tomentosa* Dennis, *C. turbinata* (Syd.) Dennis in the north and west.

Cyathicula cacaliae (Pers.) Dennis	1	2		
C. coronata (Bull.) de Not.	1			4
C. cyathoidea (Bull.) Thuem.	1	2	3	4
C. dolosella (Karst.) Dennis	1			4
C. pteridicola (Crouan) Dennis	1			
C. rubescens (Mout.) Arendh.	1			
C. subhyalina (Rehm) Dennis	1	2		

Cystopezizella Svrček
Saprophytes.

Cystopezizella conorum (Rehm) Svrček	1	4

Dematioscypha Svrček
Lignicolous saprophyte.

Dematioscypha dematiicola (B.& Br.) Svrček	Anamorph *Haplographium delicatulum* B. & Br.	1	3

Dencoeliopsis Korf
Lignicolous saprophyte, *D. johnstonii* (Berk.) Korf on *Betula* in the north.

Dennisiodiscus Svrček
Saprophytes on marsh vegetation &c.; *D. prasinus* (Quél.) Svrček on *Glyceria* in Middlesex, *D. virescentibus* (Mout.) Svrček on *Carpinus* in Essex.

Dermea Fr.
Saprophytes or facultative parasites of angiosperm trees; *D. ariae* (Pers.) Tul. on *Sorbus* in the north and west, *D. tulasnei* Groves on *Fraxinus* in the west.

Dermea cerasi (Pers.) Fr.	Anamorph *Foveostroma drupacearum* (Lev.) DiCosmo	1	3
D. prunastri (Pers.) Fr.		1	

Dibeloniella Nannf.
Saprophytes; *D. trichophoricola* Graddon in Scotland is a *Nimbomollisia* teste Nannfeldt.

Diplocarpa Massee
Lignicolous saprophyte.

Diplocarpa bloxamii (Berk.) Seaver		1	3 4

Diplocarpon Wolf
Foliar parasites of angiosperms.

Diplocarpon earlianum (Ell. & Ev.) Wolf	Anamorph *Marssonina fragariae* (Lib.) Kleb.	2	4
D. mespili (Sorauer) Sutton	Anamorph *Entomosporium mespili* (DC.) Sacc.	1	3 4
D. rosae Wolf	Anamorph *Marssonina rosae* (Lib.) Diedicke	1 2	3 4
D. saponariae (Ces.) Nannf.	Anamorph *Diplosporonema delastrei* (Delacr.) Petrak	1	3 4

Discinella Boud.
Terrestrial saprophytes.

Discinella boudieri (Quél.) Boud. var. *spadicea* Boud.	1
D. margarita Buckley	1

Discohainesia Nannf.
Saprophyte or facultative parasite.

Discohainesia oenotherae (Cooke & Ellis) Nannf.	Anamorph *Hainesia lythri* (Desm.)v. Höhn.	2 3

Drepanopeziza (Kleb.) v. Höhn.
Facultative foliar parasites, especially of Grossulariaceae and Salicaceae; *D. populi-albae* (Kleb.) Nannf. in E. Anglia.

Drepanopeziza populorum (Desm.) v. Höhn.	Anamorph *Marssonina populi-nigrae* Kleb.	1	3
D. punctiformis Gremmen	On *Populus*, anamorph *M. brunnea* (Ell.& Ev.) Magn.	1	
D. ribis (Kleb. v. Höhn.	Anamorph *Gloeosporidiella ribis* (Lib.) Petrak	1	3 4
D. salicis (Tul.) v. Höhn.	Anamorph *Monostichella salicis* (West.) v. Arx	1	

D. sphaerioides (Pers.) v. Höhn.	Anamorph *Marssonina salicicola* (Bres.) Magn.	1	
D. triandrae Rimpau	Anamorph *Marssonina kriegeriana* (Bres.) Magn.	1	4

Duebenia Fr.
Saprophyte on dead herbaceous stems; *D. compta* (Sacc.) Nannf.

Durella Tul.
Lignicolous saprophytes, *Xylogramma* Wallr. is a doubtful synonym; *D. compressa* (Pers.) Tul. anamorph only formerly in the west, *D. macropus* Fuck. & *D. suecica* (Starb.) Nanff. in the north.

Durella atrocyanea (Fr.) v. Höhn.	1		3	4
D. commutata Fuck.	1	2		4
D. connivens (Fr.) Rehm	1			

Echinula Graddon
Foliicolous saprophyte.

Echinula asteriadiformis Graddon	1

Encoelia (Fr.) Karst.
Lignicolous saprophytes; *E. glauca* Dennis on *Corylus* in the north, *E. siparia* (B. & Br.) Nannf. on *Ulmus* in E. Anglia, *E. tiliacea* (Fr.) Karst. on *Tilia* in the midlands and west.

Encoelia fascicularis (A. & S.) Karst.	On *Fraxinus* & *Populus*	1			4
E. fuckelii Dennis	On *Prunus spinosa*			3	
E. furfuracea (Roth) Karst.	On *Alnus* & *Corylus*	1	2	3	4
E. glaberrima (Rehm) Kirschstein	On *Carpinus*				4

Encoeliopsis Nannf.
Lignicolous saprophytes, *E. laricina* (Ettlinger). Groves in Hampshire.

Eriopezia (Sacc.) Rehm
Lignicolous saprophytes.

Eriopezia caesia (Pers.) Rehm	Especially on *Quercus*	1	2	3	4

Eupropolella v. Höhn.
Foliar parasites.

Eupropolella britannica Greenh. & M. Jones	On *Prunus laurocerasus*, anamorph *Cryptocline phacidiella* (Grove) v. Arx	1	4

Fabrella Kirschstein
Foliar parasite, *F. tsugae* (Farl.) Kirschst. on *Tsuga canadensis* in the north.

Geoglossum Pers.
Terrestrial saprophytes.

Geoglossum cookeianum Nannf.	1	2	3	4
G. fallax Durand	1	2	3	4
G. glutinosum Pers.	1	2	3	4
G. nigritum Cooke	1	2		4
G. starbaeckii Nannf.		2		
G. vleugelianum Nannf.				4

Gloeotinia Wilson et al.
Facultative parasite of Gramineae, distributed on infected seed.

Gloeotinia granigena (Quél.) Schumacher   Anamorph *Endoconidium temulentum* Prill. & Del. No local records but inevitably present, ubiquitous on *Lolium*.

Godronia Moug. & Lév.
Facultative parasites on woody angiosperms, *G. cassandrae* Peck, *G. spiraeae* (Rehm) Seaver, *G. uberiformis* Groves and *G. urceolus* (Schm.) Karst. in the north, *G. fuliginosa* (Fr.) Seaver and *G. ribis* (Fr.) Seaver in E. Anglia.

Godronia callunigera (Karst.) Karst.   Unconfirmed record   1

Gorgoniceps Karst.
Saprophytes; *G. boltonii* (Phill.) Dennis on *Equisetum* and *G. micrometra* (B. & Br.) Sacc. on *Juncus* &c., in the north and west.

Gorgoniceps aridula (Karst.) Karst.   On *Pinus*   1

Graddonia Dennis
Lignicolous saprophyte, *G. coracina* (Bres.) Dennis in the north and west.

Grahamiella Spooner
Foliicolous saprophyte, *G. dryadis* (Nannf.) Spooner on *Dryas* in the north.

Gremmeniella Morelet
Facultative parasite of *Pinus*, *Gremmeniella abietina* (Lagerh.) Morelet with anamorph *Brunchorstia pinea* (Karst.) v. Höhn. widespread in the north and west.

Habrostictis Fuck.
Lignicolous saprophyte, *H. rubra* Fuck. in E. Anglia.

Haglundia Nannf.
Lignicolous saprophytes akin to *Mollisia*, *H. elegantior* Graddon in the north.

Haglundia perelegans Nannf.   2

Hamatocanthoscypha Svrček
Saprophytes; *H. uncipila* (LeGal) Huhtinen in the midlands.

Hamatocanthoscypha laricionis (Vel.) Svrček On *Larix*   1   3

Heterosphaeria Grev.
Saprophytes on angiosperm stems.

Heterosphaeria patella (Tode) Grev.   Anam. *Heteropatella bonordenii* (Haszl.) Lind on Umbelliferae   1   2   3   4

Heyderia (Fr.) Link
Saprophytes on coniferous debris.

Heyderia abietis (Fr.) Link   On *Pinus* (on *Picea* elsewhere)   1   4

Hyalinia Boud.
Lignicolous saprophytes, allied to *Orbilia*.

Hyalinia rubella (Pers.) Nannf.   Record unreliable, Surrey material under the name is *Orbilia delicatula* Karst.

Hyalopeziza Fuck.

Saprophytes on dead stems and leaves, *H. corticicola* (Dennis) Raitviir on *Myrica* in the north.

Hyalopeziza ciliata Fuck.	On leaves of *Acer, Betula* &c.	1		
H. trichodea (Phill. & Plowr.) Raitviir	On *Pinus* needles	1		

Hyaloscypha Boud.

Saprophytes on woody substrata; *H. britannica* Huhtinen in Berkshire, *H. epipora* Huhtinen and *H. intacta* Svrček in the north.

Hyaloscypha albohyalina (Karst.) Boud.		1	
H. albohyalina var *spiralis* (Vel.) Huhtinen		1	3
H. aureliella (Nyl.) Huhtinen	Anamorph *Cheiromycella microscopicum*	1	4
(*H. stevensonii* (B. & Br.) Nannf;	(Karst.) Hughes)		
H. velenovskii Grad.)			
H. daedaleae Vel. (*H. hyalina* (Pers.) Boud. ss auct. angl. pro max parte)		1	
H. fuckelii Nannf.		1	
H. herbarum Vel.		1	
H. leuconica (Cooke) Nannf.		1	4
H. leuconica var *bulbopilosa* (Feltg.) Huhtinen		1	2
H. paludosa Dennis	This is a *Protounguicularia* for Huhtinen	1	
H. vitreola (Karst.) Boud.		1	

Hymenoscyphus S.F. Gray

Saprophytes on diverse plant remains; *H. equisetinus* (Vel.) Dennis in the midlands, *H. rhodoleucus* (Fr.) Phill. and *H. robustior* (Karst.) Dennis on marsh plants in E. Anglia.

Hymenoscyphus albidus (Rob. & Desm.) Phill.		1	2		4
H. caudatus (Karst.) Dennis		1	2	3	4
H. cortisedus (Karst.) Dennis		1	2		
H. epiphyllus (Pers.) Rehm		1	2		4
H. fagineus (Pers.) Dennis		1	2		
H. friesii (Weinm.) Arendholz		1			
H. fructigenus (Bull.) S.F. Gray		1	2	3	4
H. herbarum (Pers.) Dennis		1	2		
H. humuli (Lasch) Dennis	Needs confirmation	1			
H. imberbis (Bull.) Dennis		1			
H. immutabilis (Fuck.) Dennis		1			
H. laetus (Boud.) Dennis		1			
H. lutescens (Hedw.) Phill.		1			
H. phyllogenon (Rehm) O.K.		1	2		4
H. phyllophilus (Desm.) O.K.		1			4
H. pileatus (Karst.) O.K.			2		
H. repandus (Phill.) Dennis		1		3	
H. rokebyensis (Svrček) Matheis		1			
H. salicellus (Fr.) Dennis		1			
H. scutula (Pers.) Phill.		1	2	3	4
H. scutula var *consobrinum* (Boud.)		1	2		
H. serotinus (Pers.) Phill.		1			
H. splendens Abdulla, Descals & Webster	Anamorph *Tricladium splendens* Ingold	1			
H. vernus (Boud.) Dennis		1			
H. virgultorum (Vahl) Phill.		1	2	3	4
H. vitellinus (Rehm) O.K.		1		3	
H. vitigenus (de Not.) Dennis		1			

Hysteronaevia Nannf.

Saprophytes on Cyperaceae, *H. fimbriata* Dennis & Sponer, *H. lyngei* (Lind) Nannf., *H. olivacea* (Mout.) Nannf. in Scotland.

Hysteropezizella v. Höhn.
Saprophytes on Cyperaceae, Gramineae and Juncaceae; *H. diminuens* (Karst.) Nannf. in Isle of Wight, *H. hysterioides* (Desm.) Nannf. and *H. rehmii* (Jaap) Nannf. in the west, *H. pusilla* (Lib Nannf. in E. Anglia, *H. foecunda* (Phill.) Nannf. in the north.

Hysterostegiella v. Höhn.
Saprophytes on Cyperaceae and Gramineae; *H. fenestrata* (Rob.) v. Hohn. in E. Anglia, *H. valvata* (Mont.) v. Höhn. widespread on *Ammophila*.

Incrupila Raitviir
Saprophytes; *I. aspidii* (Lib.) Raitv. on *Athyrium* and *Polystichum* in Scotland.

Incrupila melatheja (Fr.) Dennis	On miscellaneous debris	1
I. viridipilosa Graddon		2

Ionomidotis Durand
Lignicolous saprophytes; *I. fulvotingens* (B. & Br.) Cash on *Corylus* in Wales.

Lachnellula Karst.
Saprophytes and facultative parasites on Gymnosperms; *L. calyciformis* (Batsch) Dharne in E. Anglia; *L. pseudofarinacea* (Crouan) Dennis and *L. resinaria* (Cooke & Phill.) Rehm in Hampshire.

Lachnellula occidentalis (Hahn & Ayers) Dharne (*L. hahniana* (Seav.) Dennis)	On *Larix*	1	2	3	4
L. subtilissima (Cooke) Dennis	On *Pinus*	1	2		
L. willkommii (Hartig) Dennis	On *Larix*				4

Lachnum Retz.
Saprophytes on diverse substrata (=*Dasyscyphus* S.F. Gray; *Lachnella* Boud. non Fr.), includes *Albotricha* Raitviir, *Belonidium*, Mont. & Dur., *Dasyscyphella* Tranz., *Erinella* Quél., and *Trichopezizella* (Dennis) Raitviir. *Dasyscyphus acerinus* (Cke. & Ell.) Cash in E. Anglia and *L. imbecille* Karst. common on *Eriophorum* in the north west, also *L. caricis.* (Desm.) v. Höhn. & *L. salicariae* (Rehm) Vel. in E. Anglia, *L. radotinense* Vel. in the midlands, *L. acutum* Vel., *L. clavigerus* (Svrček) Raitviir, *L. clavisporus* (Mout.) Haines, *L. tenue* Kirschst., and *L. trapeziformis* Vel. in the north, *L. relicinum* (Fr.) Karst. in the west.

Dasyscypha abscondita Massee	1			
Lachnum acutipilum (Karst.) Karst. (*Albotricha acutipila* (Karst.) Raitviir)	1			
L. albotestaceum (Desm.) Karst.	1			
L. apalum (B. & Br.) Nannf.	1	2	3	4
L. bicolor (Bull.) Karst.	1	2	3	4
L. brevipilosum Baral	1	2	3	
L. calyculiformis (Schum.) Karst.	1			
L. capitatum (Peck) Svrček (*Dasyscyphus scintillans* Massee)		2		4
L. cerinum (Pers.) Morgan	1	2	3	
L. ciliare (Schrad.) Rehm	1	2	3	
L. clandestinum (Bull.) Karst.	1			
L. controversum (Cooke) Rehm	1	2		
L. (Belonidium) corticale (Pers,) Nannf. (*Lachnella canescens* Cooke)	1	2	3	4
Dasyscyphus coruscatus Graddon	1			
L. crystallinum (Fuck.) Rehm	1			4
L. diminutum (Rob.) Rehm	1		3	4
L. dumorum (Rob.) Huhtinen	1	2	3	4
L. elongatisporum Baral (*Dasyscyphus carneolus* (Sacc.) Rehm v. *longisporus* D.)	1	2		4
L. fascicularis Vel.		2		
L. fuscescens (Pers.) S.F. Gray	1	2	3	4
L. fuscescens var *fagicola* Phill.	1		3	

	1	2	3	4
L. minutissimum (Crouan) Baral (*Dasyscyphus rhytismatis* (Phill.) Sacc.)	1			
L. misellum (Rob. & Desm.) Huhtinen	1			
L. mollissimum (Lasch) Karst. (*Dasyscyphus leucophaeus* (Pers.) Massee)	1	2	3	4
L. (*Trichopezizella*) *nidulum* (Schmidt & Kunze) Karst.		2		4
L. (*Dasyscyphella*) *niveum* (Hedw.) Karst.	1	2	3	4
L. nudipes (Fuck.) Nannf. (*Dasyscypha spiraeicola* (Karst.) Sacc.)	1	2	3	4
L. palearum (Desm.) Karst.	1			
L. papyraceum (Karst.) Karst.	1			
Dasyscypha pteridis (A. & S.) Massee	1	2		
L. pudibundum (Quél.) Schroet.	1	2		
L. pulverulentum (Lib.) Karst.	1			
L. pygmaeum (Fr.) Bres.	1	2		4
L. soppitii (Massee) Raitviir	1			4
L. subnudipes Baral (*Dasyscyphus nudipes* var *minor* Dennis)	1			
L. sulfureum (Pers.) Karst.	1	2	3	4
L. tenuissimum (Quél.) Korf (*Dasyscyphus pudicellus* (Quél.) Schroet.)	1			
L. virgineum (Batsch) Karst.	1	2	3	4

Laetinaevia Nannf.
Saprophytes; *L. marina* (Boyd) Spooner on rotting *Fucus* and *Ascophyllum* in E. Anglia, also *L. Luzulae* Spooner in the north and *L. minutissima* (Rostr.) Hein in the west.

Lambertella v. Höhn.
Saprophytes, mainly in the tropics.

		1	2	3	4
Lambertella tubulosa Abdullah & Webster	As anamorph *Helicodendron tubulosum* (Riess) Linder	1			4

Lanzia Sacc.
Saprophytes, *L. coracina* (Dur. & Lév.) Spooner on *Quercus ilex* in the west, *L. cuniculi* (Boud.) Dumont in the north, *L. stellariae* (Vel.) Spooner & *L. vacini* (Vel.) Spooner in E. Anglia.

		1	2	3	4
Lanzia luteovirescens (Rob.) Dumont & Korf	On *Acer pseudoplatanus*	1	2		4

Leotia Pers.
Terrestrial saprophytes.

	1	2	3	4
Leotia atrovirens Pers.				4
L. lubrica Pers.	1	2	3	4
L. lubrica f. *chlorocephala* (Schw.) Massee	1			

Leptotrochila Karst.
Facultative parasites of herbaceous dicotyledons; *L. brunellae* (Lind) Dennis and *L. radians* (Rob.) Karst. in the north or west; anamorphs in *Sporonema*.

		1	2	3	4
Leptotrochila cerastiorum (Wallr.) Schuepp	On *Cerastium*	1	2		
L. ranunculi (Fr.) Schuepp	On *Ranunculus* spp.	1		3	4
L. verrucosa (Wallr.) Schuepp	On *Asperula* and *Galium* spp.	1			4

Loramyces Weston
Saprophytes on sodden Cyperaceae and Juncaceae; *L. juncicola* Weston and *L. macrospora* Ingold & Chapman in the north and west.

Martininia Korf
Coprophilous/lignicolous saprophyte, *M. panamaensis* (Whetzel) Dumont & Korf in the north.

Melittosporiella v. Höhn.
Lignicolous saprophyte; *M. pulchella* v. Höhn. in the north and west.

Merostictis Clem.
Saprophytes on Cyperaceae and Juncaceae; *M. exigua* (Desm.) Defago on *Juncus* and *M. seriata* (Lib.) Defago on *Carex* in E. Anglia.

Microglossum Gill.
Terrestrial saprophytes.

Microglossum olivaceum (Pers.) Gill.		1		
M. viride (Pers.) Gill.		1	2	3

Micropeziza Fuck.
Saprophytes on Cyperaceae and Gramineae = *Actinoscypha* Karst.; *M. cornea* (B. & Br.) Nannf. in the west.

Micropeziza karstenii Nannf. (*Actinoscypha graminis* Karst.)			3
M. poae Fuck.		1	

Micropodia Boud.
Saprophytes ?=*Pezizella*

Micropodia pteridina (Nyl.) Boud.	1

Microscypha Syd. & Syd.
Saprophytes; *M. ellisii* Dennis and var *eriophori* Dennis in E. Anglia.

M. grisella (Rehm) Syd. & Syd.	On *Pteridium aquilinum*	1	4

Mitrula Fr.
Saprophytes and ? hyperparasites on Typhula; *M. sclerotipus* Boud. in Hertfordshire.

Mitrula paludosa Fr.	On rotting leaves especially *Quercus* and *Molinia*	1	2	3	4

Mniacea Boud.
?Parasites on Hepaticae, *M. nivea* (Crouan) Boud. in the north.

Mniacea jungermanniae (Nees) Boud.	On *Aplozia crenulata* and *Diplophyllum albicans*	1

Moellerodiscus Henn.
Saprophytes, = *Ciboriopsis* Dennis; *M. tenuistipes* (Schroet.) Dumont in E. Anglia.

Moellerodiscus advenulus (Phill.) Dumont	On fallen needles of *Larix*	1

Mollisia (Fr.) Karst.
Saprophytes on diverse substrata. The number of species is uncertain, the distinction between them ill defined and all determinations are provisional. Any separation from *Pyrenopeziza* on one hand and *Tapesia* on the other is irrational and at best traditional. Other names current in the literature include *M. typhae* (Cooke) Phill. in Hertfordshire, *M. artemisiae* (Lasch) Phill, *M. atrorufa* Sacc., *M. fuscostriata* Graddon, *M. junciseda* Karst., *M. millegrana* (Boud.) Nannf., *M. nervicola* Gill. in E. Anglia, *M. caespiticia* Karst., *M. uda* (Pers.) Gill. in the north and west.

Mollisia amenticola (Sacc.) Rehm	1	
M. benesuada (Tul.) Phill.		3

	1	2	3	4
M. caricina Fautrey	1		3	
M. cinerea (Batsch) Karst.	1	2	3	4
M. cinerella (Sacc.) Sacc.	1			4
M. clavata Gremmen	1			
M. coerulans Quél.	1			
M. discolor (Mont.) Karst. var *longispora* LeGal	1			4
M. escharodes (B. & Br.) Gremmen	1		3	4
M. fallax (Desm.) Karst.	1			
M. heterosperma Le Gal	1			
M. hydrophila (Karst.) Sacc.	1			
M. lycopi Rehm	1			
M. lycopincola Rehm	1			
M. ligni (Desm.) Karst.	1	2	3	4
M. melaleuca (Fr.) Sacc.	1			4
M. palustris (Rob.) Karst.	1			
M. phalaridis (Lib.) Rehm	1			
M. pteridina (Nyl.) Karst.	1			
M. ramealis (Karst.) Karst.	1		3	4
M. revincta (Karst.) Rehm	1			4
M. spectabilis Kirschstein	1			
M. undulatodepressa Le Gal		2		
M. ventosa (Karst.) Karst.		2		

Mollisiella Boud.
Saprophytes.

Mollisiella pallens Boud.	1			

Mollisina v. Höhn.
Saprophytes.

		1	2	3	4
Mollisina acerina (Mout.) v. Höhn.	On fallen leaves of *Acer, Quercus* &c.	1			
M. oedema (Desm.) Dennis	With *Phragmidium* pustules on *Rubus*	1			
M. rubi (Rehm) v. Höhn.		1	2		4

Mollisiopsis Rehm
Saprophytes, *Mollisia* with lanceolate paraphyses, *M. lanceolata* (Gremm.) Hawks. in W and N.

Mollisiopsis dennisii Graddon	On *Ulex europaus*	3

Monilinia Honey
Facultative parasites especially on fruit of Rosaceae and *Vaccinium*; anamorphs *Monilia* Bon.

		1	2	3	4
Monilinia baccarum (Schroet.) Whetzel	On *Vaccinium myrtillus*	1			
M. fructigena (Aderh. & Ruhl.) Honey	On *Malus* and *Prunus*, anamorph only	1			4
M. johnsonii (Ell. & Ev.) Honey	On *Crataegus*	1	2	3	4
M. laxa (Aderh. & Ruhl.) Honey	On *Malus* and *Prunus*, anamorph only	1	2	3	4
M. linhartiana (Prill. & Del.) Buchw.	On *Cydonia*	1			
M. mespili (Schell.) Whetzel	On *Mespilus*, if distinct from *M. johnsonii*	1	2		

Naevala Hein
Saprophyte, on fallen leaves of *Quercus*.

Naevala perexigua (Rob.) Holm & Holm (*Naevia minutissima* (Auersw.) Rehm)	1 ·	

Naeviopsis Hein
Saprophytes, *N. tithymalina* (Kunze) Hein on *Euphorbia* in the west.

Neobulgaria Petrak
Lignicolous saprophytes.

	1	2	3	4
Neobulgaria lilacina (Wulf.) Dennis	1	2		
N. pura (Fr.) Petrak	1	2	3	4

Nimbomollisia Nannf.
Saprophytes; *N. melatephroides* (Rehm) Nannf. on *Molinia* in Hampshire, two other species in Scotland.

Niptera Fr.
Saprophytes, most species so called belong elsewhere e.g. in *Belonopsis* or *Mollisia*.

	1		3	4
Niptera lacustris (Fr.) Fr.	1			
N. melatephra (Lasch) Rehm	1			4
N. pilosa (Crossland) Boud.			3	

Ocellaria (Tul.) Karsten
Lignicolous saprophytes hardly separable from *Pezicula*.

		1	2
Ocellaria ocellata (Pers.) Schroet.	On *Salix*, anamorph *Cryptosporiopsis scutellata* (Otth) Petrak	1	2

Ombrophila Fr.
Saprophytes; *O. ambigua* v. Höhn. on *Glyceria* in the north and west.

		1
Ombrophila violacea Fr.	On sodden vegetation	1

Orbilia Fr.
Saprophytes, lignicolous or herbicolous. The species are superficially much alike and the names have been widely misinterpreted. Also *O. acuum* Vel. and *O. epipora* (Nyl.) Karst. in the north and west, *O. comma* Graddon in E. Anglia.

		1	2	3	4
Orbilia alnea Vel.	(*O. xanthostigma* (Fr.) Fr. ss Svrček)	1			
O. auricolor (Blox.) Sacc.	(*O. curvatispora* Boud.)	1			
O. cardui Vel.	(*O. arundinacea* Vel.)		2		
O. delicatula Karst.	(*O. xanthostigma* auct. angl.)	1	2	3	4
O. eucalypti Harkn.		1			
O. inflatula Karst.		1	2		4
O. leucostigma (Fr.) Fr.		1	2		4
O. luteorubella (Nyl) Karst.					4
O. sarraziniana Boud.		1	2		
O. vinosa (A. & S.) Fr.		1			4

Ovulinia Weiss
Facultative parasite.

		1
Ovulinia azaleae Weiss Buchw.	On *Rhododendron* spp. anamorph *Ovulitis azaleae*	1

Parorbiliopsis Spooner & Dennis
Lignicolous saprophytes; *P. extumescens* (Karst.) Spooner & Dennis in E. Anglia, *P. minuta* Spooner & Dennis in the north.

Patellariopsis Dennis
Lignicolous saprophytes; *P. atrovinosa* (Bloxam) Dennis in the midlands and west, *P. clavispora* (B. & Br.) Dennis in the north and west.

Patellea (Fr.) Sacc.
Lignicolous saprophyte, *P. pallida* (Berk.) Massee in the midlands.

Patinella Sacc.
Lignicolous saprophyte, *P. diaphana* (Sow.) in Middlesex.

Patinella rubrotingens (B. & Br.) Sacc.   1

Perrotia Boud.
Saprophytes, segregate from *Lachnum*; *P. flammea* (A. & S.) Boud. in the north and west.

Perrotia phragmiticola (Henn. & Ploett.)   On *Phragmites*   1
Dennis

Pezicula Tul.
Saprophytes and facultative parasites with anamorphs in *Cryptosporiopsis* Bub. & Kab. also *P. carpinea* (Pers.) Tul. on *Carpinus* and *Fagus* in Hertfordshire and Essex; *P. scoparia* (Cooke) Dennis in E. Anglia; six other species in the north and west.

Pezicula alba Guthrie	On *Malus*, anamorph *Phlyctaena*			4
	vagabunda Desm.			
P. cinnamomea (DC.) Sacc.	On *Quercus* &c., anam. *C. quercina* Petrak	1		
P. corticola (Jorg.) Nannf.	On *Malus*, anam. *C. corticola* (Edgerton)	1		4
	Nannf.			
P. coryli (Tul.) Tul.	On *Corylus*, anam. *C. grisea* (Pers.) Petrak	1		
P. corylina Groves	On *Corylus*, anam. *C. coryli* (Peck) Sutton	1		
P. frangulae (Fr.) Fuck.	On *Frangula alnus*, anam. *C. versiformis*	1		
	(A. & S.) Woll.			
P. livida (B. & Br.) Rehm	On various conifers, anam. *Myxosporium*	1	3	4
	abietinum Rostr.			
P. malicorticis (Jackson) Nannf.	On Malus, anam. *C. malicorticis* (Cordl.)	1		4
	Nannf.			
P. myrtillina Karst.	On *Vaccinium myrtillus*	1		
P. paradoxa Dennis	On *Corylus*	1		
P. rubi (Lib.) Niessl	On *Rosa* and *Rubus*, anam. *C. phaeosora*	1	3	4
	(Sacc.) v. Arx			

Pezizella Fuck.
Saprophytes, small hymenoscyphoid apothecia on diverse substrata, cf. *Micropodia pteridina*. Many additional speices in the north and west; *P. albosanguinea* (v.H.) Graddon in E. Anglia.

Pezizella alniella (Nyl.) Dennis	On *Alnus*	1	2		
P. amenti (Batsch) Dennis	On *Salix*	1			
P. (Pachydisca) ascophanoides (Boud.)		1		3	
P. chrysostigma (Fr.) Sacc.	On *Dryopteris*	1	2		4
P. discreta (Karst.) Dennis		1		3	
P. eburnea (Rob.) Dennis	On Gramineae	1		3	
P. filicum (Phill.) Sacc.			2		4
P. minutissima Vel.	On *Rubus*	1			
P. vulgaris (Fr.) v. Höhn.		?1			

Pezizellaster v. Höhn.
Saprophytes; *P. radiostriatus* (Feltg.) v. Höhn. and *P. serratus* (Hoffm.) Dennis in the north and west.

Pezoloma Clem.
Saprophytes on marshy ground; *P. ciliifera* (Karst.) Korf in the north and west, *P. obstricta* (Karst.) Korf in E. Anglia.

Phaeangella Massee
Saprophyte; *P. ulicis* (Cooke) Sacc. in the north and west.

Phaeangellina Dennis
Saprophyte; *P. empetri* (Phill.) Dennis on dead leaves of *Empetrum* in the north.

Phaeohelotium Kanouse
Saprophytes; *P. flexuosum* (Crossland) Dennis in Hertfordshire.

Phaeohelotium geogenum (Cooke) Svrček & Matheis	1	3
P. monticola (Berk.) Dennis (*P. flavum* Kanouse)	1	
P. nobilis (Vel.) Dennis	1	2
P. trabinellum (Karst.) Dennis	2	
P. umbilicatum (Le Gal) Dennis	2	

Phaeoscypha Spooner
Saprophytes with *Chalara* anamorphs; *P. cladii* (NagRaj & Kendrick) Spooner in E. Anglia.

Phialina v. Höhn.
Saprophytes, mainly foliicolous; *P. foliicola* (Graddon) Huhtinen in E. Anglia, *P. ulmariae* (Lasch) Dennis in Hertfordshire.

Phialina flaveola (Cooke) Raitviir	On *Pteridium aquilinum*	1
P. lachnobrachya (Desm.) Raitviir	On *Acer, Aesculus, Corylus, Fraxinus, Quercus*	1
P. pseudopuberula (Graddon) Raitviir (*P. plowrightii* Arends.)	On *Quercus*	1

Phragmonaevia Rehm
Hyperparasites; *P. fuckelii* Rehm & *P. peltigerae* (Nyl.) Rehm on *Peltigera* in the north.

Pirottaea Sacc.
Saprophytes on dead herbaceous stems, eight other species in the north and west.

Pirottaea anglica Nannf.	On *Centaurea*	1
P. nigrostriata Graddon	On *Heracleum sphondylium*	1

Pithyella Boud.
British records unconfirmed.

Ploettnera Henn.
Caulicolous saprophytes; *P. solidaginis* (de Not.) Hein on *Aster tripolium* in E. Anglia.

Ploettnera exigua (Niessl) v. Höhn.	On *Rubus*	1	3

Pocillum de Not.
British status unconfirmed.

Poculum Vel. see *Rutstroemia*

Podophacidium Niessl
Terrestrial saprophyte.

Podophacidium xanthomelum (Pers.) Schroet.	Under conifers	1	4

Polydesmia Boud.
Saprophyte or ?hyperparasite on senescent Diatrypaceae, *Hypoxylon* &c.

Polydesmia pruinosa (B. & Br.) Boud.		1	2	3	4

Proliferodiscus Haines & Dumont
Lignicolous saprophytes.

Proliferodiscus pulveraceus (A. & S.) Spooner		1	2	3

Protounguicularia Raitviir
Lignicolous saprophytes.

Protounguicularia barbata (Vel.) Raitviir	On *Fagus*	1

Pseudographis Nyl.
Lignicolous saprophytes; *P. elatina* (Ach.) Nyl. in Scotland.

Pseudographis pinicola (Nyl.) Rehm	On *Pinus*	4

Pseudohelotium Fuck.
Saprophytes; *P. alaunae* Graddon On *Deschampsia caespitosa* in the north and west.

Pseudohelotium pineti (Batsch) Fuck.	On *Pinus* needles and cones	1

Pseudonaevia Dennis & Spooner
Saprophyte; *P. caricina* Dennis & Spooner on *Carex* in Orkney.

Pseudopeltis Holm & Holm
?Parasite; *L. filicum* Holm & Holm on *Athyrium* and *Dryopteris* in Scotland.

Pseudopeziza Fuck.
Facultative parasites of angiosperms.

Pseudopeziza calthae (Phill.) Karst.	On *Caltha palustris*	1	2		
P. medicaginis (Lib.) Sacc.	f. sp. *medicaginis-lupulinae* Schmied.			3	4
	f.sp. *medicaginis-sativae* Schmied.				4
P. trifolii (Biv. Bern.) Fuck.	f. sp. *trifolii-pratensis* Schuepp			3	
	f. sp. *trifolii-repentis*	1	2	3	4
	Probably a race of this on *Coronilla* sp.	1			

Psilachnum v. Höhn.
Saprophytes; *P. acutum* (Vel.) Dennis on grasses, *P. asemum* (Phill.) Dennis on *Carex, P. rubrotinctum* Graddon on *Filipendula* and *Gunnera, T. tami* (Lamy) Dennis in the north and west, *P. lateritioalbum* (Karst.) v. Höhn in Hertfordshire, *P. pteridigenum* Graddon in the midlands.

Psilachnum inquilinum (Karst.) Dennis	On *Equisetum* spp.	1	2

Psilocistella Svrček
Differs from *Hyaloscypha* in obtuse hairs and united with it by Huhtinen; *P. priapi* (Vel.) Svrček in the north, *P. quercina* (Vel.) Svrček in Middlesex.

Pycnopeziza White & Whetzel
Saprophytes with anamorphs in *Acarosporium* Bub. & Vleug.; *P. pachyderma* (Rehm) White & Whetzel on *Quercus* litter in the north.

Pyrenopeziza Fuck.
Saprophytes, in effect *Mollisia* on nonlignicolous substrata, animadversions under *Mollisia* apply; *P. arctii* (Phill.) Nannf., *P. ebuli* (Fr.) Sacc., *P. mercurialis* (Fuck.) Boud., *P. plantaginis* Fuck., *P. polygoni* (Lasch) Gremmen in East Anglia, *P. carduorum* Rehm, *P. foliicola* (Karst.) Sacc., *P. fuckelii* Nannf., *P. petiolaris* (A. & S.) Nannf. in the north and west, *P. plicata* Rehm in Hertfordshire &c. &c.

Pyrenopeziza alismatis Feltg.	On *Alisma plantago*		4
P. arenivaga (Desm.) Boud.	On *Ammophila*	2	
P. brassicae Sutton & Rawlinson	As anamorph *Cylindrosporium concentricum* Grev.		4
P. chamaenerii Nannf.	On *Chamaenerion*	1	
P. commoda (Rob.) Nannf.	On *Viburnum*	1	4
P. compressula Rehm	On *Lotus, Ononis, Trifolium*	1	4
P. digitalina (Phill.) Sacc.	On *Digitalis purpurea*	1	
P. lychnidis (Sacc.) Rehm	On *Lychnis* and *Silene*		4
P. oenanthes Graddon	On *Oenanthe crocata*	1	
P. rubi (Fr.) Rehm	On *Rubus* spp.	1	
P. urticicola (Phill.) Boud.	On *Urtica dioica*	1	4

Rodwayella Spooner
  Saprophytes.

Rodwayella citrinula (Karst.) Spooner	On *Carex, Digitalis, Oenanthe, Urtica* &c.	1

Roesleria Thuem. & Pass.
  Lignicolous saprophytes.

Roesleria pallida (Fr.) Sacc.	On roots of *Fagus, Malus, Pinus, Rosa, Ulmus* &c.	1	4

Rutstroemia Karst.
Saprophytes, see also *Ciboria*; *R. calopus* (Fr.) Rehm in E. Anglia, *R. hercynica* (Kirschst.) Dennis, *R. juniperi* Holm, *R. rhenana* (Kirschst.) Dennis in the north.

Rutstroemia echinophila (Bull.) v. Höhn.	On *Castanea*	1	2	3	4
R. firma (Pers.) Karst.	On *Quercus*	1	2	3	4
R. fruticeti Rehm	On *Rubus*	1			
R. lindaviana (Kirscht.) Dennis	On *Gramineae*		2		
R. petiolorum (Rob.) White	On *Fagus*	1			
R. sydowiana (Rehm) White	On *Fagus* and *Quercus*	1			4

Sarcoleotia Imai
  Terrestrial saprophyte.

Sarcoleotia turficola (Boud.) Dennis	1

Sarcotrochila v. Höhn.
  Saprophytes, *S. alpina* (Fuck.) v. Höhn. on *Larix* in the north.

Sclerotinia Fuck.
Facultative parasites, including *Botryotinia* Whetzel, *Ciborinia* Whetzel, *Dumontinia* Kohn, *Myriosclerotinia* Buchwald; *S. bulborum* (Wakker) Sacc. in the midlands, *S. (M.) scirpicola* Rehm & *S. (M.) sulcata* Whetzel in East Anglia, *S. gregoriana* Palmer, *S. (M.) juncifida* (Nyl.) Palmer, and *S. vahliana* Rostrup in the north *S. (B.) sphaerosperma* Gregory in the west, *S. draytoni* Buddin & Wakef. and *S. minor* Jagger on cultivated plants with undisclosed distributions.

Sclerotinia (Ciborinia) bresadolae Rick (*S. hirtella* Boud.)	On *Castanea* and *Quercus*	1

		1	2	3	4
S. (Ciborinia) candolleana (Lév.) Fuck.	On *Quercus*	1		3	
S. (Myriosclerotinia) curreyana (Berk.) Karst.	On *Juncus*	1	2	3	4
S. (Botryotinia) fuckeliana (de Bary) Fuck.	Anamorph *Botrytis cinerea* Pers.	1	2	3	4
S. (B.) globosa (Buchwald) Webster	On *Allium ursinum*. Anam. *B. globosa* Raabe		2		
S. (B.) narcissicola Gregory	On *Narcissus*. Anam. *B. narcissicola* Kleb.	1			
S. (B.) polyblastis Gregory	On *Narcissus*. Anam. *B. polyblastis* Dowson	1			4
S. sclerotiorum (Lib.) de Bary	Polyphagous	1	2	3	4
S. (B.) squamosa (Viennot-Bourgin) Dennis	On *Allium*. Anam. *B. squamosa* Walker	1			
S. trifoliorum Eriksson	On Leguminosae	1	2	3	4
S. (D.) tuberosa (Hedw.) Fuck.	On *Anemone nemorosa*	1		3	4
S. sp.	On *Endymion nonscriptum*	1			

Scutomollisia Nannf.

Saprophytes on Monocotyledons, *S. stenospora* Nannf. in E. Anglia, four more species in the north and west.

Scutoscypha Graddon

Saprophyte.

		1			
Scutoscypha fagi Graddon	On leaves of *Fagus*	1			

Skyathea Spooner & Dennis

Saprophyte, *S. hederae* Spooner & Dennis on *Hedera helix* in the north.

Spathularia Pers.

Terrestrial saprophyte.

		1			4
Spathularia flavida Pers.		1			4

Spilopodia Boud.

Saprophytes; *S. melanogramma* Boud. on *Mercurialis* in Hertfordshire, *S. nervisequa* Boud. on *Plantago* in the west.

Stamnaria Fuck.

Facultative parasite, *S. persoonii* (Pers.) Fuck. on *Equisetum* spp. in E. Anglia.

Stegopeziza v. Höhn.

Saprophytes, akin to *Hysterostegiella* but on leaves of Dicotyledons; *S. dumeti* (Sacc. & Speg.) Spooner on *Rubus* in the north and west, *S. lauri* (Cald.) v. Höhn and *S. quercea* (Fautr. & Lamb.) Spooner in the west.

Stromatinia Boud.

Akin to *Sclerotinia* but with microsclerotia only.

	1
Stromatinia gladioli (Drayton) Whetzel	1

Strossmayeria Schulzer

Lignicolous saprophytes.

	1
Strossmayeria basitricha (Sacc.) Dennis	1

Symphyosirinia Ellis

Facultative parasites; *S. angelicae* Ellis on fruit of *Angelica* in Essex; *S. galii* Ellis in E. Anglia, *S. heraclei* Ellis in the north.

Tapesia (Pers.) Fuck.
See *Mollisia*, retained here for species with subiculum on wood.

Tapesia cinerella Rehm	Needs confirmation	1		
T. fusca (Pers.) Fuck.		1	2	4
T. lividofusca (Fr.) Rehm		1		4
T. rosae (Pers.) Fuck.		1		

Thuemenidium Kuntze
Terrestrial saprophytes; *T. atropurpureum* (Batsch) Kuntze formerly in Berkshire, now only in the north and west, *T. arenarium* (Rostr.) in Dorset.

Torrendiella Boud. & Torr.
Saprophytes; *T. ciliata* Boud. on *Rubus* in the west, *T. eucalypti* (Berk.) Boud. in the north.

Triblidium Reb.
Corticolous saprophytes; *T. octosporum* Hawksw. & Coppins on *Fraxinus* in Hampshire.

Triblidium caliciiforme Reb.	On *Quercus*	3

Trichobelonium (Sacc.) Rehm
Saprophytes.

Trichobelonium asteroma (Fuck.) Rehm	On *Carex*		3
T. obscurum (Rehm) Rehm	On *Calluna*	1	

Trichoglossum Boud.
Terrestrial saprophytes, *Geoglossum* with setose hymenium; *T. rasum* Pat. in the west, *T. variabile* (Durand) Nannf. and *T. walteri* (Berk.) Durand in the north.

Trichoglossum hirsutum (Fr.) Boud.		1	2	3	4

Trochila Fr.
Facultative foliicolous parasites.

Trochila craterium (DC.) Fr.	On *Hedera helix* Anam. *Cryptocline paradoxa* (de Not.) v. Arx	1			4
T. ilicina (Nees) Greenhaugh	On *Ilex aquifolium*	1	2	3	4
T. laurocerasi (Desm.) Fr.	On *Prunus laurocerasus* and *P. lusitanica*	1		3	4

Tympanis Tode
Lignicolous saprophytes with anamorphs in *Sirodothis; T. alnea* (Pers.) Fr. & *T. saligna* Fr. in the west, *T. confusa* Nyl., *T. hypopodia* Nyl. and *T. spermatiospora* (Nyl.) Nyl. in the north.

Tympanis conspersa (Fr.) Fr.	On *Malus* and *Sorbus*	1	4
T. laricina (Fuck.) Sacc.	On *Larix*	1	
T. ligustri Tul.	On *Ligustrum* (?& *Fraxinus*)	1	

Unguicularia v. Höhn.
Saprophytes; *U. aspera* (Fr.) Nannf. on *Osmunda regalis* in the north and west, *U. costata* (Boud.) Dennis on *Juncus* in E. Anglia, *U. dilatopilosa* Graddon on *Epilobium* in the west, *U. raripila* v. Höhn in Hertfordshire and *U. ulmariae* (Vel.) Dennis in the north.

Unguicularia incarnatina (Quél.) Nannf.	On dead *Cirsium*	1
U. millepunctata (Lib.) Dennis (*U. cirrhata* (Crouan) Le Gal; *U. scrupulosa* (Karst.) v. Höhn.)		1

Unguiculella v. Höhn.
Saprophytes; *U. eurotioides* (Karst.) Nannf. in E. Anglia, *U. robergei* (Desm.) Dennis in the north and west on *Rosa* and *Rubus*.

Unguiculella hamulata (Feltg.) v. Höhn.	On *Heracleum* and *Urtica*	1

Urceolella Boud.
Saprophytes; *U. carestiana* (Rab.) Dennis on *Athyrium* and *Dryopteris* in the north and west, *U. crispula* (Karst.) Boud. (*U. spirotricha* (Oud.) Boud.) on dead stems in the north and west, *U. salicicola* Raschle on dead leaves in the west.

Velutarina Korf
Lignicolous saprophyte.

Velutarina rufoolivacea (A. & S.) Korf	On *Acer, Fagus, Fraxinus, Rubus, Ulex*	1

Verpatinia Whetzel
Saprophytes; *V. calthicola* Whetzel on *Iris*, *V. spiraeicola* Dennis in East Anglia.

Vibrissea Fr.
Lignicolous on sodden wood, *V. truncorum* (A. & S.) Fr. in the north and west.

Voralbergia Grumm.
Saprophytes or associate of algae, *V. medioincrassata* Grumm. and *V. renitens* Grumm. in north.

Zoellneria Vel.
Saprophyte, distinguished from *Torrendiella* by the anamorph *Amerosporium patellarioides* Sm. & Ramsb.; *Z. rosarum* Vel. on fallen leaves of *Rosa* in the north and west.

PHACIDIALES/RHYTISMATALES

Arwidssonia B. Eriksson
Facultative parasite, *A. empetri* (Rehm) B. Eriksson on *Empetrum* in the north.

Ascodichaena Butin
Corticolous saprophyte.

Ascodichaena rugosa Butin	On *Fagus* and *Quercus*, anamorph *Polymorphum quercinum* (Pers.) Chev.	1	2	3	4

Coccomyces de Not.
Saprophytes; *C. arctostaphyli* (Rehm) Erikss.; *C. boydii* A.L. Sm., *C. juniperi* Karst., *C. leptideus* (Fr.) B. Erikss. in the north. Reputed anamorphs in *Leptothyrium*.

Coccomyces coronatus (Schum.) de Not.	On *Betula, Fagus, Quercus*	1		3	
C. dentatus (Kunze & Schm.) Sacc.	On *Castanea* and *Quercus*	1			4
C. tumidus (Fr.) de Not.	On *Castanea* and *Quercus*	1			

Colpoma Wallr.
Facultative parasites.

Colpoma crispum (Pers.) Sacc.	On coniferae	1			
C. quercinum Pers.) Wallr.		1	2	3	4

Cryptomyces Grev.
Facultative parasite, *C. maximus* (Fr.) Rehm on *Salix* spp. in East Anglia.

Cryptomycina v. Höhn.
Facultative parasite.

Cryptomycina pteridis (Rebent.) v. Höhn.   On *Pteridium aquilinum*   4

Cycloneusma Di Cosmo, Peredo & Minter
Facultative parasite?

Cycloneusma minus (Butin) Di Cosmo   On *Pinus* spp.   1   3
et al (*Naemacyclus minor* Butin)

Didymascella Maire & Sacc.
Facultative parasites of conifers, *D. tetraspora* (Phill. & Keith) Maire on *Juniperus* in the north.

Didymascella thujina (Durand) Maire   On *Thuja*. Alien introduced with the host 1   2   3   4

Hypoderma DC.
Saprophytes; *H. alpinum* Spooner on *Carex* in the north, *H. ilicinum* de Not. on *Quercus* in the west, *H. scirpinum* DC. in East Anglia.

Hypoderma commune (Fr.) Duby	On stems, anamorph *Leptothyrium vulgare* Sacc.	1			4
H. hedera (Mart.) de Not.,	On *Hedera*, anamorph *L. hederae* Starb.	1			
H. rubi DC. (*H. virgultorum* DC.)	Probably synonym of *H. commune*, anamorph *L. rubi* Sacc.	1	2	3	4

Karstenia Fr.
Saprophytes; *K. clematidis* (Phill.) Sherwood & *K. lonicerae* (Vel.) Sherwood in the north, *K. idaei* (Fuck.) Sherwood in the west.

Laqueria Fr.
Doubtfully British.

Leptopeltis v. Höhn.
Pteridicolous saprophytes formerly erroneously referred to Dothideales, *L. filicina* (Lib.) v. Höhn and *L. nebulosa* (Petrak) Holm & Holm in East Anglia.

Leptopeltis litigiosa (Desm.) Holm & Holm On *Pteridium aquilinum*   1   4

Lirula Darker
Parasites of conifer foliage; *L. macrospora* (Hartis) Darker on *Picea abies* in the north.

Lophodermella v. Höhn.
Parasitic of foliage of conifers; *L. conjuncta* (Darker) Darker and *L. sulcigena* (Rostr.) v. Höhn on *Pinus* in the north.

Lophodermium Chev.
Saprophytes and facultative parasites; *L. alpinum* (Rehm) Terrier, *L. apiculatum* (Wormsk.) Sacc., *L. aucupariae* (Schleich.) Darker, *L. gramineum* (Fr.) Chev., *L. laricinum* Duby, *L. L. maculare* (Fr.) de Not., *L. melaleucum* (Fr.) de Not., *L. oxycocci* (Fr.) Karst., *L. petiolicola* Fuck., *L. sieglingiae* Hilitzer, *L. vagulum* Wils. & Rob., *L. versicolor* (Wahlenb.) Rehm in the north, *L. festucae* (Roum.) Terrier in Essex, *L. hedericola* Ahmad, *L. nanakii* Cannon & Minter in the west.

Lophodermium arundinaceam (Schrad.)   On *Ammophila* & *Phragmites*   1   2   4
Chev.

		1	2	3	4
L. caricinum (Rob.) Duby	On Carex spp.		2		
L. conigenum (Brun.) Hilitzer	On Pinus spp.	1			4
L. culmigenum (Fr.) de Not.		1			
L. foliicola (Fr.) Cannon & Minter	On Crataegus (L. hysterioides (P.) Sacc)	1	2		
L. juniperinum (Fr.) de Not.		1	2	3	4
L. piceae (Fuck.) v. Höhn.		1			
L. pinastri (Schrad.) Chev.		1	2	3	4
L. piniexcelsae Ahmad					4
L. seditiosum Minter et al	On Pinus sylvestris	1			4
L. typhinum (Fr.) Lamb.		1			4

Lophomerum Oellette & Magasi
 Facultative parasite.

Lophomerum ponticum Minter	On *Rhododendron ponticum*	1	2

Meloderma Darker
 Foliar parasites of coniferae.

Meloderma desmazierii (Duby) Darker (*Hypoderma brachysporum* Rostr.)	On *Pinus* spp.	3

Micraspis Darker
 Saprophytes, *M. strobilina* Dennis on *Pinus* in the north.

Naemacyclus Fuck.
 Saprophytes, *N. caulium* v. Höhn. on *Urtica*, *N. fimbriatus* (Schw.) DiCosmo et al on *Pinus* in the north (=*Lasiostictis fimbriata* (Schw.) Bäumler).

Phacidium Fr.
 Facultative parasites or saprophytes; *P. abietinum* Schm., *P. coniferarum* (Hahn) DiCosmo, *P. infestans* Karst, *P. vaccinii* Fr. in the north, *P. aquifolii* (DC.) Rehm in the west.

Phacidium lacerum Fr.	On *Pinus* anamorph *Ceuthospora pinastri* (Fr.) v. Höhn.	1		
P. (Phacidiostroma) multivalve (DC.) Schm.	On *Ilex* anam. *C. phacidioides* Grev.	1	2	4

Potebniamyces Smerlis
 Facultative parasite, *P. pyri* (B.& Br.) Dennis widespread as bark canker of rosaceous trees.

Potebniamyces coniferarum (v. Höhn.) Smerlis (*Phacidiopycnis tseudotsugae* (Wils.) Hahn)	3

Propolina Sacc.
 Lignicolous saprophytes, *Propolomyces* with polysporous asci.

Propolina cervina Sacc.	On *Quercus*	1

Propolis (Fr.) Corda
 Saprophytes, *P. emarginata* (Cooke & Massee) Sherwood on *Eucalyptus*, *P. phacidioides* (Fr.) Corda on *Arctostaphylos* in the north.

Propolomyces Sherwood
 Lignicolous saprophytes, *P. betulae* (Fuck.) Dennis in the north and west.

Propolomyces versicolor (Fr.) Dennis		1	2	3	4

Pseudophacidium Karst.
Facultative parasites with anamorphs in *Myxofusicoccum*, *P. callunae* (Karst.) Karst., *P. microspermum* Rehm and *P. piceae* Müller in the north.

Rhabdocline Syd.
Parasite on needles of *Pseudotsuga*.

Rhabdocline pseudotsugae Syd.	Anamorph *Rhabdogloeum hypophyllum* Ellis and Gill	1	4

Rhytisma Fr.
Facultative foliar parasites of dicotyledons, *R. andromedae* (Pers.) Fr. in the north and west. Also, if distinct, *R. symmetricum* Müller on *Salix* in East Anglia. Anamorphs in *Melasmia*.

Rhytisma acerinum (Pers.) Fr.	On *Acer campestre* and *A. pseudoplatanus*	1	2	3	4
R. punctatum (Pers.) Fr.	Probably merely a state of the preceding	1			
R. salicinum (Pers.) Fr.	On *Salix caprea*, *S. cinerea* and many other species	1	2		4

Sporomega Corda
Parasite of *Vaccinium uliginosum*, *S. degenerans* (Fr.) Corda in the north.

Terriera B. Eriksson
Facultative parasite on *Vaccinium*.

Terriera cladophilum (Lév.) B. Eriksson	On *Vaccinium myrtillus*	1

Therrya Sacc. & Penz.
Saprophyte on coniferae, *Coccophacidium* Rehm is a synonym.

Therrya pini (A.& S.) v. Höhn.	On *Pinus sylvestris* anamorph *Colpomella pini* v. Höhn.	1

ARTHONIALES

Arthonia Ach.
Principally lignicolous and lichenised, a few hyperparasites.= *Celidium* Tul. and *Conida* Massal. *A. glaucomaria* (Nyl.) Nyl. on *Lecanora rupicola* in the north and west (*Cel. varians* (Dav.) Arn.).

Arthonia clemens (Tul.) Th. Fr. (*A. subvarians* Nyl.)	On *Lecanora dispersa* &c.	3
A. epiphyscia Nyl.	On *Phaeophyscia orbicularis* and *Xanthoria*	4

Plectocarpon Fée
Hyperparasites = *Lichenomyces* Sant., *P. lichenum* (Somm.) Hawksw. on *Lobaria* in north and west.

Seuratia Pat.
Honeydew saprophyte.

Seuratia millardetii (Racib.) Meeker	Anam. *Atichia glomerulosa* (Ach.) Stein	3

Tarbertia Dennis
Saprophyte, *T. juncina* Dennis on Juncus in the north.

LECANORALES (nonlichenised)

Agyrium Fr.
Lignicolous, *A. rufum* (Pers.) Fr. in the north.

Arthrorhaphis Th. Fr.
Hyperparasites on *Baeomyces, A. citrinella*) (Ach.) Poelt & *A. grisea* Th. Fr. in the north.

Carbonea (Hertel) Hertel
Hyperparasites, *C. supersparsa* (Nyl.) Hertel on *Lecanora* in the north, *C. vitellinaria* (Nyl.). Hertel on *Candellariella* in the west.

Dactylospora Körber
Hyperparasites or lignicolous, *D. athallina* (M. Arg.) Haf. on *Baeomyces* in the west, *D. bloxami* (Berk.) Haf. in midlands, *D. caledonica* Haf., *D. lamyi* (Rich.) Arn. *D. saxatilis* (Schaer.) Haf., *D. urceolata* (Th. Fr.) Arn. in the north, *D. lobariella* (Nyl.) Haf. in the West.

Dactylospora parasitica (Floerke) Zopf	On *Pertusaria* and *Ochrolechia*	3
D. parellaria (Nyl.) Arn.	On *Ochrolechia parella*	3
(*Leciographa parellaria* (Nyl.) Sacc.)		
D. stygia (Berk. & Curt.) Haf.		2

Mycobacidia Rehm
Hyperparasites, *M. killiasii* (Hepp) Rehm on *Peltigera, M. plumbina* (Anzi) Sant. on *Parmeliella* in the north.

Mycobilimbia Rehm
Hyperparasites, *M. endocarpicola* (Lindsay) Vouaux on *Catapyrenium* in the west.

Mycomelaspilea Reinke
Hyperparasites, *M. leciographoides* (Vouaux) Keissler on *Verrucaria* in the north and west, *Melaspilea lentiginosa* (Lyell) Mull. Arg. on *Phaeographis*.

Nesolechia Massal.
Hyperparasites, *N. cetrariicola* (Lindsay) Arn., *L. cladoniaria* (Nyl.) Arh., *N. leptostigma* (Nyl.) Sacc., *L. oxyspora* (Tul.) Massal. in the north and west.

Sarea Fr.
Resinicolous saprophytes, segregate from *Biatorella*, = *Retinocyclus* Fuck., *S. difformis* (Fr.) Fr. in the north.

Sarea resinae (Fr.) Kuntze (*Biatorella*	On resin of conifers, anam. *Pycnidiella*	1	2
resinae (Fr.) Th. Fr.)	*resinae* (Fr.) v. Höhn.		

Scutula Tul.
Hyperparasites, *S. cristata* (Leighton) Sacc. & D. Sacc., *S. epicladonia* (Nyl.) Sacc. & D. Sacc. *S. (Catillaria) episema* (Nyl.) Zopf in the west, *S. epiblematica* (Wallr.) Rehm, *S. krempelhuberi* Körber and *S. stereocaulorum* (Anzi) Körber in the north.

Xylographa (Fr.) Fr.
Lignicolous saprophytes, *X. abietina* (Pers.) Zahlbr. (*X. parallela* Nyl.), *X. minutula* Körber and *X. trunciseda* (Th. Fr.) Minks in the north, *X. vitiligo* (Ach.) Laundon in the midlands.

OPEGRAPHALES

Bactrospora Massal.
Lignicolous saprophytes.

Bactrospora dryina (Ach.) Massal   3

Opegrapha Ach.
Predominantly lichenised, a few hyperparasites, *O. parasitica* (Massal.) Olivier, *O. pertusariicola* Coppins & James, *O. pulvinata* Rehm in the north and west.

Ptychographa Nyl.
Lignicolous saprophytes, = *Placographa, P. flexella* (Ach.) Coppina and *P. xylographoides* Nyl in the north.

OSTROPALES

Ostropa Fr.
Lignicolous saprophytes, *O. barbara* (Fr.) Nannf. perhaps not British.

Paschelkiella Sherwood
Lignicolous saprophytes, *P. pini* (Rommell) Sherwood (*Odontotrema pini* Romm.) in the north.

Plejobolus (Bomm., Rouss., & Sacc.) O. Eriksson
Saprophyte, *P. arenarius* (B.R.& S.) O. Erikss. On *Ammophila arenaria* in the north. As this was described from the vicinity of Ostend it surely only awaits record from Kent and Sussex.

Pleospilis Clements
Hyperparasite = *Spilomela* (Sacc.) Keissler, *P. ascaridiella* (Nyl.) Hawksw. on *Pertusaria* and *Huilia* in the north and west.

Robergea Desm.
Lignicolous saprophytes.

Robergea cubicularia (Fr.) Rehm   On *Fraxinus*   3 4

Schizoxylon Pers.
Saprophytes, *S. berkeleyanum* (Dur. & Lév.) Fuck. on *Epilobium* &c. in East Anglia, *S. atro-album* Rehm in the north, *S. ligustri* (Schw.) Sherwood on *Rubus* in the west.

Skyttea Sherwood et al
Hyperparasites of lichens, *S. nitschkei* (Korb.) Sherw. et al on *Thelotrema* in Hampshire, others in the north and west.

Stictis Pers.
Saprophytes, *S. elevata* (Karst.) Karst. on *Clematis* and *Juncus* in north and west, *S. elongatispora* Graddon on *Carex* in East Anglia.

Stictis arundinacea Pers.	On *Carex, Gramineae, Luzula*		3
S. friabilis (Phill. & Plowr.) Sacc. & Trav.	On dicot wood	1	
S. pusilla Speg.	On *Carex*		3

S. radiata (L.) Pers.	On wood of angiosperms and gymnosperms	3
S. stellata Wallr.	On dead stems of angiosperms and *Pteridophyta*	2

Xerotrema Sherwood & Coppins
Lignicolous saprophyte, on *Pinus* in the north, *X. megalospora* Sherw. & Copp.

Xylopezia Höhn.
Lignicolous saprophytes (*Odontotrema* auct. angl.), *X. hemispherica* (Fr.) Sherwood in the north, *X. inclusa* (Pers.) Sherwood in the north and west.

CLAVICIPITALES

Acrospermum Tode
Saprophytes, perhaps dothideal, *A. compressum* Tode on *Angelica* in Hertfordshire.

Acrospermum graminum Lib.	On Gramineae	4
A. pallidum Kirschst.	On *Galium*, doubtfully distinct from *A. compressum*	2

Apiocrea Syd.
Hyperparasites of Agaricales (Boletes), *A. tulasneana* (*Plowr.*) Petch in East Anglia.

Apiocrea chrysosperma (Tul.) Syd.	Anamorph *Sepedonium chrysospermum* (Bull.) Link	1 2 3 4

Arachnocrea Moravec
Lignicolous saprophyte.

Arachnocrea stipata (Fuckel) Moravec	1

Barya Fuck.
Hyperparasite, *B. aurantiaca* Plowr. & Wilson on sclerotia of *Claviceps*, perhaps in E. Anglia.

Byssostilbe Petch
Hyperparasites of Myxomycetes.

Byssostilbe stilbigera (B. & Br.) Petch	As anamorph *Blistum tomentosum* (Schrad.) Sutton	1 2 3 4

Claviceps Tul.
Parasites of Cyperaceae and Gramineae.

Claviceps nigricans Tul.	On *Eleocharis*, anamorph *Sphacelia nigricans* Sacc.	1
C. purpurea (Fr.) Tul.	Collective species with races on many Gramineae, anamorph *Sphacelia segetum* Lév.	1 2 3 4

Cordyceps (Fr.) Link
Obligate parasites of insects and *Elaphomyces*, *C. forquignoni* Quél. common on diptera in the north and west, *C. memorabilis* Ces., *C. sphecocephala* (Klotz.) B.& C. and *C. tuberculata* (Lebert) Maire in East Anglia.

Cordyceps capitata (Holmsk.) Link	On *Elaphomyces*	1 2 4

C. entomorrhiza (Holmsk.) Link		1	2		4
C. gracilis Mont. & Dur.	On lepidoptera	1	2		4
C. longisegmentis Ginns (*C. canadensis* auct.)	On *Elaphomyces*	1			4
C. militaris (L.) Link	On lepidoptera, anam. *Cephalosporium militare* Kob.	1	2		4
C. ophioglossoides (Ehrhart) Link	On *Elaphomyces*	1	2		4

Epichloe (Fr.) Tul.
  Parasite of Gramineae.

Epichloe typhina (Pers.) Tul.	Anamorph *Sphacelia typhina* Sacc.	1	2		4

Hypomyces Tul.
  Mainly hyperparasites, *H. broomeanus* Tul. on *Heterobasidion* in E. Anglia, *H. papulasporae* Rogerson & Samuels in Hampshire.

Hypomyces aurantius (Pers.) Tul.	On *Polyporus*, anam. *Cladobotryum varium* Nees	1	2	3	
H. ochraceus (Pers.) Tul.	On Russulaceae anam. *Verticillium agaricinum* Corda	1	2		
H. odoratus Arnold		1	2	3	
H. rosellus (A.& S.) Tul.	On *Stereum*, anam. *Cladobotryum dendroides* (Bull.) Gams	1	2	3	4

Oomyces B. & Br.
  Saprophyte, perhaps dothideal, *O. carneoalbus* (Lib.) B.& Br. on *Deschampsia* in Hampshire.

Ophiocordyceps Petch
  Parasite of scale insects, *O. clavulata* (Schwein.) Petch in the west, perhaps extinct.

Peckiella Sacc.
  Hyperparasites of Russulaceae, synonym *Byssonectria*.

Peckiella lateritia (Fr.) Maire	On *Lactarius*	1	
P. viridis (A. & S.) Sacc.	On *Lactarius*	1	

Torrubiella Boud.
  Parasites of Arachnida and Diptera, *T. albolanata* Petch.

T. albotomentosa Petch	On Diptera			4
Torrubiella arachnophila (Johnston) Mains v.*pleiopus* Mains	As the anamorph *Gibellula pleiopus* (Vuill.) Mains	1	2	
T. aranicida Boud.		1		

HYPOCREALES

Battarrina (Sacc.) Clem. & Shear
  Hyperparasites, *B. inclusa* (B.& Br.) Clem. & Shear in *Tuber puberulum* in the west.

Calonectria de Not.
  Saprophytes, see also under *Nectria*.

Calonectria kyotensis Terashita	As anamorph *Cylindocladium scoparium* Morgan	4

			1	2	3	4
C. pyrochroa (Desm.) Sacc. (*C. hederae* Arn.)	Anam. *C. ilicicola* (Hawley Boed. & Reitsma)					4

Cesatiella Sacc.
Lignicolous saprophyte, *C. lancastriensis* Grove in the north.

Chromocrea Seaver = *Creopus* Link
Lignicolous saprophytes, *Hypocrea* with coloured ascospores, *C. cupularis* (Fr.) Petch in E. Anglia, *C. spinulosum* (Fuck.) Petch in E. Anglia.

		1	2	3	4
Chromocrea aureoviridis (Plowr. & Cooke) Petch			2		
C. gelatinosus (Tode) Seaver (*Creopus gelatinosus* (Tode) Link)		1	2	3	4

Epigloea Zukal
Saprophytes, *E. bactrospora* Zukal on peat in the north.

Gibberella Sacc.
Facultative parasites, anamorphs in *Fusarium*, *G. acervalis* (Moug.) Sacc. in the west Midlands.

		1	2	3	4
Gibberella acuminata Booth	As *Fusarium acuminatum* Ell. & Ev.		2		
G. avenacea R.J. Cooke	Anamorph *F. avenaceum* (Corda) Sacc.	1	2		4
G. baccata (Wallr.) Sacc. (*G. moricola* (de Not.) Sacc)	Anam. *F. lateritium* Nees	1			4
G. buxi (Fuck.) Wint.	Anam. *F. lateritium* var *buxi* Booth	1			
G. cyanogena (Desm.) Sacc. (*G. saubinetii* (Mont.) Sacc.)	Anam. *F. sulphureum* Schlecht.	1			
G. fujikuroi (Saw.) Ito	Anam. *F. moniliforme* Sheldon		2		
G. pulicaris (Fr.) Sacc.	Anam. *F. sambucinum* Fuck.	1	2	3	4
G. zeae (Schw.) Petch	Anam. *F. graminearum* Schwäbe	1			

Halonectria Gareth Jones
Marine lignicolous saprophyte, *H. milfordensis* Gareth Jones.

Heleococcum Jorgensen
Saprophyte, *H. aurantiacum* Jorgensen on mushroom compost in Hampshire.

Hypocrea Fr.
Saprophytes and hyperparasites, *H. placentula* Grove in Midlands, *H. splendens* Phill. & Plowr. in East Anglia.

		1	2	3	4
Hypocrea argillacea Phill. & Plowr.		1			
H. citrina (Pers.) Fr. (*H. lactea* (Fr.) Fr.)		1			4
H. lutea (Tode) Petch					4
H. pallida Ell. & Ev.	On Polypores	1	2		
H. pilulifera Webster & Rifai		1			
H. pulvinata Fuck.	On effete carpophores of *Piptoporus*	1	2	3	
H. rufa (Pers.) Fr.		1	2	3	4
H. schweinitzii (Fr.) Sacc.					4
H. tremelloides (Schum.) Fr.			2		

Hypocreopsis Karst.
Saprophytes, *H. lichenoides* (Tode) Seaver and *H. rhododendri* Thaxter in the north.

Nectria (Fr.) Fr.
Saprophytes and facultative parasites, including *Calonectria* pro max. part., *Dialonectria* (Sacc.) Cooke, *Lasionectria* (Sacc.) Cooke, *Nectriopsis* Maire, *Sphaerostilbe* Tul., additional

species in E. Anglia include *N. aurantiaca* (Tul) Jacz., *N. brassicae* Ell. & Sacc., *N. funicola* (B. & Br.) Berk., *N. keithii* B. & Br., *N. lecanodes* (Ces.) Ces. & de Not., *N. rishbethii* Booth, in Hampshire *N. berkeleyana* (Plowr. & Cooke) Dingley, in the west *N. boothii* Hawksw., *N. citrinoaurantia* Del., *N. flavoviridis* (Fuck.) Woll., *N. peristomialis* (B. & Br.) Samuels, in the north *N. candicans* (Plowr.) Samuels, *N. sylvana* Mout., *N. ventricosa* Booth and *N. aureola* Wint.

Nectria aquifolii (Fr.) Berk.	On *Ilex*	1		3	4
N. arenula (B. & Br.) Berk (*Nectriella bloxami* (B. & Br.) Fuck		1	2		
N. cinnabarina (Tode) Fr.	Anamorph *Tubercularia vulgaris* Tode	1	2	3	4
N. coccinea (Pers.) Fr.	Anamorph *Cylindrocarpon candidum* (Lk.) Woll.	1	2	3	4
N. coryli Fuck.	On *Corylus* and *Salix*	1			
N. desmazierii Becc. & de Not.	Anamorph *Fusarium buxicola* Sacc. on *Buxus*	1			
N. ditissima Tul.	Anamorph *Cylindrocarpon willkommii* (Lindau) Woll. on *Fagus*	1	2		4
N. ellisii Booth		1			
N. episphaeria (Tode) Fr.	Anamorph *Fusarium aqueductum* Lag. v. *medium* Woll.	1	2	3	4
N. flammea (Tul.) Dingley	Anamorph *Fusarium coccophilum* (Desm.) Woll. & Reinke	1			
N. fuckeliana Booth	Anamorph *Cylindrocarpon cylindroides* Woll. v. *tenue* Woll.	1			
N. galligena Bres.	Anamorph *Cylindrocarpon heteronemum* (B. & Br.) Woll.	1	2	3	4
N. gliocladioides Smalley & Hansen	Anamorph *Gliocladium roseum* (Lk.) Bainier	1	2		4
N. haematococca B. & Br.	Anamorph *Fusarium solani* (Mart.) Sacc.	1	2		
N. hederae Booth	Anamorph *Cylindrocarpon hederae* Booth	1			
N. inventa Pethybridge	Anamorph *Verticillium cinnabarinum* (Corda) Reinke & Berth.	1	2	3	4
N. leptosphaeriae Niessl		1			
N. lugdunensis Webster	As anamorph *Heliscus lugdunensis* Webster		2		
N. magnusiana Rehm	Anamorph *Fusarium epistromum* (V. Höhn.) Booth, on *Diatrypella*	1			
N. mammoidea Phill. & Phill. & Plowr.	Anamorph *Cylindrocarpon ianthothele* Woll. v. *Majus* Woll.	1	2	3	
N. mammoidea var. *rubi* (Osterw.) Weese	Anamorph *C. ianthothele* v. *ianthothele*	1			
N. modesta v. Höhn.			2		
N. myxomyceticola Samuels	As anamorph *Verticillium rexianum* (Sacc.) Sacc.	1		3	
N. ochroleuca (Schw.) Berk.	Alien	1			
N. pallidula Cooke	Anamorph *Gliocladium penicillioides* Corda		2		
N. peziza (Tode) Fr.		1	2	3	4
N. pinea Dingley	Anamorph *Cylindrocarpon pineum* Booth		2		
N. pseudopeziza (Desm.) Rossman (*Calonectria ochraceopallida* (B. & Br.) Sacc.)					4
N. punicea (Schm.) Rabenh.	Anamorph *Cylindrocarpon album* (Sacc.) Woll.		2		
N. punicea var. *ilicis* Booth	On *Ilex*	1	2	3	4
N. purtonii (Grev.) Berk.		1			
N. radicicola Gerlach & Nilsson	As anamorph *Cylindrocarpon destructans* (Zins) Scholt.	1			4
N. ralfsii B. & Br.		1			
N. sinopica (Fr.) Fr.	Anamorph *Zythiostroma mougeotii* (Fr.) v. Höhn on *Hedera helix*	1	2		4
N. suffulta Berk. & Curt. (*Neohenningsia suffulta* (B. & C.) v. Höhn.)	?Alien	1			

N. veuillotiana Roum. & Sacc.   Anamorph *Cylindrocarpon candidum*   4
                                      (Sacc.) Woll.
N. viridescens Booth   Anamorph *Acremonium butyri* (v. Beyma)   1   2   3
                                      Gams

Nectriella Nits.
Saprophytes and hyperparasites of hepaticae and lichens, *N. dacrymycella* (Nyl.) Rehm, *N. robergei* (Mont. & Desm.) Weese and *N. umbelliferarum* (Crouan) Sacc. in E. Anglia, *N. exigua* Dennis, *N. luteola* (Rob.) Weese, *N. paludosa* Fuck., *N. santessonii* Lowen & Hawks. and *N. tincta* (Fuck.) Sant. in the west, *N. sambuci* (v. Höhn.) Weese, *N. laminariae* O. Erikss. and *N. tenuispora* Hawks. in the north.

Nectriella consolationis (Sacc.) Müller   On *Laurus nobilis*   2

Nectriopsis Maire see *Nectria* and *Sphaerostilbella*

Neocosmospora F. Smith
Facultative parasite, *N. vasinfecta* E.F. Smith casual on imported fruit.

Orcadia Sutherland
Saprophyte on Phaeophyceae, *O. ascophylli* Suth. on *Ascophyllum* and *Pelvetia* in the north and west.

Paranectria Sacc.
Lichenicolous parasites, *P. affinis* (Ces.) Sacc. on *Ephebe*, *P. oropensis* (Ces.) Hawksw. & Piroz. on *Parmeliella* in the north, *P. superba* Hawksw. on *Peltigera* in the midlands.

Phaeonectriella Eaton & Gareth Jones
Marine lignicolous saprophyte, *P. lignicola* Eaton & Gareth Jones in the west.

Podostroma Karst.
Lignicolous saprophyte.

Podostroma alutaceum (Pers.) Atk.   1   2

Protocrea Petch
Lignicolous saprophytes, *P. delicatula* (Tul.) Petch in the west.

Protocrea farinosa (B. & Br.) Petch   On *Fagus, Fraxinus, Quercus*   1   2   4

Pseudonectria Seaver
Facultative parasites with anamorphs in *Volutella*, for hepaticolous species see *Octosporella* Dobb.; *P. pachysandricola* Dodge is an alien.

Pseudonectria rousseliana (Mont.) Woll.   On *Buxus*, anamorph *Volutella buxi* (Cda)   1
                                      Berk.

Pyxidiophora Bref. & v. Tavel
Saprophytes = *Mycorhynchus* Sacc., *P. asterophora* (Tul.) Lindau and *P. petchii* (Breton & Faur.) Lundq. in East Anglia, three others in the north.

Scoleconectria Seaver
?Facultative parasite.

Scoleconectria cucurbitula (Tode) Booth   On *Pinus*, anamorph *Zythiostroma pinastri*   1
                                      (Karst.) v. Höhn.

Selinia Karst.
Coprophilous saprophyte, *S. pulchra* (Wint.) Sacc. in East Anglia.

Thuemenella Penzig & Sacc.
Lignicolous saprophyte, *T. britannica* Rifai & Webster on *Fagus* in Hampshire.

Thyronectria Sacc.
Lignicolous saprophytes, *T. berolinensis* (Sacc.) Seaver on *Euonymus* and *Ribis* in the north.

Trailia Sutherland
?Parasite on Phaeophyceae, *T. ascophylli* Sutherland on *Ascophyllum* and *Fucus* in the west.

Trichonectria Kirschst.
Saprophytes, or lichenicolous, *T. hirta* (Blox.) Petch in the midlands, *T. hyalocristata* Scheuer on *Carex* in the west.

POLYSTIGMATALES

Gibellina Pass.
Facultative parasite on *Triticum*, *G. cerealis* Pass. in Hertfordshire.

Glomerella Spaulding & v. Schrenk
Facultative parasites, *G. montana* (Sacc.) v. Arx & Müll. on *Sesleria* in the north.

Glomerella cincta (Stonem.) Spaulding & v. Schrenk	Glasshouse alien on orchids, anamorph *Colletotrichum cinctum* (Berk. & Curt.) Stonem.	1
G. cingulata (Stonem.) Spaulding & v. Schrenk	Anamorph *C. fructigenum* (Berk.) Vassil. (*G. miyabeana* Fukushi) v. Arx; *G. phacidiomorpha* (Ces.) Petrak)	1 2 3 4

Isothea Fr.
Foliar parasite, *I. rhytismoides* (Bab.) Fr. on *Dryas octopetala* in the north.

Phycomelaina Kohlmeyer
?Parasite on Phaeophyceae, *P. laminariae* (Rostrup) Kohlm. on *Laminaria* in the north.

Phyllachora Nits.
Parasites, *P. sylvatica* Sacc. & Speg. on *Festuca ovina*, *F. rubra* and *F. vivipara* in North and West.

Phyllachora dactylidis Delacr.	On *Dactylis glomerata*	1 2 3 4
P. graminis (Pers.) Fuck.	On *Agropyron caninum*, *A. repens* and *Bromus ramosus*. Also reported on *Agrostis*, *Brachypodium*, *Arrhenatherum* & *Deschampsia* in N. & W.	1
P. junci (A. & S.) Fuck.	On *Juncus*	1 2 3 4

Plagiosphaera Petrak
Saprophyte, *P. immersa* (Trail) Petrak on *Urtica dioica* in the north.

Polystigma DC.
Foliicolous parasites, *P. fulvum* DC. on *Prunus padus* in the north.

Polystigma rubrum (Pers.) DC.   Anamorph *Polystigmina rubra* Sacc., on   1 2 3 4
                                  Prunus spinosa

Teliminella Petrak
   Foliar parasites, *T. gangraena* (Fr.) Petrak on *Poa annua, P. nemoralis & P. trivialis* in the
north.

Thamnogalla Hawksw.
   Lichenicolous parasite, *T. crombei* (Mudd) Hawksw. on *Thamnolia* in the north and
west.

SPHAERIALES

Adelococcus Theissen & Sydow
   Lichenparasites, a few species in the north and west.

Amphisphaerella (Sacc.) Kirschstein
   Lignicolous saprophytes, *A. xylostei* (Pers.) Munk on *Lonicera* in East Anglia.

Amphisphaeria Ces. & de Not.
   Saprophytes, *A. culmicola* Sacc. on *Spartina* in Hampshire, *A. paedida* (B. & Br.) Sacc. on
Fagus in the west, *A. umbrina* (Fr.) de Not. on *Quercus* in the north, *A. vibratilis* (Fuck.)
Müller on *Prunus* in E. Anglia.

Anisostomula v. Hohn. see *Hyponectria*

Anthostoma Nits.
   Lignicolous saprophytes.

Anthostoma decipiens (DC.) Nits	On *Carpinus*	1		
A. dryophilum (Currey) Sacc.	On *Quercus*	1		
A. melanotes (B. & Br.) Sacc.	On *Ulmus*	1	2	
A. plowrightii (Niessl) Sacc.	On *Ulex*			3

Anthostomella Sacc.
   Herbicolous and graminicolous saprophytes; *A. alchemillae* (Sm. & Ram.) Francis, *A.
arenaria* O. Erikss., *A. chionostoma* (Dur. & Mont.) Sacc., *A. conorum* (Fuck.) Sacc., *A.
pedemontana* Ferr. & Sacc., *A. sabiniana* Francis in the north, *A. lugubris* (Rob.) Sacc. in the
north and west, *A. caricis* Francis, *A. fuegiana* Speg., *A. leptospora* (Sacc.) Francis, *A. scotina*
(Dur. & Mont.) Sacc. in East Anglia, *A. miscanthea* Sacc. in the midlands.

Anthostomella appendiculosa (B. & Br.) Sacc.	On *Rubus* needs confirmation	1		3
A. clypeoides Rehm	On *Chamaenerion & Rubus*	1		3
A. formosa Kirschst.	On *Pinus* needles		2	
A. limitata Sacc.	On *Carex*	1		
A. phaeosticta (Berk.) Sacc.	On *Ammophila*			3
A. punctulata (Rob.) Sacc.	On *Carex*			3
A. rubicola (Speg.) Sacc.	On *Rubus*	1		3
A. tomicoides Sacc.		1		3

Apiorhynchostoma Petrak
   Lignicolous sparophyte.

Apiorhynchostoma curreyi (Rab.) Müller   1

Apiospora Sacc.
Graminicolous saprophytes with *Arthrinium* or *Scyphospora* anamorphs, *A. bambusae* (Turc.) Siv. in the north.

Apiospora montagnei Sacc.	As the anamorph *Papularia arundinis* (Cda) Fr.	1

Ascotricha Berk.
Saprophytes, *A. erinacea* Zambett. on *Ulmus* in the west midlands.

Ascotricha amphitricha (Corda) Hughes	1
A. chartarum Berk.	1
A. lusitanica R. Kenneth	1

Biscogniauxia O. Kuntze.
Lignicolous saprophytes = *Nummariella*, *B. marginata* (Fr.) Pouzar., on *Pyrus* in the north.

Biscogniauxia mediterranea (De Not.)		1		
B. nummularia (Bull.) O.K.	Mainly on *Fagus* (*Nummularia bulliardii* Tul.)	1	2	4

Blogiascospora Shoemaker et al.
  Saprophyte.

Blogiascospora marginata (Fuck.) Shoemaker et al.	On *Rosa*, anam. *Seiridium marginatum* Nees	3

Broomella Sacc.
Saprophytes on *Clematis*, *B. montaniensis* (Ell. & Ev.) Müller & Ahmad in the Midlands, *Broomella vitalbae* (B. & Br.) Sacc. in E. Anglia.

Cainiella E. Müller
  Saprophytes, *C. johansonii* (Rehm) Müller. on *Dryas octopetala* in the north.

Calosphaeria Tul.
Lignicolous saprophytes, *C. cyclospora* (Kirschst.) Petrak & *C. dryina* (Curr.) Nits. in E. Anglia.

Calosphaeria parasitica Fuck.	On *Eutypella quaternata* on *Fagus*	1
C. pulchella (Pers.) Schroet.	On *Prunus*	1
C. wahlenbergii (Desm.) Nits.	On *Betula*	1

Camarops Karst.
Lignicolous saprophytes, *C.* (*Bolinia*) *tubulina* (A. & S.) Shear on *Abies* in Wiltshire.

Camarops (*Bolinia*) *lutea* (A. & S.) Nannf.	On *Alnus, Betula, Buxus, Corylus, Fagus, Ilex, Quercus, Salix, Ulex*	1	2	4
C. (*Anthostoma*) *microspora* (Karst.) Shear	On *Alnus* and *Corylus*	1		
C. polysperma (Mont.) Miller	On *Alnus*	1		

Ceriophora v. Höhn.
  Saprophyte.

Ceriophora palustris(B. & Br.) v. Höhn.	On *Carex*	3

Ceriospora Niessl
Saprophytes, *C. caudaesuis* Ingold in the north, *C. polygonacearum* (Petr.) Piroz. & M. Jones in the Midlands.

Ceriosporopsis Linder
  Marine lignicolous saprophytes, five other British species.

Ceriosporopsis halima Linder		3 4

Chaetomastia (Sacc.) Berl.
  Saprophytes, *C. hispidula* (Sacc.) Berl. in E. Anglia, *C. canescens* (Speg.) Berl. in the north.

Chaetosphaerella Müller & Booth
  Lignicolous saprophytes.

Chaetosphaerella fusca (Fuck.) Müll. & Booth	Anamorph *Oedemium didymum* (Schm.) Hughes	1 2 4
C. fusispora Siv.		1
C. phaeostroma (Dur. & Mont.) Müll. & Booth	Anamorph *Oedemium minus* (Link) Hughes	1 2 3 4

Chaetosphaeria Tul.
  Lignicolous saprophytes, *C. anglica* Fisher & Petrini in the west, *C. bramleyi* Booth in north and west.

Chaetosphaeria callimorpha(Mont.) Sacc.		1 3 4
C. cupulifera (Br. & Br.) Sacc.	Anamorph *Catenularia cuneiformis* (Richon) Mason	1
C. inaequalis (Grove) Gams & Hol. Jech.	Anamorph *Gonytrichum caesium* Nees	1 4
C. innumera Tul.	Anamorph *Chloridium botryoideum* (Corda) Hughes	1 2
C. lentomita Gams & Hol. Jech.	Anamorph *Chloridium pachytrachelum* Gams & Hol. Jech.	1
C. myriocarpa (Fr.) Booth	Anamorph *Chloridium clavaeforme* (Preuss) Gams. & H.Jech.	1 2
C. preussii Gams & Hol. Jech.	Anamorph *Chloridium preussii* Gams & Hol. Jech.	1
C. pulviscula (Currey) Booth	Anamorph *Menispora caesia* Preuss	1 3
C. vermicularioides (Sacc. & Roum.) Gams. & Hol. Jech.		1

Clypeosphaeria Fuck.
  Lignicolous saprophytes.

Clypeosphaeria mamillana (Fr.) Lamb.		1
C. notarisii Fuck.		1 3 4

Coniochaeta (Sacc.) Cooke
  Coprophilous or lignicolous saprophytes, many additional species in the west and north.

Coniochaeta ambigua (Sacc.) Popuscho		2	
C. hansenii (Oud.) Cain	1		
C. ligniaria (Grev.) Massee (*C. discospora* (Auersw.) Cain)	1		4
C. pulveracea (Ehrh.) Munk	1		
C. scatigena (B. & Br.) Cain		2	
C. subcorticalis (Fuck.) Munk			3
C. velutina (Fuck.) Cooke	1		3

Coniochaetidium Mallock & Cain
  Saprophyte, *C. savoryi* (Booth) Mallock & Cain on *Juniperus* in the west.

Corollospora Werdermann
  Marine lignicolous saprophytes.

Cryptovalsa Ces. & de Not.
Lignicolous saprophytes, *C. suaedicola* Spooner on *Suaeda fruticosa* in Essex.

Cryptovalsa protracta (Pers.) Ces. & de Not.	On *Fraxinus*	1

Daldinia Ces. & de Not.
Lignicolous saprophytes and facultative parasites.

Daldinia concentrica (Bolt.) Ces. & de Not.	1 2 3 4
D. vernicosa (Schw.) Ces. & de Not.	3

Diapleella Munk
Saprophyte.

Diapleella clivensis (B. & Br.) Munk	4

Discostroma Clem.
Saprophytes or facultative parasites, *D. strobiligenum* (Müll. & Loeff.) Brockm. in the north.

Discostroma corticola (Fuck.) Brockman (*Griphosphaeria corticola* (Fuck.) v. Höhn)	Anamorph *Seimatosporium lichenicola* (Cda) Shoem. & Müller	1

Dothivalsaria Petrak
Saprophyte, *D. megalospora* (Auersw.) Petrak on *Alnus* in the north.

Endoxylina Romell
Lignicolous saprophytes, *E. pini* Siv. in the north.

Eriosphaeria Sacc.
Lignicolous saprophytes, *E. membranacea* (B. & Br.) Sacc. in the west.

Eriosphaeria aggregata Müller & Munk	On *Larix*, as anamorph *Trimmatostroma scutellare* (B. & Br.) Ellis	1

Haligena Kohlmeyer
Marine lignicolous saprophytes, *H. elaterophora* Kohlm. on *Spartina*, Lincolnshire.

Halosarpheia Kohlm. & Kohlm.
Marine lignicolous saprophyte, *H. spartinae* (Gareth Jones) Shearer & Crane on E. & W. coasts.

H. unicaudata Gareth Jones et al.

Halosphaeria Linder
Marine lignicolous saprophytes, 8 other species no doubt generally distributed.

Halosphaeria appendiculata Linder	3 4

Hypocopra (Fr.) Kickx
Coprophilous saprophytes, no doubt less common through decline in equine population, *B. brefeldii* (Zopf) Zopf, *H. equorum* (Fuck.) Wint., *H. merdaria* (Fr.) Fr., *H. stercoraria* (Sow.) Sacc. in East Anglia, *H. stephanophora* Krug & Cain in the midlands.

Hyponectria Sacc.
Saprophytes or facultative parasites.

Hyponectria buxi (DC.) Sacc.	On *Buxus sempervirens*	1	2		4
H. cookeana (Auersw.) Barr (*Anisostomula* cookeana (Auersw.) v. Höhn.	On *Quercus*	1			

Hypoxylon Bull.
Lignicolous saprophytes, see also *Biscogniauxia, Nemania, Ustulina.*

Hypoxylon cohaerens (Pers.) Fr.	1	2	3	4
H. confluens (Tode) West. (*H. semiimmersum* Nits.)	1	2	3	4
H. fragiforme (Scop.) Kickx	1	2	3	4
H. fraxinophilum Pouzar (*H. argillaceum* (Pers.) Berk.)	1	2		4
H. fuscum (Pers.) Fr.	1	2	3	4
H. howeanum Peck	1	2		4
H. mammatum (Wahlenb.) Miller	1	2		
H. multiforme (Fr.) Fr.	1	2	3	4
H. rubiginosum (Pers.) Fr.	1	2	3	4
H. rutilum Tul.	1	2		
H. udum (Pers.) Fr.	1			4

Leiosphaerella v. Höhn.
Saprophytes, see also *Paradidymella, L. vexans* (Sacc.) Müller on *Cornus* in Dorset.

Leiosphaerella tosta (B. & Br.) Müller	Anamorph *Seimatosporium passerinii* (Sacc.) Brock.	1		3	4

Lentomita Niessl
Lignicolous saprophytes, *L. stylophora* (B. & Br.) Sacc. of uncertain status.

Lentomita hirsutula Bres. Ss. Munk	On *Prunus*	1
L. stylophora (B. & Br.) Sacc.	On *Acer*, doubtful record.	1

Lepteutypa Petrak
Lignicolous saprophyte, *L. hippophaes* (Sollm.) v. Arx in the west.

Lignincola Höhnk
Maritime saprophyte, *L. laevis* Höhnk in Lincolnshire.

Lindra I.M. Wilson
Marine lignicolous saprophyte, *L. inflata* I.M. Wilson in the west.

Linostomella Petrak
Lignicolous saprophyte, *L. sphaerosperma* (Fuck.) Petrak in the west midlands.

Litschaueria Petrak
?Hyperparasite.

Litschaueria corticiorum (v. Höhn.) Petrak	On *Phanerochaete sordida*	2

Lopadostoma (Nits.) Trav.
Lignicolous saprophytes, segregates from *Anthostoma.*

Lopadostoma gastrinum (Fr.) Trav.	On *Ulmus*	1	2		
L. turgidum (Pers.) Trav.	On *Fagus*	1	2	3	4

Lulworthia Sutherland
Marine lignicolous saprophytes, *L. fucicola* Suth. & *L. medusa* (Ell. & Ev.) Cribb widespread.

Lulworthia purpurea(Wilson) Johnson	3

Melanopsamma Niessl
Lignicolous saprophytes.

Melanopsamma pomiformis (Pers.) Sacc.	Anamorph *Stachybotrys socia* (Sacc.) Sacc.	1
M. pustula (Currey) Sacc.		1

Melomastia Nits.
Lignicolous saprophyte.

Melomastia mastoidea (Fr.) Schroet.   1   2

Monographella Petrak
Facultative parasite of Gramineae.

Monographella nivalis (Schaffnit) Müller   Anamorph *Fusarium nivale* (Fr.) Ces.   1

Nais Kohlmeyer
Marine lignicolous saprophyte, *N. inornata* Kohlmeyer.

Nautosphaeria Gareth Jones
Marine lignicolous saprophyte, *N. cristaminuta* Gareth Jones in the west.

Nemania S.F. Gray
Lignicolous saprophytes, segregate from *Hypoxylon*, *N. effusa* (Nits.) Pouzar in Hampshire, *N. gwyneddi* (Walley et al) Pouzar in the west.

Nemania aenea (Nits.) Pouzar	1			
N. bipapillata (Berk. & Curt.) Pouzar	1			
N. chestersii (Rogers & Walley) Pouzar	1			
N. serpens (Pers.) S.F. Gray	1	2	3	4
N. subannulata (Henn. & Nym.)	1			

Neolamya Theiss. & Syd.
Parasite on lichens, *N. peltigerae* (Mont.) Theiss. & Syd. in the west.

Niesslia Auersw.
Saprophytes, *N. cladoniicola* Hawksw. & Gams in the north, *N. exilis* (A. & S.) Wint. in north and west.

Niesslia exosporioides (Desm.) Wint.	Anamorph *Monocillium granulatum* (Fuck.) Gams	2
N. ilicifolia (Cooke) Wint.		1

Paradidymella Petrak
Saprophytes, = the older *Leiosphaerella* but combinations not made. *Paradidymella clarkii* Hawks. & Siv., *P. holci* Hawksw. & Siv. and var. *moliniae* Dennis & Spooner in the north and west.

Phaeotrichosphaeria Siv.
Lignicolous saprophyte.

Phaeotrichosphaeria britannica Siv.	On *Fagus* & *Quercus*, anamorph *Endophragmiella uniseptata* (Ellis) Hughes	1

Phomatospora Sacc.
Saprophytes, *P. arenaria* Sacc. Bomm. & Rouss., *P. claraebonae* (Speg.) Barr and *P. endopteris* (Plowr.) Phill. & Plowr. in north and west.

143

Phomatospora berkeleyi Sacc.	On herbaceous stems	1	4
P. coprophila Richardson	On dung of herbivores	1	
P. dinemasporium Webster	On Gramineae, anamorph *Dinemasporium graminum* (Lib.) Lév.	1	4
P. gelatinospora Barr	On *Rhododendron ponticum*	1	
P. ribesia (Cooke & Massee) Sacc.	On *Ribes*		3
P. therophila (Desm.) Sacc.	On *Juncus*		4

Physalospora Niessl
  Saprophytes, much confused with *Botryosphaeria, Glomerella, Guignardia*; eight other species reported from the north and west.

Physalospora lonicerae Grove	On *Lonicera periclymenum*. Clarification desirable	3

Podosordaria Ellis & Holway
  Coprophilous saprophytes.

Podosordaria pedunculata (Dickson) Dennis		3 4
P. tulasnei (Nits.) Dennis		3

Poikiloderma Fuisting
  Lignicolous saprophyte.

Poikiloderma bufonia (B. & Br.) Fuisting	On *Quercus*	1

Poronia Willd.
  Coprophilous saprophytes, *P. erici* Lohmeyer & Benkert in East Anglia.

Poronia oedipus Mont.	Presumably a casual alien	"Sussex"
P. punctata (L.) Fr.	On horse dung	1 3 4

Pseudoguignardia Gutner
  Saprophyte, *P. scirpi* Gutner in the west midlands.

Pseudomassaria Jacz.
  Saprophytes, including *Chaetapiospora* Petrak, the setae are inconstant; *P. islandica* (Joh.) Barr, *P. lycopodina* (Karst.) v. Arx, *P. vaccinii* Dennis in the north.

Pseudomassaria chondrospora (Ces.) Jacz.	On *Tilia*	2
P. corni (Sow.) v. Arx	On *Cornus*	1
P. sepincoliformia (de Not.) v. Arx	On *Rosa*	1
P. thistletonia (Cooke) v. Arx	On *Rhododendron*	1

Rosellinia de Not.
  Lignicolous saprophytes, *R. necatrix* Berl. records may have originated in confusion.

Rosellinia aquila (Fr.) de Not.		1 2 4
R. britannica Petrini et al		2
R. buxi Fabre		1
R. mammaeformis (Pers.) Ces. & de Not.	Needs confirmation but likely to occur	1 2
R. thelena (Fr.) Rab.		1

Savoryella Gareth Jones & Eaton
  Marine lignicolous saprophyte, *S. lignicola* Gareth Jones & Eaton in the west.

Scopinella Lév.
  Saprophytes, *S. caulincola* (Fuck.) Malloch in the west, *S. solani* (Zukal) Malloch in north.

Scotiosphaeria Siv.
Hyperparasite *S. endoxylinae* Siv. on *Pinus* in the north.

Spumatoria Massee & Salmon
Coprophilous saprophyte, *S. longicollis* Massee & Salmon. on dung in Essex.

Synaptospora Cain
Hyperparasite, *S. tartaricola* (Nyle.) Cain. on *Ochrolechia* & *Pertusaria* in the west.

Torpedospora Meyers
Saprophyte, *T. radiata* Meyers in the west.

Trichosphaerella Bomm., Rouss., & Sacc.
Lignicolous saprophytes.

Trichosphaerella decipiens Bomm., Rouss., & Sacc.	3

Trichosphaeria Fuck.
Lignicolous saprophytes, *T. barbula* (B. & Br.) Wint., *T. crassipila* Grove, *T. superficialis* (Currey) Sacc. in the west, *T. melanostigmoides* (Feltg.) Munk in the north.

Trichosphaeria notabilis Mouton	1
T. pilosa (Pers.) Fuck.	1

Ustulina Tul.
Saprophyte or facultative parasite especially of *Tilia*.

Ustulina deusta (Hoffm.) Lind (*U. vulgaris* Tul., *Hypoxylon deustum* (Hoffm.) Grev.)	1	2	3	4

Valsaria Ces. & de Not.
Lignicolous saprophytes, *V. foedans* (Karst.) Sacc. & *V. rubricosa* (Fr.) Sacc. in East Anglia, *V. niesslii* (Wint.) Sacc. in the north.

Valsaria anserina (Pers.) Sacc.	1
V. capronii (Cooke) Berl. & Vogl.	1
V. cincta (Currey) Sacc.	1
V. exasperans (Gerard) Sacc.	1
V. insitiva (Tode) Ces. & de Not.	1

Vialaea Sacc.
Saprophyte.

Vialaea insculpta (Fr.) Sacc.	On *Ilex aquifolium*	1

Wawelia Namyslowski
Saprophytes, *W. octospora* Minter & Webster in the west.

Xylaria Hill.
Lignicolous saprophytes, *X. bulbosa* (Pers.) B. & Br. & *X. friesii* Laessøe in the west, *X. guepini* (Fr.) Fr. in the north, *X. filiformis* (Fr.) Fr. in East Anglia, *X. oxyacanthae* Tul. in the north and west.

Xylaria carpophila (Pers.) Fr.	On *Fagus* cupules	1	2	3	4
X. digitata (L.) Grev. ss. auct. angl.		1	2	3	4
X. hippotrichoides (Sow.) Sacc.		1			
X. hypoxylon (L.) Grev.		1	2	3	4
X. longipes Nits (*X. corniformis* (Fr.) Fr.?)	On *Acer pseudoplatanus*	1	2	3	4
X. mellisii (Berk.) Cooke	Glasshouse alien	1			
X. polymorpha (Pers.) Grev.	In Britain characteristic of *Fagus*	1	2	3	4

Zignoella Sacc.
Lignicolous saprophytes, *Z. eutypoides* Sacc., *Z. morthieri* (Fuck.) Sacc., *Z. seriata* (Curr.
Sacc. in the north, *Z. rhytidodes* (Br. & Br.) Sacc. in the west, *Z. pachyspora* in the north and
west.

Zignoella cf. *aterrima* (Feld.) Sacc.	1	
Z. collabens (Currey) Sacc.	1	
Z. cf. *dolichospora* Sacc.	1	
Z. fallax (Sacc.) Sacc.	1	
Z. hysterioides (Currey) Sacc.		4
Z. macrospora Sacc.	1	
Z. ovoidea (Fr.) Sacc.	1	
Z. rhodobapha (B. & Br.) Sacc. "South Kensington"		

SORDARIALES

Apiosordaria v. Arx & Gams
Terrestrial saprophytes, *A. verruculosa* (Jensen) v. Arx & Gams Midlands and west.

Arnium Nits.
Predominantly coprophilous saprophytes; *A. caballinum* Lundq. & *A. macrothecum*
(Crouan) Lundq. in Essex, *A. tomentosum* (Speg.) Lundq. in East Anglia, *A. apiculatum*
Griff.) Lundq. in the west.

Arnium cervinum Lundq.	1	
A. hirtum (Hansen) Lundq. & Krug	1	
A. leporinum (Cain) Lundq. & Krug	1	
A. mendax Lundq.	1	
A. olerum (Fr.) Lundq. & Krug (*Podospora brassicae* (Klotzsch) Wint.)	1	2

Arxiomyces Cannon & Hawksw.
Lignicolous saprophyte, *A. (Phaeostoma) vitis* (Fuck.) Cannon & Hawksw. in the west
midlands.

Bombardia (Fr.) Fuck.
Lignicolous saprophyte.

Bombardia bombarda (Batsch) Schroet. (*B. fasciculata* (Fr.) Fuck.)	1	4

Bombardioidea Moreau
Coprophilous saprophytes; records of two other British species also need
confirmation.

Bombardioidea bombardioides (Auersw.) Moreau & Lundq. needs confirmation	1

Cercophora Fuck.
Saprophytes on various substrata, (*Lasiosordaria* Chen.); *C. arenicola* Hilber in the west
midlands, *C. mirabilis* Fuck. & *C. silvatica* Lundq. in the north. A British record of *C.
sulphurella* (Sacc.) Hilber is unlocalised.

Cercophora caudata (Currey) Lundq.	1	2	
C. coprophila (Fr.) Lundq.	1	2	3
Sordaria sparganicola Plowr. Has been referred here by Lundq. in herb	1		

Chaetomidium (Zopf) Sacc.
Saprophytes, *C. unciatum* Dennis in the west midlands.

Chaetomium Kunze
Saprophytes with cosmopolitan distribution; at least 30 more British species, of which
C. aureum Chivers & *C. warcupii* Saxena & Mukerj. are in E. Anglia, *C. megalocarpum* Bain.
in Hertfordshire, *C. erectum* Skolko & Groves and *C. gelasinosporum* v. Arx & Müller in
Hampshire, *C. reflexum* Skolko & Groves in Berkshire and *C. crispatum* Fuck. in Essex.

Chaetomium bostrychodes Zopf	1		3	4
C. britannicum Ames		2		
C. caprinum Bain.	1			
C. cochliodes Pall.	1			
C. dolichotrichum Ames	1			
C. elatum Schm. & Kunze	1	2		4
C. funicola Cooke	1			
C. globosum Kunze	1	2	3	4
C. indicum Corda				4
C. murorum Corda	1			
C. olivaceum Cooke & Ellis				4
C. simile Massee & Salmon	1			
C. spirale Zopf	"Sussex"			

Corynascus v. Arx
Saprophytes, *C. sepedonium* (Emmons) v. Arx in the midlands, *C. thermophilus* (Fergus &
Sinden) Klapotek in Hertfordshire.

Erostrotheca Martin & Charles
Facultative parasite, connection with the reputed anamorph has been denied and if so
the genus is not British.

Erostrotheca multiformis Martin & Charles	On *Lathyrus odoratus,* as the putative anamorph *Ramularia alba* (Dowson) Nannf.	1

Gelasinospora Dowding
Saprophytes.

Gelasinospora adjuncta Cain		2	
G. cerealis Dowding	1		4
G. reticulata (Booth & Ebben) Cailleux		2	
G. reticulospora (Greis & Greis-Dengler) Moreau (*G. retispora* Cain)	1		4
G. tetrasperma Dowding	1		

Helminthosphaeria Fuck.
Hyperparasite on *Clavulina*.

Helminthosphaeria clavariarum (Tul.) Fuck.	Anam. *Spadicoides clavariarum* (Desm) Hughes	1	3	4

Jugulospora Lundq.
Terrestrial saprophyte.

Jugulospora rotula (Cooke) Lundq.	On burnt ground	1

Lasiosphaeria Ces. & de Not.
Lignicolous saprophytes; *L. dactylina* Webster, *L. helicoma* (Phill. & Plowr.) Cooke &
Plowr., *L. mutablis* (Pers.) Fuck. in East Anglia, *L. felina* (Fuck.) Cooke & Plowr., *L.
phyllophila* Mouton and *L. sorbina* (Nyl.) Karst, in the west.

Lasiosphaeria canescens (Pers.) Karst.	1	2	3
L. caudata (Fuck.) Sacc.	1	2	

L. hirsuta (Fr.) Ces. & de Not.	1	2		4
L. hispida (Tode) Fuck.	1			
L. ovina (Pers.) Ces. & De Not.	1	2	3	4
L. rhacodium (Pers.) Ces. & de Not.	1			
L. spermoides (Hoffm.) Ces. & de Not.	1	2	3	4

Melanocarpus v. Arx
Thermophilic saprophyte, *M. albomyces* (Cooney & Emerson) v. Arx.

Melanospora Corda
Saprophytes and hyperparasites, *M. fusispora* (Petch) Doguet in East Anglia.

Melanospora brevirostris (Fuck.) v. Höhn.	On *Sepultaria* and debris	1		4
M. caprina (Fr.) Sacc.		1	
M. damnosa (Sacc. & Berl.) Lindau		1	2
M. chionea (Fr.) Corda	On needles of *Pinus*	1	
M. fallax Zukal		1	
M. lagenaria (Pers.) Fuck.		1	
M. longisetosa Cannon & Hawksw.		1	
M. zamiae Corda		1	2

Neurospora Shear & Dodge
Saprophytes.

Neurospora sitophila Shear & Dodge	Anamamorph *Chrysonilia sitophila* (Mont.) v. Arx	1 2 3 4
N. tetrasperma Shear & Dodge	Anamorph *Chrysonilia tetrasperma* (Shear. & Dodge) v. Arx	4

Ophioceras Sacc.
Saprophytes, *O. leptosporum* (Iqbal) Walker in the west.

Persiciospora Cannon & Hawksw.
Lignicolous saprophyte.

Persiciospora masonii (Kirschst.) Cannon & Hawksw. (*Ceratostoma masonii* Kirschst.)	1

Podospora Ces.
Saprophytes, predominantly coprophilous; *P. excentrica* Lundq. in Hampshire, *P. intestinaceae* Lundq. in Middlesex, *P. myriospora* (Crouan) Niessl in East Anglia.

Podospora appendiculata (Auersw.) Niessl		1		
P. curvicolla (Wint.) Niessl		1	2	
P. decipiens (Wint.) Niessl				3	
P. ellisiana (Griff.) Mirza					4
P. fimbriata (Bayer) Cain		1			4
P. fimiseda (Ces. & de Not.) Niessl		1	2	
P. globosa (Massee & Salmon) Cain		1		
P. granulostriata Lundq.		1		
P. gwynnevaughaniae (Page) Cain				3	
P. pauciseta (Ces.) Trav.	(*P. anserina* (Rab.) Niessl)	1		
P. perplexans (Cain) Cain				
P. pilosa (Mout.) Cain					4
P. pleiospora (Wint.) Niessl		1	2	3
P. pyriformis (Bayer) Cain		1		
P. setosa (Wint.) Niessl		1	2	3

Pustulipora Cannon
Saprophyte, *P. corticola* Cannon in the midlands.

Rhamphoria Niessl
Lignicolous saprophytes; *R. bevanii* Siv. in the north.

Rhamphoria pyriformis (Pers.) v. Höhn.	1
R. tympanidispora Rehm	1

Schizothecium Corda
Coprophilous saprophytes; *S. aloides* (Fuck.) Lundq., *S. glutinans* (Cain) Lundq., *S. hispidulum* (Speg.) Lundq. *S. nanum* Lundq. in East Anglia, *S. dubium* Hansen) Lundq. in midlands.

Schizothecium conicum (Fuck.) Lundq.	(*Podospora curvula* (De Bary) Niessl)	1	2		4
S. squamulosum (Crouan) Lundq.		1			
S. tetrasporum (Wint.) Lundq.		1	2	3	
S. vesticola (B. & Br.) Lundq.	(*Podospora minuta* (Fuck.) Niessl)	1		3	

Sordaria Ces. & de Not.
Coprophilous saprophytes; *S. alcina* Lundq. in the midlands, *S. minima* Sacc. & Speg. in the north.

Sordaria fimicola (Rob.) Ces. & de Not.	1	2		4
S. humana (Fuck.) Wint.	1	2		4
S. lappoe Potebnia	1			
S. macrospora Auersw.	1		3	4
S. superba de Not.	"Sussex"			

Sphaerodes Clem.
Hyperparasites or saprophytes, *S. episphaeria* (Phill. & Plowr.) Cannon & Hawksw. in E. Anglia.

Sphaerodes fimicola (Hansen) Cannon & Hawsk.	1	4

Strattonia Ciferri
Saprophytes; *S. minor* Lundq. in the midlands.

Strattonia carbonaria (Phill. & Plowr.) Lundq.	On burnt ground	1

Syspastospora Cannon & Hawksw.
Hyperparasites, *S. parasitica* (Tul.) Cannon & Hawksw. in Berkshire & East Anglia.

Thielavia Zopf
Saprophytes, *T. fimeti* (Fuck.) Malloch & Cain in the west midlands, *T. terrestris* (Apinis) Malloch & Cain in the midlands, *T. wareingii* Seth in the west. All records of *T. basicola* Zopf are highly suspect and to be rejected.

Thielavia terricola (Gilman & Abbott) Emmons	1

Viennotidia Negru & Verona
Coprophilous saprophyte.

Viennotidia fimicola (Marchal) Cannon & Hawksw.	1

Zopfiella Wint.
Saprophytes, *Z. inermis* (Cailleux) Malloch & Cain in the west, *S. (Tripterospora) erostrata* (Griffiths) Udagawa & Furuya in Warwickshire.

Zygospermella Cain
Coprophyllous saprophyte.

Zygospermella insignis (Mout.) Cain 1

CORONOPHORALES

Acanthonitschkea Sacc.
Lignicolous saprophyte.

Acanthonitschkea tristis (Pers.) Nannf. 4

Bertia de Not.
Lignicolous saprophyte.

Bertia moriformis (Tode) de Not. 1 2 3 4

Coronophora Fuck.
Lignicolous saprophytes, *C. gregaria* (Lib.) Fuck. in East Anglia, *C. annexa* (Nits.) Fuck. in the north.

Coronophora angustata Fuck. On *Betula* and *Fagus* 1

Lasiosphaeriopsis Hawksw. & Siv.
Hyperparasite, *L. salisburyi* Hawksw. & Siv. on *Peltigera* in the north.

Nitschkia Otth
Lignicolous saprophytes.

Nitschkia collapsa (Romell) Chenant. 1 2
N. confertula (Schw.) Nannf. (*Tympanopsis euomphala* (B. & C.) Starb.) 1
N. cupularis (Pers.) Karst. 1 2
N. grevillei (Rehm) Nannf ("*Calyculosphaeria tristis*" auct angl. non Pers.) 1 2 3 4
N. parasitans (Schw.) Nannf. On *Nectria cinnabarina* 1

Rhagadostoma Körber
Hyperparasite, *R. (Bertia) lichenicola* (de Not.) Keissler on *Solorina crocea* in the north.

DIATRYPALES

Cryptosphaeria Grev.
Lignicolous saprophytes, *C. lignyota* (Fr.) Auersw. (*C. populina* (Pers.) Sacc.) in the north.

Cryptosphaeria eunomia (Fr.) Fuck. 1 2 4
C. eunomia var *fraxini* (Richon) Rappaz (*Cryptosphaerina fraxini* Fautr. & Lamb.) 1

Diatrype Fr.
Lignicolous saprophytes.

Diatrype bullata (Hoffm.) Fr. 1 2 3
D. disciformis (Hoffm.) Fr. Anamorph *Libertella disciformis* v. Höhn. 1 2 3 4
 On *Fagus*
D. stigma (Hoffm.) Fr. Anamorph *Libertella betulina* Desm. 1 2 3 4

Diatrypella (Ces. & de Not.) de Not.
Lignicolous saprophytes.

Diatrypella quercina (Pers.) Cooke Grove	Anamorph *Libertella quercina* (Sacc.)	1	2	3	4
D. verruciformis (Ehrh.) Nits (*D. favacea* Fr.) de Not.)	Anamorph *L. favacea* Trav.	1	2	3	4

Eutypa Tul.
Lignicolous saprophytes.

Eutypa flavovirens (Hoffm.) Tul. (*Diatrype flavovirens* (Hoffm.) Fr.)	1	2	3	4
E. lata (Pers.) Tul.	1	2	3	4
E. lejoplaca (Fr.) Cooke	1			
E. maura (Fr.) Fuck. (*E. acharii* Tul.)	1	2		4
E. polycocca (Fr.) Karst. (*E. aspera* (Nits.) Fuck.)	1			
E. spinosa (Pers.) Tul.	1	2		4

Eutypella (Nits.) Sacc.
Lignicolous saprophytes; *E. sorbi* (Schmidt) Sacc. in the north.

Eutypella dissepta (Fr.) Rappaz		1	2		4
Eutypella leprosa (Pers.) Berl. (*Diatrype berberidis* Cooke, *E. acericola* (de Not.) Berl.)			2		
E. prunastri (Pers.) Sacc.		1		3	
E. quaternata (Pers.) Rappaz	(*Quaternaria quaternata* (Pers.) Schroet.)	1	2	3	4
E. scoparia (Schw.) Ell. & Ev.	(*Peroneutypa heteracantha* (Sacc.) Berl.)	1	2	3	4
E. stellulata (Fr.) Sacc.		1	2	3	4

DIAPORTHALES

Anisogramma Theiss. & Syd. facultative parasite of Betula, *A. virgultorum* (Fr.) Theiss. & Syd. in the north.

Apiognomonia v. Höhn.
Facultative parasites of angiosperm trees.

Apiognomonia errabunda (Rob.) v. Höhn	On *Fagus* & *Quercus*, anamorph *Discula umbrinella* (B. & Br.) Sutton	1		
A. erythrostoma (Pers.) v. Höhn	On *Prunus*, anamorph *Libertina effusa* (Lib.) v. Höhn.			4
A. petiolicola (Fuck.) Monod	On *Tilia*		2	4
A. veneta (Sacc. & Speg.) v. Höhn. (Peck) Sacc.	On *Platanus*, anamorph *Discula platani*	1	2	4

Apioplagiostoma Barr
Saprophyte.

Apioplagiostoma aceriferum (Cooke) Barr	*On Acer campestre*	4

Apioporthe v. Höhn.
Saprophyte.

Apioporthe vepris (Del.) Wehmeyer	On *Rubus*	1

Aporhytisma v. Höhn.
Saprophyte.

Aporhytisma urticae (Wallr.) v. Höhn	Anamorph *Apomelasmia urticae* (Fr.) Grove	1	4

Calospora Sacc.
Saprophyte.

Calospora platanoides (Pers.) Niessl (*C. innesii* (Currey) Sacc.)	On *Acer*	1			

Caudospora Starb.
Saprophyte or weak facultative parasite, segregate from *Diaporthe*.

Caudospora taleola (Fr.) Starb.	On *Quercus*, anamorph *Endogloea taleola* v. Höhn.	1	2	3	4

Ceratosphaeria Niessl
Lignicolous saprophytes.

Ceratosphaeria crinigera (Cooke) Sacc.		1			
C. fragilis Wilberforce	On *Diatrype disciformis*				4
C. lampadophora (B. & Br.) Niessl		1	2		
C. mycophila Wint.		1	2		
C. rhenana (Auersw.) Wint.		2			

Ceratostomella Sacc. see *Endoxyla*

Clypeocarpus Kirschstein
Saprophyte, segregate from *Diaporthe*.

Clypeocarpus lirella (Moug. & Nestl.) Kirschst.	On *Filipendula ulmaria*	1			

Cryptodiaporthe Petrak
Saprophytes or facultative parasites of woody dicotyledons; *C. aubertii* (West.) Wehm. on *Myrica* in the north and west, *C. pyrrhocystis* (B. & Br.) Wehm. in the west.

Cryptodiaporthe aesculi (Fuck.) Petrak	Anamorph *Discella aesculi* (Cda.) Oud.	1			
C. castanea (Tul.) Wehm.	Anamorph *Fusicoccum castaneum* Sacc.	1	2		4
C. galericulata (Tul.) Wehm.	On *Fagus*, anamorph *Malacostroma carneum* (Thüm.) v. Höhn				
C. hranicensis (Petrak) Wehm.	On *Tilia*	1			
C. hystrix (Tode) Petrak	On *Acer*, anamorph *Diplodina acerina* (Pass.) Sutton	1		3	4
C. lebiseyi (Desm.) Wehm.	On *Acer*, anamorph *Phomopsis lebiseyi* (Sacc.) Died.	1			
C. populea (Sacc.) Butin	On *Populus*, anamorph *Chondroplea populea* (Sacc. & Briard) Kleb.	1	2		
C. robergeana (Desm.) Wehm.	On *Staphylea*	1			
C. salicella (Fr.) Petrak	On *Salix*, anamorph *Discella salicis* (West.) Boerema	1		3	4
C. salicina (Pers.) Wehm.	On *Salix*, anamorph *Discella carbonacea* (Fr.) B. & Br.	1		3	4

Cryptospora Tul.
Lignicolous saprophytes, see *Ophiovalsa*.

Cryptospora intexta (Currey) Sacc.	On *Quercus*	1			

Cryptosporella Sacc. see *Wuestnia*

Debaryella v. Höhn.
Lignicolous saprophytes or ?hyperparasites.

Debaryella gracilis Munk					4
D. hyalina Syd.		1			

Diaporthe Nits.
Saprophytes or facultative parasites on woody plants, with anamorphs in *Phomopsis*.

Species	Host / notes	1	2	3	4
Diaporthe arctii (Lasch) Nits.	A collective species on herbaceous stems	1	2	3	4
D. beckhausii Nits.	On *Viburnum*	1			4
D. chailletii Nits.	On *Atropa belladonna*	1	2		
D. circumscripta (Fr.) Otth	On *Sambucus*	1			4
D. crataegi (Currey) Nits.		1		3	
D. decedens (Fr.) Fuck.	On *Corylus avellana*	1			
D. decorticans (Lib.) Sacc. & Roum.	On *Prunus cerasus*	1			
D. eres Nits. Typically on *Ulmus* but a collective species including races called:		1	2	3	4
D. conorum (Desm.) Niessl	On coniferae	1	2	3	
D. crustosa Sacc. & Roum.	On *Ilex aquifolium*		2		4
D. incarcerata (B. & Br.) Nits.	On *Rosa*	1			
D. laschii Nits.	On *Euonymus europaeus*	1			
D. nucleata (Currey) Cooke	On *Ulex europeus*	1	2		4
D. occulta (Fuck.) Nits.	On *Pinus*			3	
D. ophites Sacc.	On *Hibiscus*	1			
D. perniciosa Marchal	On *Prunus domestica*	1			
D. radula Nits.	On *Prunus laurocerasus*	1			
D. resecans Nits.	On *Syringa vulgaris*	1			
D. revellens Nits.	On *Corylus avellana*	1			
D. rhois Nits.	On *Rhus*				4
D. ryckholtii (West.) Nits.	On *Symphoricarpos*	1			
D. scobina Nits.	On *Fraxinus excelsior*	1	2		4
D. velata (Pers.) Nits.	On *Tilia*	1			
D. fibrosa (Pers.) Fuck.	On *Rhamnus cartharticus*	1	2		4
D. hederae Weymeyer	On *Hedera helix*	1			
D. ilicis Weymeyer	On *Ilex*	1			
D. impulsa (Cooke & Peck) Sacc.	On *Sorbus aucuparia*				4
D. inaequalis (Currey) Nits.	On *Ulex europeus*	1		3	4
D. leiphamia (Fr.) Sacc.	On *Quercus*	1	2	3	4
D. medusae Nits.	On *Fagus*	1			4
D. oncostoma (Duby) Fuck.	On *Robinia pseudacacia*	1		3	4
D. pardalota (Mont.) Nits.	On *Convallaria*	1		3	4
D. pulla Nits.	On *Hedera helix*	1	2	3	4
D. rudis (Fr.) Nits.	On *Laburnum anagyroides*	1			
D. sarothamni Auersw.	On *Sarothamnus scoparius*				4
D. sarothamni var *dulcamarae* (Nits.) Wehm.	On *Solanum dulcamara*	1			
D. sociabilis Nits.	On *Morus*	1			
D. sophorae Sacc.	On *Sophora japonica*	1	2		
D. strumella (Fr.) Fuck.	On *Ribes*			3	4
D. syngenesia (Fr.) Nits.	On *Frangula alnus*				4
D. varians (Currey) Sacc.	On *Acer campestre*	1			4

Diaporthopsis Fabre
Saprophytes, see also *Aporhytisma*

Species	Host / notes	1	2	3	4
Diaporthopsis angelicae (Berk.) Wehm.	On dead stems of Umbelliferae	1			4
D. pantherina (Berk.) Wehm.	On *Pteridium aquilinum*	1			

Ditopella de Not.
Saprophyte.

Species	Host / notes	1	2	3	4
Ditopella ditopa (Fr.) Schroet.	On *Alnus glutinosa*	1		3	

Enchnoa Fr.
Lignicolous saprophytes; *E. lanata* (Fr.) Fr. on *Betula* in the north.

Species	Host / notes	1	2	3	4
Enchnoa infernalis (Kunze) Fuck.	On *Quercus*	1			4

Endoxyla Fuck.
Lignicolous saprophytes; *E. operculata* (A. & S.) Sacc. & *E. parallela* (Fr.) Sacc. in the north.

Endoxyla cirrhosa (Pers.) Müller & v. Arx	1		3
E. laevirostris Munk	1	2	3
E. vestita (Sacc.) Munk	1		

Gaeumannomyces v. Arx & Olivier
Facultative parasites; *G. caricis* Walker in E. Anglia, *G. cylindrosporum* Hornby in Herts.

Gaeumannomyces graminis (Sacc.) v. Arx. & Olivier v. *tritici* Walker		3	4
v. *avenae* (Turner) Dennis	Probably widespread on native grasses but overlooked		4

Gnomonia Ces. & de Not.

Saprophytes and facultative parasites; *G. alni-viridis* Podlahova & Svrček on *Betula* in the north and west, *G. graphis* Fuck. on *Rosa & Rubus* in E. Anglia, *G. tetraspora* Wint. on *Euphorbia* in the west.

Diaporthe acus (Blox.) Cooke	On Rumex is *Gnomonia* teste Wehmeyer	1	3	4
Gnomonia cerastis (Riess) Ces. & de Not.	On *Acer pseudoplatanus*	1		4
G. comari Karst. (*G. fructicola* (Arn.) Fall)	On *Agrimonia, Fragaria, Geum, Potentilla*	1		4
G. geranii Hollos	On *Geranium* (cult.)	1		
G. gnomon (Tode) Schroet. (*H. vulgaris* Ces. & de Not)	On *Corylus*	1		4
G. leptostyla (Fr.) Ces. & De Not.	Anamorph *Marssoniella juglandis* (Lib.) v. Höhn.		3	4
G. longirostris Cribb & Cribb.	On submerged wood			4
G. rostellata (Fr.) Brefeld (*G. rubi* (Rehm) Wint.)	On *Rubus*		3	
G. setacea (Pers.) Ces. & de Not.	On *Quercus*		2	

Gnomoniella Sacc.
Saprophytes; *G. rubicola* Pass. in East Anglia.

Gnomoniella carpinea (Fr.) Monod (*Sphaerognomonia carpinea*) (Fr.) Pot.)	On *Carpinus*	4
G. tubaeformis (Tode) Sacc.	On *Alnus*, anamorph *Cylindrosporella alnea* (Pers.) v. Höhn.	1

Hapalocystis Auersw.
Lignicolous saprophytes; *H. bicaudata* Fuck. on *Ulmus* in the west.

Hapalocystis berkeleyi Auersw.	On *Platanus*, anamorph *Fusicoccum hapalocystis* Sacc.	1

Hercospora Fr.
Lignicolous saprophyte.

Hercospora tiliae (Pers.) Fr.	Anamorph *Rabenhorstia tiliae* Fr.	1	4

Hypospilina (Sacc.) Trav.
Foliar saprophytes, see also *Plagiostoma*

Hypospilina bifrons (DC.) Trav.	On *Quercus*	1

Linospora Fuck.
Foliar saprophytes.

Linospora capreae (DC.) Fuck.　　　　On *Salix*　　　　　　　　　　1

Mamiana Ces. & de Not.
Foliar parasite.

Mamiana fimbriata (Pers.) Ces. & de Not.　On *Carpinus*　　　　　1　　3　4

Mamianella v. Höhn.
Foliar parasite.

Mamianella coryli (Batsch) v. Höhn.　　　　　　　　　　　　　　　　4

Mazzantia Mont.
Saprophytes, *M. galii* (Fr.) Mont. in the north.

Melanamphora La Flamme
Lignicolous saprophyte.

Melanamphora spinifera (Wallr)　　　　On *Fagus*　　　　　　1　2　3　4

Melanconis Tul.
Lignicolous saprophytes or facultative parasites (including *Melanconiella* Sacc.) with anamorphs in *Melanconium*.

Melanconis alni Tul.　　　　　　On *Alnus*, anamorph *Melanconium*　1
　　　　　　　　　　　　　　　　　　apiocarpum Link
M. aucta (B. & Br.) Wehm.　　　On *Alnus*　　　　　　　　　　　　1
M. carthusiana Tul.　　　　　　On *Juglans*, anamorph *Melanconium*　1
　　　　　　　　　　　　　　　　　　juglandinum Kunze
M. chrysostroma (Fr.) Tul.　　　On *Carpinus* anamorph *Melanconium*　1　　3
　　　　　　　　　　　　　　　　　　microsporum Nees
M. flavovirens (Otth) Wehm.　　On *Corylus*　　　　　　　　　　　2
　　(*Discodiaporthe sulphurea* (Fuck.) Petr.)
M. modonia Tul.　　　　　　　　On *Castanea*, anamorph *Coryneum*　1　　　　4
　　　　　　　　　　　　　　　　　　modonium (Sacc.) Griff. & Maubl.
M. (Melanconiella) spodiaea Tul.　On *Carpinus*　　　　　　　　　　1
M. stilbostoma (Fr.) Tul.　　　On *Betula*, anamorph *Melanconium*　1　2　3　4
　　　　　　　　　　　　　　　　　　betulinum Schm. & Kunze (*M. bicolor*
　　　　　　　　　　　　　　　　　　Nees)
M. thelebola (Fr.) Sacc.　　　　On *Alnus*, anamorph *Cytosporopsis*　1　　3　4
　　　　　　　　　　　　　　　　　　umbrinus (Bon.) v. Höhn.

Melogramma Fr.
Lignicolous saprophytes, see also *Melanamphora*; *M. elongatum* A.L. Smith in the north.

Melogramma campylosporum Fr.　　On *Carpinus* needs confirmation　　　　　4
　　(*M. vagans* de Not.)

Obryzum Wallr.
Hyperparasite, *O. corniculatum* Wallr. on *Leptogium* in the southwest.

Ophiognomonia (Sacc.) Sacc.
Foliar saprophytes; *O. melanostyla* (DC.) Sacc. on *Tilia* in East Anglia, *O. padicola* (Lib.) Monod on *Prunus padus* in the north.

Ophiovalsa Petrak
   Lignicolous saprophytes = *Cryptospora* Tul. non Karelin & Kirilow.

Ophiovalsa betulae (Tul.) Petrak	Anamorph *Cryptosporium betulinum* Jaap	1		3 4
O. corylina (Tul.) Petrak		1		
O. suffusa (Fr.) Petrak	Anamorph *Cryptosporium neesii* Corda	1		3 4

Phragmoporthe Petrak
   Lignicolous saprophyte.

Phragmoporthe conformis (B. & Br.) Petrak	On *Alnus*	1	4

Plagiostoma Fuck.
   Saprophytes, *P.devexum* (Desm.) Fuck. On *Polygonum* in E. Anglia, *P. alnea* v. *betulina* Barr, *P. lugubre* (Karst.) Bolay & *P. tormentillae* (Lind) Bolay in the north.

P. inclinatum (Desm.) Barr	On Acer	1		
P. pustula (Pers.a) v. Arx	On *Quercus*	1	2	4

Pleuroceras Riess
   Foliar saprophytes, *P. groenlandicum* (Rostr.) Barr on *Salix* in the north, *P. pseudoplatani* (v. Tub.) Monod on *Acer* in the midlands.

Prosthecium Fr.
   Lignicolous saprophytes, see also *Calospora*, *Hapalocystis*, *Melanconis*.

Pseudovalsa Ces. & de Not.
   Lignicolous saprophytes with anamorphs in *Coryneum*.

Pseudovalsa lanciformis (Fr.) Ces. & de Not.	Anamorph *Coryneum brachyurum* Link on *Betula*	1	2	3	4
P. longipes (Tul.) Sacc.	Anamorph *Coryneum umbonatum* Nees on *Quercus*	1	2	3	4
P. umbonata (Tul.) Sacc.	Anamorph *Coryneum depressum* Schm. on *Quercus*	1		3	4

Septomazzantia Theiss. & Syd.
   Saprophyte, *S. epitypha* (Cooke) Theiss. & Syd. in East Anglia.

Sillia Karst.
   Lignicolous saprophyte.

Sillia ferruginea (Pers.) Karst.	On *Corylus*	1	2	4

Stioclettia Dennis
   Saprophyte, *S. luzulina* Dennis on *Luzula* in the north.

Sydowiella Petrak
   Saprophytes, *S. depressula* (Karst.) Barr & *S. juncina* Spooner in the north.

Sydowiella fenestrans (Ruby) Petrak	On *Chamaenerion*	1	3 4

Valsa Fr.
   Saprophytes or weak facultative parasites of woody plants with anamorphs in *Cytospora*, includes *Leucostoma* (Nits.) v. Höhn.; *V. abietis* (Fr.) Fr.) & *V. cenisia* de Not. in E. Anglia, *V. mulleriana* Cooke in Essex.

Valsa ambiens (Pers.) Fr.	Anamorph *Cytospora ambiens* Sacc., plurivorous	1	2	3	4

V. (Leucostoma) cincta (Fr.) Fr.	Anamorph *C. cincta* Sacc. On *Prunus*	1			4
V. cornicola Cooke	On *Cornus*				4
V. coronata (Hoffm.) Fr. (*V. ceratophora* Tul.)		1		3	4
V. (Leucostoma) curreyi Nits.	Anamorph *C. curreyi* Sacc on *Larix*	1			
V. diatrypa (Fr.) Fr.	*On Alnus*			3	
V. germanica Nits.	Anamorph *C. germanica* Sacc. On *Betula*			3	
V. kunzei (Fr.) Fr.	Anamorph *C. kunzei* Sacc. on *Picea*	1	2		
V. laurocerasi Tul.	Anamorph *C. laurocerasi* Fuck.	1			
V. leucostoma (Pers.) Fr. (*Leucostoma persoonii* v. Höhn.)	Anamorph *C. leucostoma* Sacc. On *Prunus*	1			4
V. microstoma (Pers.) Fr.	On *Prunus*, anam. *C. microstoma* Sacc.	1			
V. nivea (Hoffm.) Fr. (*Leucostoma niveum* (Hoffm.) v. Höhn.	Anamorph *C. nivea* Sacc. on *Populus*	1	2		
V. opulina Sacc. & Sacc.	On *Viburnum lantana*	1			
V. oxystoma Rehm	On *Alnus*	1			
V. pini (A. & S.) Fr.	Anamorph *C. pini* Desm.			3	
V. pustulata Auersw.	On *Crataegus*	1		3	
V. rhodophila B. & Br.	Anamorph *C. rhodophila* Sacc. on *Rosa*	1			
V. rosarum de Not.	Anamorph *C. rosarum* Grev. on *Rosa*	1			
V. sordida Nits.	Anamorph *C. chrysosperma* Fr. on *Populus*	1			4

Valsella Fuck.
Valsa with polysporous asci, *V. clypeata* Fuck. in E. Anglia, *V. salicis* Fuck. in the north.

Valsella adhaerena Fuck.	On *Betula*	1	
V. amphoraria (Nits.) Sacc.	On *Fagus*	1	

Wuestnia Auersw.
Lignicolous saprophytes = *Cryptosporella* Sacc.

Wuestnia compta Reed & Boroth		1	2	
W. hypodermia (Fr.) Ananthapadmanan		1		4

OPHIOSTOMATALES

Ceratocystiopsis Upadhyay & Kendrick
Saprophyte, *C. falcata* (Wright & Cain) Upadhyay on timber, Hampshire and E. Anglia.

Ceratocystis Ellis & Halsted
Saprophytes, associates of bark beetles or plant pathogens; *C. coerulescens* (Munch) Bakshi, *C. minor* (Hedgc.) Hunt & *C. penicillata* (Grosm.) Moreau in E. Anglia, *C. narcissi* (Limber) Hunt in Hertfordshire, several more species in the north.

Ceratocystis moniliformis (Hedgc.) Moreau					4
C. perparvispora Hunt		1			
C. piceae (Münch) Bakshi		1			
C. pilifera (Fr.) Moreau		1			
C. pluriannulata (Hedgc.) Moreau					4
C. ulmi (Buisman) Moreau	Anamorph *Pesotum ulmi* (Schwartz) Crane & Schoknecht	1	2	3	4

Klasterskya Petrak
Saprophyte, *K. acuum* (Mouton) Petrak on needles of *Pinus* in East Anglia.

ELAPHOMYCETALES

Elaphomyces Nees
Mycorrhizal associates of trees; *E. anthracinus* Vitt. in the west.

Elaphomyces aculeatus Vitt.		2		
E. granulatus Fr. (*E. cervinus* (L.) Schlecht.)	1	2	3	4
E. muricatus Fr. (*E. variegatus* Vitt.)	1	2	3	4

ERYSIPHALES

Erysiphe Hedw. f.
Obligate parasites of phanerogams; *E. knautiae* Duby in E. Anglia, *E. valerianae* (Jacz). Blumer in Hertfordshire.

Erysiphe aquilegiae DC	On *Caltha palustris*	1	
	On *Aquilegia*	1	
E. artemisiae Grev.	On *Artemisia vulgaris*	1	4
E. betae (Vanha) Weltzien	On *Beta vulgaris* & *Spinacea oleracea*		4
E. biocellata Ehrenb. (*E. salviae* (Jacz.) Blumer)	On *Lycopus europeus*	1	
	On *Melissa officinalis*	1	
	On *Mentha aquatica* & *M.* sp.	1	
	On *Salvia* spp.	1	
E. catalpae Simonian	On *Catalpa bignonioides*	1	
E. cichoracearum DC.	On *Achillea ptarmica*	1	
	On *Aster noviangliae, A. novobelgii, A. pygmaeus*	1	
	On *Catananche caerulea*	1	
	On *Centaurea montana* & *C. nigra*	1 2	
	On *Cirsium arvense* & *C. vulgare*	1	
	On *Crepis capillaris*	1	
	On *Eupatorium cannabinum*	2	
	On *Lactuca serriola*	1	
	On *Sonchus* spp.	1	
	On *Tanacetum vulgare*	1	
	On *Tragopogon pratensis*	1	
E. cichoracearum var *latispora* Braun	On *Helianthus annuus*	1	
E. circaeae Junell	On *Circaea lutetiana*	1 2	
E. convolvuli DC.	On *Convolvulus arvensis*	1	
E. cruciferarum Opiz	On *Alliaria petiolata*	1	
	On *Brassica* spp.	1	
	On *Lunaria biennis*	1	
	On *Meconopsis* spp.	1	
	On *Rhododendron ponticum* oidia said to belong here	1	
	On *Sisymbrium officinale*	1	
E. cynoglossi (Wallr.) Braun	On *Myosotis arvensis*	1	
	On *Pulmonaria officinalis*	1	
	On *Symphytum*	1	
E. depressa (Wallr.) Schlecht.	On *Arctium* spp.	1	4
E. fischeri Blumer	On *Senecio squalidus* & *S. vulgaris*	1	
E. galeopsidis DC.	On *Galeopsis tetrahit*	1	
	On *Lamium album*	1	
	On *Stachys silvatica*	1	4
E. galii Blumer	On *Galium aparine*	1	4

		1	2	3	4
E. (Blumeria) graminis DC.	On *Agrostis* spp.	1			
	On *Agropyron caninum* & *A. repens*	1			
	On *Arrhenatherum elatius*	1			
	On *Avena sativa*	1			
	On *Bromus mollis*	1			
	On *Dactylis glomerata*	1			
	On *Festuca* spp.	1			
	On *Hordeum murinum, H. vulgare*	1	2		
	On *Holcus lanatus* & *H. mollis*	1			
	On *Lolium perenne* & *L. multiflorum*	1			
	On *Poa pratensis*	1			
	On *Triticum aestivum*	1	2		4
E. heraclei Schleich.	On *Angelica sylvestris*	1	2		
	On *Anthriscus sylvestris*	1			
	On *Heracleum sphondylium*	1	2	3	4
	On *Pastinaca sativa*	1			
	On *Petroselinum crispum*				4
E. howeana Braun	On *Oenothera erythrosepala*	1			
E. hyperici (Wallr.) Blumer	On *Hypericum perfoliatum*	1			4
E. limonii Junell	On *Limonium vulgare*				4
E. lycopsidis Zheng & Chen	On *Lycopsis arvensis*	1			4
	On *Pentaglottis sempervirens*	1			
E. magnicellulata Braun	Oidia only on *Phlox paniculata*	1			
E. mayorii Blumer	On *Carduus* & *Cirsium*	1		3	
E. orontii Cast. (*E. polyphagi* Hamm.)	On *Acanthus mollis*	1			
	On *Antirrhinum orontium*	1			
	On *Eucalyptus nutans* & *E. woodwardii*	1			
	On *Lycopersicum esculentum*		2		
	On *Solidago petiolaris*	1			
	On *Viola* cult.	1			
E. pisi DC.	On *Genista tinctoria*	1			
	On *Medicago arabica* & *M. lupulina*		2		
	On *Pisum sativum*	1	2	3	4
	On *Vicia cracca*	1			
E. polygoni DC.	On *Polygonum aviculare*	1			
E. ranunculi Grev.	On *Anemone* sp. cult.		2		
	On *Clematis jackmanii*	1			
	On *Delphinium ajacis* & *D. elatum*	1			
	On *Ranunculus acris, R. repens*	1			
E. sordida Junell	On *Plantago major*	1			
E. tortilis (Wallr.) Fr.	On *Cornus sanguinea*	1	2		4
E. trifolii Grev.	On *Lathyrus odoratus* & *L. pratensis*	1			
	On *Lotus corniculatus* & *L. uliginosus*	1			
	On *Lupinus* sp.	1			
	On *Melilotus officinalis*	1			
	On *Onobrychis viciaefolia*	1			
	On *Trifolium incarnatum, T. pratense*	1	2		4
E. urticae (Wallr.) Klotzsch	On *Urtica dioica*	1		3	
E. verbasci (Jacz.) Blumer	On *Verbascum thapsus*	1		3	4

Leveillula Arnaud

Obligate parasites; *L. cistacearum* Golovin on *Helianthemum* in Hampshire, Berkshire and Hertfordshire.

Leveillula compositarum Golovin	On *Cynara*		1

Microsphaera Lév.

Obligate parasites, *Erysiphe tortilis* & *E. trifolii* are sometimes transferred here; *M. astragali* (DC.) Trev. & *M. baumleri* Magn. in the north, *M. lonicerae* (DC.) Wint. in the west.

		1	2	3	4
Microsphaera alphitoides Griffon & Maubl.	On *Quercus*	1	2	3	4

		1	2	3	4
M. begoniae Siv.	On *Begonia macdougalii, B. peltata* etc	1			
M. berberidis (DC.) Lev.	On *Berberis vulgaris*	1			4
	On *Mahonia aquifolium*	1			
M. divaricata (Wallr.) Lév.	On *Frangula alnus*	1			
M. euonymi (DC.) Sacc.	On *Euonymus europaeus*	1	2		
M. euonymi-japonici Vienn. Bourg.	On *Euonymus japonicus*	1		3	
M. friesii Lév.	On *Rhamnus catharticus, R. oleoides*	1			
M. grossulariae (Wallr.) Lév.	On *Ribes uvacrispa*			3	4
	On *R. rubrum*				4
M. hedwigii Lév.	On *Viburnum lantana*	1		3	4
M. mougeotii Lév.	On *Lyceum barbarum*	1			4
M. penicillata (Wallr.) Lév.	On *Alnus glutinosa*		2		
M. platani Howe	On *Platanus orientalis*	1			
M. polonica Siem. (*Oidium hortensiae*)	On *Hydrangea macrophylla*	1			
M. pseudacaciae (Marcz.) Braun	On *Robinia pseudacacia*			3	
M. sparsa Howe	On *Viburnum opulus* & *V. tinus*	1			
M. syringae (Schw.) Magn.	On *Syringa vulgaris*	1			

Phyllactinia Lév.
Obligate parasites.

		1	2	3	4
Phyllactinia guttata (Wallr.) Lév.	On *Corylus, Fraxinus, Greyia* etc.	1			4

Podosphaera Kunze
Obligate parasites.

		1	2	3	4
Podosphaera aucupariae Eriksson	On *Sorbus aucuparia*	1			4
P. clandestina (Wallr.) Lév.	On *Crataegus* & *Mespilus*	1	2		4
	On *Pyracantha coccinea*	1			
P. leucotricha (Ell. & Ev.) Salmon	On *Malus* spp. & *Pyrus communis*	1	2		4
P. myrtillina (Schubert) Kunze	On *Vaccinium myrtillus*	1			
P. tridactyla (Wallr.) de Bary	On *Prunus domestica*	1			
	On *P. laurocerasus*	1			
	On *P. spinosa*	1	2		

Sawadaea Miyabe
Obligate parasites, segregate from *Uncinula*.

		1	2	3	4
Sawadaea bicornis (Wallr.) Homma	On *Acer campestre, A. negundo,* *A. pseudoplatanus*	1	2	3	4

Sphaerotheca Lév.
Obligate parasites; *S. ferruginea* (Schlecht.) Junell on *Sanguisorba* in north and west.

		1	2	3	4
Sphaerotheca aphanis (Wallr.) Braun	On *Fragaria* × *ananassa*	1			4
	On *Geum urbanum*	1			
	On *Potentilla arbuscula* & *P. reptans*	1			
	On *Rubus fruticosus*	1			
S. balsaminae (Wallr.) Kari	On *Impatiens capensis*	1			
	On *I. nolitangere*	1	2		
S. dipsacearum (Tul.) Junell	On *Dipsacus fullonum*		2		
S. epilobii (Wallr.) Sacc.	On *Epilobium hirsutum* & *E. roseum*	1			4
S. euphorbiae (Cast.) Salmon	On *Euphorbia amygdaloides, E. helioscopia* & *E. peplus*	1			
S. fugax Penz. & Sacc.	On *Geranium molle* & *G. sanguineum*	1			
S. fuliginea (Schlecht.) Poll.	On *Veronica exaltata, V. longifolia* & *V. teucrium*	1			

S. fusca (Fr.) Blumer	On *Bidens cernua*	1			
	On *Calendula officinalis*	1			4
	On *Conyza canadensis*	1			
	On *Lapsana communis*	1			
	On *Matricaria matricarioides*	1	2		
	On *Pulicaria dysenterica*	1			
	On *Senecio jacobaea*	1			
	On *Taraxacum officinale*	1			4
S. macularis (Wallr.) Lind	On *Humulus lupulus*	1			4
S. morsuvae (Schw.) Berk. & Curt.	On *Ribes nigrum* & *R. uvacrispa*		2	3	4
S. pannosa (Wallr.) Lév.	On *Rosa canina* & *Rosa* cult.	1	2	3	4
S. pannosa var *persicae* Woron.	On *Prunus persica*	1			4
S. plantaginis (Cast.) Junell	On *Plantago lanceolata*	1			4
S. spiraeae Sawada	On *Filipendula ulmaria*	1			4
S. verbenae Savul. & Negru	On *Verbena* cult.	1			

Uncinula Lév.
   Obligate parasites, see also *Sawadaea*; also *U. clandestina* (Bev Bern.) Schroet. on *Ulmus*.

Uncinula adunca (Wallr.) Lév.	On *Salix* spp.	1			
U. necator (Schw.) Burrill	On *Ampelopsis brevipedunculata* & *Vitis vinifera* etc	1		3	4
U. prunastri (DC.) Sacc.	On *Prunus spinosa*	1			

EUROTIALES

Amylocarpus Currey
   Saprophyte.

Amylocarpus encephaloides Currey	2

Aporothielavia Malloch & Cain
   Terrestrial saprophyte.

Aporothielavia leptoderma (Booth) Malloch & Cain	1

Byssochlamys Westling
   Saprophytes in soil and plant products; *B. nivea* Westling on stored *Hordeum* in Hampshire.

Byssochlamyc fulva Olliver & G. Smith	Anamorph *Paecilomyces fulvus* Stolk & Samson	4

Cephalotheca Fuck.
   Saprophytes, mainly lignicolous, see also *Fragosphaeria*.

Cephalotheca savoryi Booth	1
C. sulfurea Fuck.	1

Connersia Malloch
   Lignicolous saprophyte, *C. rilstonii* (Booth) Malloch in the west.

Dactylomyces Sopp
   Saprophytes, *D. thermophilus* Sopp in the Midlands.

Dactylomyces crustaceus Apinis & Chesters	Anamorph *Paecilomyces crustaceus* Apinis & Stolk	4

Edyuillia Subram.
Saprophyte, *E. athecia* (Raper & Fennell) Subram. apparently casual alien.

Emericella B. & Br.
Saprophytes in soil and plant products with anamorphs in *Aspergillus,* doubtless ubiquitous.

Emericella nidulans (Eidam) Vuill.	Anamorph *A. nidulans* (Eidam) Wint.	1

Emericellopsis v. Beyma
Terrestrial saprophytes doubtless ubiquitous, *E. minima* Stolk & *E. mirabilis* (Malan) Stolk in Hampshire.

Eupenicillium Ludwig
Teleomorphs of *Penicillum,* presumably ubiquitous.

Eupenicillium javanicum (v. Beyma) Stolk & Scott	Anamorph *P. simplicissimum* (Oud.) Thom	1	4
E. lapidosum Scott. & Stolk	Anamorph *P. lapidosum* Raper & Fennell	1	
E. meridianum Scott	Anamorph *P. decumbens* Thom	1	
E. pinetorum Stolk	Anamorph *P. fuscum* (Sopp) Biourge	1	
E. shearii Stolk & Scott	Anamorph *P. shearii* Stolk & Scott	1	

Eurotium Link
Teleomorphs of *Aspergillus,* doubtless ubiquitous, others in adjacent counties.

Eurotium herbariorum (Wiggers) Link	1	4
E. niveoglaucum (Thom & Raper) Malloch & Cain	1	
E. repens de Bary	1	

Fragosphaeria Shear
Lignicolous saprophytes.

Fragosphaeria purpurea Shear	1
F. reniformis (Sacc. & Therry) Mallock & Cain	1

Hemicarpenteles Sarbhoy & Elphick
Saprophyte.

Hemicarpenteles paradoxus Sarbhoy & Elphick	4

Neosartorya Malloch & Cain
Teleomorphs of *Aspergillus, N. fischeri* Malloch & Cain var. and *N. quadricincta* (Udagawa & Kawasaki) Malloch & Cain in the north.

Nigrosabulum Malloch & Cain
Saprophyte.

Nigrosabulum globosum Malloch & Cain	4

Pleuroascus Massee & Salmon
Coprophilous saprophyte.

Pleuroascus nicholsonii Massee & Salmon	1

Pseudeurotium v. Beyma
Saprophytes, *P. bakeri* Booth associate of bark beetles, *P. ovale* Stolk with nematode cysts.

Pseudeurotium zonatum v. Beyma 1

Roumegueriella Speg.
Ubiquitous saprophyte = *Lilliputia* Boud. & Pat.

Roumegueriella rufula (B. & Br.) Mallock & Cain 1 2

Sartorya Vuill.
Teleomorphs of *Aspergillus*.

Sartorya fumigata Vuill. Anamorph *Aspergillus fischeri* Wehmer 1

Syncleistostroma Subram.
Teleomorph of *Aspergillus, S. alliacea* (Thom & Church) Subram.

Talaromyces C. Benj.
Teleomorphs of *Penicillium, T. emersonii* Stolk in Berkshire, others in the midlands.

Talaromyces flavus (Klöcker) Stolk & Samson	Anamorph *P. dangeardii* Pitt	1
T. intermedius (Apinis) Stolk & Samson	Anamorph *P. intermedius* Stolk & Samson	1
T. thermophilus Stolk	Anamorph *P. dupontii* Griffon & Maublanc	1
T. wortmannii (Klöcker) C. Benj.	Anamorph *P. kloeckeri* Pitt	1

Thermoascus Miehe
Terrestrial saprophytes.

Thermoascus aurantiacus Miehe 4

MICROASCALES

Kernia Nieuwl.
Coprophilous saprophyte.

Kernia nitida (Sacc.) Nieuwl. 1

Lophotrichus R. Benj.
Coprophilous saprophytes, *L. ampullus* R. Benj. in the north.

Lophotrichus bartlettii (Massee & Salmon) Malloch & Cain (*Magnusia bartlettii*) 1

Microascus Zukal
Teleomorphs of *Scopulariopsis, M. trigonosporus* Emmons & Dodge in Hampshire.

Microascus cinereus (Emile-Weil & Gaudin) Curzi	Anamorph *S. cinerea* E-Weil & Gaudin	1
M. desmosporus (Lechmere) Curzi (*M. cirrhosus* Curzi)		1 2
M. longirostris Zukal (*M. variabilis* Massee & Salmon)		1
M. manginii (Loub.) Curzi	Anamorph *S. candida* (Guegin) Vuill.	1

Petriella Curzi
    Saprophytes.

Petriella sordida (Zukal) Barron & Gilman   "Sussex"

Pithoascus v. Arx
    Saprophytes.

Pithoascus nidicola (Massee & Salmon) v. Arx   1

Pseudallescheria Negroni & Fischer
    Terrestrial saprophyte and facultative human pathogen, *P. boydii* (Shear) McGinnis et al.

ASCOSPHAERIALES

Ascosphaera Olive & Spiltoir
    ?Pathogen of *Apis*.

Ascosphaera apis (Maassen) Olive & Spiltoir   4

Bettsia Skou
    ?Pathogen of *Apis*.

Bettsia alvei (Betts) Skou   1

GYMNOASCALES

Actinodendron Orr & Kuehn
    Saprophyte.

Actinodendron verticillatum (Smith) Orr & Kuehn   1

Amauroascus Schroet.
    Terrestrial saprophyte, *A. verrucosus* (Eidam) Schroet. in the Midlands.

Anixiopsis Hansen
    Saprophytes = *Aphanoascus* Zukal.

Anixiopsis fulvescens (Cooke) de Vries var *stercoraria* (Hansen) de Vries   1

Apinisia La Touche
    Saprophyte, *A. graminicola* La Touche in the north.

Arachniotus Schroet.
    Saprophytes.

Arachniotus candidus (Eidam) Schroet.   1
A. citrinus Massee & Salmon   1
A. confluens (Sartory & Bainier) Apinis   1
A. ruber (v. Tieghem) Schroet.   1

Arachnomyces Massee & Salmon
  Saprophytes.

Arachnomyces nitidus Massee & Salmon   1
A. sulphureus Massee & Salmon   1

Arthroderma Currey
  Saprophytes, often keratinophilic.

Arthroderma curreyi Berk.   1
A. quadrifidum Dawson & Gentles   Anamorph *Trichophyton terrestre* Durie   1
        & Frey

Ctenomyces Eidam
  Saprophyte.

Ctenomyces serratus Eidam   1

Diehliomyces Gilkey
  Saprophyte or ?hyperparasite.

Diehliomyces (Pseudobalsamia) microsporus (Diehl & Lambert) Gilkey   4

Gymnoascus Baranetzky
  Saprophytes, *G. umbrinus* Boud. in Middlesex and Hampshire, *G. zuffianus* Morini in E. Anglia.

Gymnoascus californiensis (Orr & Kuehn) Apinis   1
G. reessii Baranetzky   1   3   4
G. uncinatus Eidam   1

Myxotrichum Kunze
  Saprophytes, *M. cancellatum* Phill. in the West Midlands, *M. stipitatum* (Lindfors) Orr & Kuehn in E. Anglia, *M. ochraceum* B. & Br. in the west.

Myxotrichum aeruginosum Mont.   1
M. chartarum Kunze   1
M. deflexum Berk.   1
M. harbariense (Orr & Kuehn) Apinis   1
M. setosum (Eidam) Orr & Plunket   1

Nannizzia Stockdale
  Saprophytes or facultative parasites of animals causing ringworm, *N. fulva* Stockdale in the west.

Nannizzia cajetani Ajello   Anamorph *Microsporum cookei* Ajello   1
N. gypsea (Nannizzi) Stockdale   Anamorph *M. gypseum* (Bodin) Guart   1
        & Grig.
N. incurvata Stockdale   1
N. ossicola (Rostrup) Apinis   2
N. persicolor Stockdale   Needs confirmation.

Onygena Pers.
  Saprophytes, *O. equina* (Willd.) Pers. in the north and west.

Onygena corvina A. & S.   1

Pseudogymnoascus Raillo
  Saprophyte, *P. roseus* Raillo in the E. Midlands & North.

Shanorella Benj.
Saprophyte, *S. spirotricha* R. Benj. in Wiltshire.

Xanthothecium v. Arx & Samson
Saprophyte, *X. peruvianus* (Cain) v. Arx & Samson in Hampshire.

ENDOMYCETALES

Ascoidea Brefeld & Lindau
Saprophyte, *A. rubescens* Brefeld & Lindau in the West Midlands.

Debaryomyces Klöcker
Saprophytes, or facultative parasites, two other species.

Debaryomyces kloecheri Guill. & Peju ?1

Dekkera v.d. Walt
Saprophytes, two species of yeasts in beer.

Dipodascus Lagerh.
Saprophytes, *D. macrosporus* Madelin & Feest in the west.

Endomyces Reess
Saprophytes, *E. decipiens* Reess reputedly widespread, *E. lactis* (Fres.) Wind. in Midlands.

Eramascus Eidam
Saprophyte.

Eramascus fertilis Stoppel ?1

Hansenula Syd. & Syd.
Saprophytes or facultative parasites, *H. anomala* (Hansen) Syx. & Syd. perhaps widespread.

Hansenula polymorpha de Morais & Maia (*H. angusta* Teunisson et al) ?1

Hormoascus v. Arx
Saprophyte; *H. platypodis* (Baker & Kruger van Rij) v. Arx on *Quercus cerris* in England.

Hyphopichia v. Arx & v.d. Walt
Saprophyte?, *H. burtonii* (Boidin et al) v. Arx & v.d. Walt (*Sporotrichum anglicum* Castellani)

Kluyveromyces v.d. Walt
Saprophytes and facultative parasites.

Kluyveromyces marxianus (Hansen) v.d. Walt (*Saccharomyces fragilis* Jörg.) ?1

Lipomyces Lodder & Kreger van Rij
Saprophytic yeasts, *L. lipofer* Lodder &Kreger van Rij and *L. starkeyi* Lodder & Kreger van Rij in soils.

Metschnikowia Kamienski
Saprophytes, *M. bicuspidata* (Metschnikow) Kamienski in the west.

Pichia Hansen
Saprophytes and facultative parasites.

Pichia farinosa (Linder) Hansen   ?1
P. membranifaciens (Hansen) Hansen   ?1

Saccharomyces Meyer
Saprophytes and facultative parasites, *S. diastaticus* Andr. & Gill. Associated with *S. cerevisiae.*

Saccharomyces bisporus (Naganishi) Lodder & van Rij   ?1
S. cerevisiae Hansen   Ubiquitous in domestic use, bakeries
                                                and breweries

LABOULBENIALES

Arthrorhynchus Kolenati
Ectoparasite, *A. eucampsipodae* Thaxter in Hampshire.

Cantharomyces Thaxter
Ectoparasites, three species in Hampshire.

Cantharomyces orientalis Speg.   4

Chitonmyces Peyritsch
Ectoparasites, two species unlocalised.

Compsomyces Thaxter
Ectoparasites.

Compomyces lestevae Thaxter   1

Dichomyces Thaxter
Ectoparasites, two species in Middlesex, one in the midlands.

Dichomyces furcifer Thaxter   3

Dimeromyces Thaxter
Ectoparasite, *D. corynetis*, Thaxter in Essex.

Euhaplomyces Thaxter
Ectoparasite, *E. ancyrophori* Thaxter.

Euphoriomyces Thaxter
Ectoparasite.

Euphoriomyces liodivorus (Huggert.) Tav.   1

Euzodiomyces Thaxter
Ectoparasite, *E. lathrobii* Thaxter in London.

Haplomyces Thaxter
Ectoparasite, *H. texanus* Thaxter in the Isle of Wight.

167

Helodiomyces Picard
　Ectoparasite, *H. elegans* Picard in Lancashire.

Hydrophilomyces Thaxter
　Ectoparasite.

Hydrophilomyces hamatus T. Majewski　　　　　　　　　　　　　　　　4

Idiomyces Thaxter
　Ectoparasite, *I. peyritschii* Thaxter.

Laboulbenia Mont. & Robin
　Ectoparasites, *L. dubia* Thaxter & *L. rougetii* Mont. & Robin in Hampshire, 9 others elsewhere.

Laboulbenia gyrinidarum Thaxter　　　　　　　　　　　　　　　　3

Misgomyces Thaxter
　Ectoparasites, *W. dyschirii* Thaxter.

Monoicomyces Thaxter
　Ectoparasite, *M. athetae* Thaxter in Hampshire.

Monoicomyces nigrescens Thaxter　　　　　　　　　　　　　　　　1

Peyritschiella Thaxter
　Ectoparasite, *P. protea* Thaxter in Hampshire.

Polyascomyces Thaxter
　Ectoparasite.

Polyascomyces trichophyae Thaxter　　　　　　　　　　　　　　　　1

Rhacomyces Thaxter
　Ectoparasites, *R. philonthinus* Thaxter widespread.

Rhacomyces furcatus Thaxter　　　　　　　　　　　　　　　　3

Rhadinomyces Thaxter
　Ectoparasites.

Rhadinomyces cristatus Thaxter　　　　　　　　　　　　　　　　4
R. pallidus Thaxter　　　　　　　　　　　　　　　　　　　3

Sphaleromyces Thaxter
　Ectoparasite, *S. lathrobii* Thaxter.

Stigmatomyces Karsten
　Ectoparasites of Diptera, *S. entomophilus* (Peck) Thaxter & *S. purpureus* Thaxter.

Symplectromyces Thaxter
　Ectoparasites, *S. vulgaris* (Thaxter) Thaxter in the north.

Teratomyces Thaxter
　Ectoparasite.

Teratomyces actobii Thaxter　　　　　　　　　　　　　　　　1

VERRUCARIALES

Predomiantly lichenised, a few hyperparasites.

Muellerella Hepp
Hyperparasites, four species in the north and west.

Phaeospora Hepp
Hyperparasites, four other species in the north and west.

Phaeospora hetairizans (Leighton) Arnold   3

Stigmidium Trevisan
Hyperparasites, about ten species in the north and west. (= *Pharcidia* Körber)

LECANIDIALES

Epilichen Clem.
Hyperparasite, *E. scabrosus* (Ach.) Clem. on *Baeomyces rufus* in the north and west.

Lecanidion Endl.
Lignicolous saprophyte. = *Patellaria* Fr.

Lecanidion atratum (Hedw.) Rab.   1   3   4

Poetschia Körber
Lignicolous saprophyte, *P. cratincola* (Rehm) Hafellner in the north and west.

DOTHIDIALES

Abrothallus de Not.
Hyperparasites of lichens with anamorphs in *Vouauxiomyces*, four or five species in the north and west.

Acanthophiobolus Berlese
Saprophyte.

Acanthophiobolus helicosporus (B. & Br.) Walker   1

Acanthostigmella v. Höhn.
Saprophytes, *A. genuflexa* v. Höhn. in E. Anglia, *A. pallida* Dennis & Barr in the west.

Actidium Fr.
Saprophytes, *A. nitidum* (Ellis) Zogg on *Juniperus* in the north.

Actidium hysterioides Fr.   1

Actinopeltis v. Höhn.
Saprophytes, *A. palustris* Ellis in East Anglia.

Amarenomyces O. Eriksson
  Saprophyte.

Amarenomyces ammophilae (Lasch) O. Eriksson	Anam. *Tiarospora perforans* (Rob.) v. Höhn.	2

Anomalemma Siv.
  Saprophyte or hyperparasite, *A. epochnii* (B. & Br.) Siv. in East Anglia.

Appendiculella v. Höhn.
  Obligate parasite.

Appendiculella calostroma (Desm.) v. Höhn.	On *Rubus "fruticosus"*	3

Arnaudiella Petrak
  Parasite, *A. genistae* (Petrak) Müller on *Sarothamnus* in the north.

Asteridium (Sacc.) Speg.
  Parasite, *A. juniperinum* (Cooke) Sacc. in the north.

Asterina Lév.
  Parasite.

Asterina (Dimerosporium) veronicae (Lib.) Cooke	On *Veronica officinalis*	1

Asteromassaria v. Höhn.
  Lignicolous saprophyte.

Asteromassaria macrospora (Desm.) v. Höhn.	Anam. *Scoliciosporium macrosporium* (Berk.) Sutton	1

Astrosphaeriella Syd. see *Kirschteiniothelia*

Atopospora Petrak
  Saprophyte.

Atopospora betulina (Fr.) Petrak	Anam. *Didymochora betulina* v. Höhn.	1

Aulographina v. Arx. & Müller
  ?Saprophyte, *A. eucalypti* (Cooke & Massee) v. Arx & Müller in the north and west.

Aulographum Lib.
  Saprophytes.

Aulographum hederae Lib. (*A. vagum* Desm.)	On fallen leaves, *Hedera, Ilex* etc.	1

Berlesiella Sacc.
  ?Hyperparasite.

Berlesiella nigerrima (Bloxam) Sacc.	On effete diatrypaceous stromata	1	2	3

Botryosphaeria Ces. & de Not.
Saprophytes or facultative parasites, *B. abietina* (Prill. & Del.) Maubl. in the west, *B. festucae* (Lib.) v. Arx & Müller, *B. hyperborea* Barr and *B. philoprina* (B. & C.) v. Arx & Müller in the north and west.

Botryosphaeria dothidea (Moug.) Ces. & de Not.	On *Rosa*	1		
B. foliorum (Sacc.) v. Arx & Müller	On *Taxus*, anam. *Phyllostictina hysterella* (Sacc.) Petr.	1	2	3
B. melanops (Tul.) Wint.	On *Fagus*, anam. *Dothiorella advena* Sacc.	1		
B. obtusa (Schw.) Shoemaker	On *Pyrus*, anam. *Sphaeropsis malorum* Peck			4
B. quercuum (Schw.) Sacc.	On *Quercus* etc., anam. *S. sumachi* (Schw.) Cooke & Ellis	1	2	
B. rhodorae (Cooke) Barr (*Laestadia rhodorae* (Cooke) Berl. & Vogl.)	On *Rhododendron*	1		
B. stevensii Shoemaker	Anam. *Diplodia mutila* (Fr.) Mont.	1	3	4

Bryochiton Döbbler & Poelt
Parasites of musci, as yet two british species.

Bryomyces Döbbler
Parasites of bryophyta, *B. microcarpa* Döbbler.

Buergenerula Syd.
Saprophytes, *B. septata* (Rostr.) Syd. on *Carex* & *B. typhae* (Fabre) v. Arx on *Phragmites* in E. Anglia, *B. spartinae* Kohlm. & Gessner in Hampshire.

Byssolophis Clements
Lignicolous saprophyte, *B. ampla* (B. & Br.) Holm in the midlands.

Byssothecium Fuck.
Lignicolous saprophyte.

Byssothecium circinans Fuck. (*Trematosphaeria circinans* (Fuck.) Wint.	1	4

Capnodium Mont.
Saprophyte.

Capnodium salicinum Mont.	On *Salix*, anam. *Fumagospora capnodioides* Arnaud	3

Capronia Sacc.
Saprophytes, including *Dictyotrichiella* Munk, *Didymotrichiella* Munk & *Herpotrichiella* Petr. *C. pleiospora* (Mout.) Sacc. in the north.

Capronia (*Didymotrichiella*) *inconspicua* (Munk) Müller et al.		1	
C. (*Dictyotrichiella*) *mansonii* (Schol.-Schwarz) Müller et al.	Anam. *Exophiala mansonii* (Castell.) de Hoog	1	
C. moravica (Petrak) Müller et al.		1	
C. parasitica (Ell. & Ev.) Müller et al.		1	
C. (*Herpotrichiella*) *pilosella* (Karst.) Müller et al.		1	
C. (*Dictyotrichiella*) *pulcherrima* (Munk) et al.		1	2
C. sexdecemspora (Cooke) Sacc.		1	

Caryospora de Not.
Lignicolous saprophyte.

Caryospora callicarpa (Currey) Nits.	On *Quercus*	4

Cercidospora Körber
Hyperparasite, *C. epipolytropa* (Mudd) Arnold on apothecia of *Lecanora* in the north and west.

Chaetoscutula Müller
?Saprophyte, *C. juniperi* Müller on *Juniperus* in the north.

Chitonospora Bomm., Rouss. & Sacc.
Saprophyte, *C. ammophilae* Bomm., Rouss. & Sacc. in the north and west but surely to be expected in 2, 3, & 4 since it was described from the Belgian coast.

Cilioplea Munk
Saprophyte, *C. coronata* (Niessl) Munk in the west.

Clypeococcum Hawksw.
Hyperparasites, two lichenicolous species.

Cochliobolus Drechsler
Facultative parasites of Gramineae.

Cochliobolus sativus (Ito & Kuribay) Drechsler	As anam. *Drechslera sorokiniana* (Sacc.) Subram, & Jain	4

Coleroa (Fr.) Rab.
Foliar parasites, *C. alchemillae* (Grev.) Wint. in the north and west; see also *Hormotheca*.

Coleroa chaetomium (Kunze) Rab.	On *Rubus*	1 2	
C. circinans (Fr.) Wint.	On *Geranium dissectum* etc.		4
C. potentillae (Wallr.) Wint.	On *Potentilla anserina*		4

Cucurbidothis Petrak
Facultative parasite.

Cucurbidothis pithyophila (Schm. & Kunz.) Petrak var *cembrae* (Rehm) Holm	1

Cucurbitaria Gray
Saprophytes or facultative parasites, *D. acervata* (Fr.) Ces. & de Not. in the midlands, *C. dulcamarae* (Schm.) Fuck. in E. Anglia, *C. nauceosa* (Fr.) Fuck. on *Ulmus* in the west.

Cucurbitaria aspegrenii (Fr.) Ces. & de Not.	On *Prunus*	1		
C. berberidis (Pers.) Gray	On *Berberis* and *Mahonia*	1		4
C. elongata (Fr.) Grev. (reported from 2 on *Prunus*)	On *Robinia*	1		
C. euonymi Cooke	On *Euonymus*	1		
C. laburni (Pers.) de Not.	On *Laburnum anagyroides*	1	3	4
C. rhamni (Nees) Ces. & de Not.	As anamorph *Diplodia frangulae* Fuck.	1		
C. spartii (Nees) Ces. & de Not.	On *Sarothamnus* and *Ulex*	1		4

Cymadothea Wolf, see *Mycosphaerella*

Dacampia Massal.
Hyperparasite, *D. hookeri* (Borrer) Massal. in the north.

Dangeardiella Sacc. & Syd.
Saprophyte, *D. fusiformis* Obrist on *Dryopteris dilatata* in the north.

Daruvedia Dennis
Saprophyte, *D. bacillatum* (Cooke) Dennis in the north.

Delitschia Auersw.
Coprophilous saprophytes.

Delitschia didyma Auersw.		3
D. furfuracea Niessl	1	
D. marchalii Berl. & Vogl.	2	
D. winteri (Phill. & Plowr.) Sacc.	2 3	

Delphinella (Sacc.) O.K.
Parasite, *D.* (*Rehmiellopsis*) *abietis* (Rostrup) Müller on *Abies* in the north.

Dennisiella Bat. & Cif.
Honeydew saprophyte.

Dennisiella babingtonii (Berk.) Bat. & Cif.　Anam. *Microxiphium fagi* (Pers.) Hughes　　3

Didymella Sacc.
Saprophytes and facultative parasites of herbaceous plants, many other reported species elsewhere; *D. exitialis* (Morini) Müller on *Triticum* in Hertfordshire.

Didymella applanata (Niessl) Sacc.	Spurblight of *Rubus idaeus*	1	4
D. bryoniae (Auersw.) Rehm	On cucurbits	1	
(*Mycosphaerella citrullina* Gross.)			
D. caricis Syd.	On *Carex otrubae* etc.	1	
D. chrysanthemi (Tassi) Garibaldi	Anam. *Phoma chrysanthemi* Vogl.	2	
& Gullino			
D. exigua (Niessl) Sacc.		1	
D. holosteae Syd.	On *Stellaria holostea*	1	
D. lycopersici Kleb.	Anam. *Ascochyta lycopersici* (Plowr.) Brun.	1 2	
D. pinodes (Berk. & Blox.) Petrak	On *Pisum*, anam. *Ascochyta pinodes*	2	
	L.K. Jones		
D. proximella (Karst.) Sacc.	On *Carex*	1	
D. urticicola van der Aa & Boerema	As anamorph *Phoma urticicola* v.d. Aa	1	
	& Boer.		

Didymopleella Munk
Saprophyte, *D. cladii* Munk on *Cladium mariscus* in East Anglia.

Didymosphaeria Fuck.
Saprophytes, many additional names listed and in need of revision.

Didymosphaeria celata (Currey) Sacc.			4
D. epidermidis (Fr.) Fuck.		1	4
D. oblitescens (B. & Br.) Fuck.		1	3
D. superapplanata Siv.			4

Dimerium Sacc. & Syd.
Hyperparasite, *D. meliolicola* (Petrak) Hansf. on *Appendiculella calostroma* in the west.

Discosphaerina v. Höhn.
Saprophytes, *D. cytisi* (Fuck.) Siv. in E. Anglia, *D. fagi* (Hudson) Barr in E. Midlands, *D. fulvida* (Sanderson) Siv. in the north.

Discosphaerina mirbelii (v.d. Aa) Siv.　On *Buxus*, anam. *Sarcophoma mirbelii* (Fr.)　1　2　　4
　　　　　　　　　　　　　　　　　　　　v. Höhn.

Dothidea Fr.
  Lignicolous saprophytes, *D. puccinioides* (DC.) Fr. on *Ulex* in the north and west.

Dothidea sambuci (Pers.) Fr.	On *Sambucus* etc.	1

Dothiora Fr.
  Saprophytes or weak facultative parasites of woody plants, *D. pyrenophora* (Fr.) Fr. on *Sorbus* in the north, *D. taxicola* (Peck) Barr in the north and west.

Dothiora ribesia (Pers.) Barr	On *Ribes*, anam. *Rabenhorstia ribesia* Cooke & Massee	1	4

Elsinoe Racib.
  Foliar parasites with anamorphs in *Sphaceloma*.

Elsinoe ampelina Shear	On *Vitis*, anamorph *Sphaceloma ampelinum* de Bary				4
E. rosarum Jenkins & Bitancourt	On *Rosa*, anam. *S. rosarum* (Pass.) Jenkins	1	2	3	4
E. veneta (Burkh.) Jenkins	On *Rubus*, anam. *S. necator* (Ell. & Ev.) Jenkins & Shear	1			4

Endococcus Nyl.
  Hyperparasites, six species on lichens.

Epibelonium Müller
  Saprophyte, *E. gaeumannii* Müller on *Quercus ilex* in the west.

Epibryon Döbbeler
  Parasites of Bryophyta, two British species.

Eudarluca Speg.
  Hyperparasite, reputed teliomorph of the ubiquitous *Sphaerellopsis filum* (Biv. Bern.) Sutton.

Eudarluca caricis (Fr.) O. Eriksson	Anamorph only on uedosori of Uredinales	1	2	3	4

Farlowiella Sacc.
  Lignicolous saprophyte.

Farlowiella carmichaeliana (Berk.) Sacc.	Anam. *Monotospora megalospora* B. & Br.	1	2

Fenestella Tul.
  Lignicolous saprophytes.

Fenestella fenestrata (B. & Br.) Schroet.	On *Quercus* etc.	3	4
F. salicis (Rehm) Sacc.		1	
F. vestita (Fr.) Sacc.	On *Fagus* and *Ulmus*	1 3	4

Gemmamyces Casagrande
  Facultative parasite, *G. piceae* (Borthwick) Casagrande on *Picea* in the north.

Gibbera Fr.
  Parasites of Ericaceae, *G. elegantula* (Rehm) Petrak in E. Anglia, *G. vaccinii* (Sow.) Fr. in the north.

Gibbera myrtilli (Cooke) Petrak	On *Vaccinium myrtillus* 1
G. ramicola Eriksson	1

Gloniella Sacc.
Foliar parasite, *G. adianti* (Kunze) Petrak on *Trichomanes*.

Gloniopsis de Not.
Lignicolous saprophytes.

Gloniopsis curvata (Fr.) Sacc.	1			
G. praelonga (Schw.) Zogg (*G. levantica* Rehm)	1	2	3	4

Glonium Muhlenb.
Lignicolous saprophyte, *G. lineare* (Fr.) de Not. in the Midlands.

Glyphium Nits.
Lignicolous saprophyte.

Glyphium elatum (Grev.) Zogg	1	2

Guignardia Viala & Ravaz
Facultative parasites, mainly tropical; *G. istriaca* Bubak on *Ruscus* in the west, *G. cytisi* (Fuck.) v. Arx & Müller on *Spartium* in E. Anglia.

G. aesculi (Peck) V.B. Stewart	Anam. *Leptodothiorella aesculicola* (Sacc.) Siv.	1
G. punctoidea (Cooke) Schroet.	On *Quercus*	1
G. philoprina (B. & Curt.) v.d. Aa	Anam. *Phyllosticta concentrica* Sacc.	1

Herpotrichia Fuck.
Saprophytes, *H. herpotrichioides* (Fuck.) Cannon in the west, *H. juniperi* (Duby) Petr. in East Anglia.

Herpotrichia macrotricha (B. & Br.) Sacc.	1			4

Homostegia Fuck.
Hyperparasite, *H. piggotii* (B. & Br.) Karst. on *Parmelia* spp. in the west.

Hormotheca Bon.
Foliar parasite.

Hormotheca robertiani (Fr.) v. Höhn.	On *Geranium robertianum*	1	2	3	4

Hypobryon Döbbeler
Parasites of Bryophyta, *H. validum* Döbbeler in the north.

Hysterium Pers.
Lignicolous saprophytes.

Hysterium angustatum A. & S. (*H. acuminatum* Fr.)	1	2	3	4
H. pulicare Pers.	1		3	4

Hysterographium Corda
Lignicolous saprophytes.

Hysterographium fraxini (Pers.) de Not.	1			4
H. mori (Schw.) Rehm (*H. rousselii* (de Not.) Sacc.)		2		4

Kalmusia Niessl
Lignicolous saprophytes, *K. sarothamni* Feltg. in East Anglia.

Karstenula Speg.
Saprophyte, *K. rhodostoma* (A. & S.) Speg. in East Anglia on *Rhamnus frangula*.

Keissleriella v. Höhn.
Saprophytes = *Trichometasphaeria* Munk, *K. culmifida* (Karst.) Bose & *K. pinicola* Hawks. & Siv. in East Anglia, others in the north and west.

Kirschsteiniothelia Hawks.
Lignicolous saprophytes.

Kirschsteiniothelia aethiops (B. & Curt.) Hawks.   1

Kriegeriella v. Höhn.
Saprophytes = *Extrawettsteinina* Barr; *K. minuta* (Barr) v. Arx & Müller on *Juniperus* in the north, *K. mirabilis* v. Höhn. on *Pinus* in East Anglia.

Lasiobotrys Kunze
Foliar parasite, *L. lonicerae* (Fr.) Kunze, reported from East Anglia.

Lautitia Schatz
L. danica (Berl.) Schatz parasite of *Chondrus crispus* in the west.

Lembosina Theiss.
Parasites, *L. gontardii* Müller on *Arctostaphylos uvaursi* and *Lembosina (Echidnodes) aulographoides* (Bomm., Rouss. & Sacc.) Theiss. on *Rhododendron* in the north.

Leptosphaeria Ces. & de Not.
Mainly saprophytes, see also *Nodulosphaeria* and *Phaeosphaeria*; *L. galiorum* (Rob.) Ces. de Not. in Hertfordshire, *L. caricicola* Fautr., *L. caricis* (Schroet.) Leucht., *L. cladii* Cruch. *L. gloeospora* (B. & C.) Sacc., *L. grandispora* Sacc., *L. macrospora* (Fuck.) Thuem., *L. maritima* Sacc., *L. marram* (Cooke) Sacc., *L. pontiformis* (Fuck.) Sacc., in East Anglia, *L. marina* Ell. & Ev. in Hampshire, many other old or confused names in need of revision.

Leptosphaeria acuta (Fuck.) Karst.	On *Urtica dioica*	1	2	3	4
L. agnita (Desm.) Ces. & de Not.	On *Eupatorium*	1			
L. arundinacea (Sow.) Sacc.	*On Phragmites*	1			4
L. cesatiana (Mont.) Holm				3	
L. fuscella (B. & Br.) Ces. & de Not.	On *Rosa*	1		3	
L. doliolum (Pers.) Ces. & de Not.		1	2	3	4
L. maculans (Desm.) Ces. & de Not.	On *Brassica*, anam. *Plenodomus lingam* (Tode) v. Höhn.	1			4
L. niessleana Rab.	On *Lathyrus*	1			
L. ogilviensis (B. & Br.) Ces. & de Not.	On *Senecio*	1			
L. oraemaris Linder				3	4
L. pelagica Gareth Jones					4
L. purpurea Rehm		1			
L. rubicunda Rehm	*On Dipsacus*	1			
L. triglochinicola (Curr.) Sacc.	On *Triglochin*			3	
L. vectis (B. & Br.) Ces. & de Not.	On *Iris foetidissima*	1			4

Leptosphaerulina McAlp.
Facultative parasites, *L. myrtillina* (Sacc. & Fautr.) Petrak in the north, *L. personata* (Niessl) Barr on *Deschampsia* & *Glyceria* in East Anglia.

Leptosphaerulina trifolii (Rostrup) Petrak	*On Medicago & Trifolium*	4

Leptospora Rab.
Saprophyte.

Leptospora rubella (Pers.) Rab.	On dead herbaceous stems	1	2	3	4

Letendraea Sacc.
Hyperparasite of *Helminthosporium velutinum*, *L. helminthicola* (B. & Br.) Müller & v. Arx in East Anglia.

Lidophia Walker & Sutton
Parasite of many genera of Gramineae, all records are of the anamorph.

Lidophia graminis (Sacc.) Walker & Sutton	Reputed teleomorph of *Dilophospora alopecuri* (Fr.) Fr.	1	2	4

Lizonia (Ces. & de Not.) de Not.
Parasite, *L. emperigonia* (Ces. & de Not.) de Not. on *Polytrichum* in the north.

Lophiosphaera Trev.
Saprophyte.

Lophiosphaera ulicis (Pat.) Müller	On *Ulex*	4

Lophiostoma (Fr.) Ces. & de Not.
Saprophytes, including *Lophiotrema* Sacc., *L. alpigenum* (Fuck.) Sacc. v. *juncinum* Mout., *L. macrostomoides* (de Not.) Ces. & de Not., *L. quadrinucleatum* Karst. in East Anglia, *L. hysterioides* (Schw.) Sacc. & *L. pileatum* (Tode) Fuck. in the north, *L. viridarium* Cooke in the west.

Lophiostoma angustilabrum (B. & Br.) Cooke		1	2	3	4
L. arundinis (Pers.) Ces. & de Not.	On *Phragmites*	1	2	3	
L. bicuspidatum Cooke		1			4
L. caudatum Fabre				3	
L. caulium (Fr.) Ces. & de Not.		1	2	3	
L. fuckelii Sacc.		1	2	3	4
L. nucula (Fr.) Ces. & de Not.				3	
L. origani (Kunze) Wint. var *rubidum* (Sacc.) Chest. & Bell	On *Chamaenerion* and *Epilobium*		2	3	
L. semiliberum (Desm.) Ces. & de Not.		1	2	3	4
L. vagabundum (Sacc.) Chesters & Bell		1	2	3	

Lophium Fr.
Saprophyte, see also *Glyphium*.

Lophium mytilinum (Pers.) Fr.	Anam. *Papulaspora mytilina* (Pers.) Lohman	1

Massarina Sacc.
Saprophytes.

Massarina aquatica Webster	Anamorph *Dactylella aquatica* (Ingold) Ranzoni	4
M. eburnea (Tul.) Sacc.		1

Massariosphaeria (Müller) Crivelli
Saprophytes, segregated from *Pleospora*, *M. multiseptata* (Starb.) Crivelli in the north, *M. rubelloides* (Plowr.) Crivelli & *M. straminis* (Sacc. & Speg.) Crivelli in E. Anglia.

Massariosphaeria rubicunda (Niessl) Crivelli		4

Melanomma Nits.
Saprophytes, *M. jenysii* (B. & Br.) Sacc. & *M. longicolle* Sacc. in the midlands, *M. scitulum* in East Anglia, *M. rhododendri* Rehm in the west.

Melanomma brachythele (B. & Br.) Sacc.					4
M. coniothyrium (Fuck.) Holm	On *Rosa, Rubus* etc., anamorph *Coniothyrium fuckelii* Sacc.	1		3	4
M. fuscidulum Sacc.		1	2	3	
M. obliterans Berk					?4
M. pulvispyrius (Pers.) Fuck.		1	2	3	4
M. subdispersum (Karst.) Berl. & Vogl.	Anam. *Pseudospiropes longipilus* (Cda.) H. Jech.	1	2	3	4

Meliola Fr.
Obligate parasite, *M. ellisii* Roum. on *Vaccinium vitis-idaea* in the north.

Melittosporium Corda
Saprophyte, *M. propolidoides* (Rehm) Rehm on *Pinus* in the north.

Merismatium Zopf
Hyperparasite, *M. lopadii* (Anzi) Zopf on lichens in the north.

Metacapnodium Speg.
Saprophyte, *M. juniperi* (Phil. & Plowr.) Speg. on *Juniperus communis* in the north.

Metacapnodium dingleyae Hughes	On *Taxus*, as anam. *Capnobotrya dingleyae* Hughes	1	2		4

Metacoleroa Petrak
Parasite, *M. dickei* (B. & Br.) Petrak on *Linnaea borealis* in the north.

Metasphaeria Sacc.
Saprophytes, *M. cumana* (Sacc. & Speg.) Sacc. on *Carex* in E. Anglia, a few other species in the north and west.

Micronectriella v. Höhn.
Saprophytes, *M. agropyri* Apinis & Chesters in the midlands, *Micronectriella cucumeris* (Kleb.) Booth is apparently widespread, as the anamorph *Fusarium tabacinum* (v. Beyma) Gams.

Microthyrium Desm.
Obligate parasites? or saprophytes, *M. ciliatum* Gremm. & de Kam in E. Anglia, v. *hederae* Ellis in Essex, *M. fagi* Ellis in E. Anglia and *M. lauri* v. Höhn. in the west.

Microthyrium cytisi Fuck. v. *ulicis* Ellis	1			
M. cytisi v. *ulicis-gallii* Ellis	1			
M. gramineum Bomm., Rouss. & Sacc.				4
M. ilicinum de Not.	1			4
M. inconspicuum Ellis	1			
M. macrosporum (Sacc.) v. Höhn	1			4
M. microscopicum Desm.	1			
M. pinophyllum (v. Höhn.) Petrak	1			
M. versicolor (Desm.) v. Höhn.	1	2	3	

Monascostroma v. Höhn. Saprophyte.

Monascostroma innumerosa (Desm.) v. Höhn.	On *Juncus*	3

Morenoina Theiss.
Saprophytes, *M. arundinariae* Ellis, *M. fimbriata* Ellis, *M. paludosa* Ellis, *M. phragmitis* Ellis, *M. rhododendri* Ellis in East Anglia, five others in the west.

Monographos Fuck.
Saprophyte, *M. fuckelii* Holm on *Pteridium* in East Anglia.

Mycosphaerella Johanson
Saprophytes and facultative parasites = *Sphaerella* (Fr.) Rab. non Sommerf., *M. elodis* (Sm. & Rams) Tomilin in Hampshire, *M. equisiti* (Fuck.) Schroet., *M. lineolata* (Rob.) Schroet., *M. rhododendri* Lindau, *M. scirpi-lacustris* (Auersw.) Lindau & *M. typhae* (Lasch) Lindau in East Anglia, *M. sagedioides* (Wint.) Lindau in the midlands, several others in the north and west.

		1	2	3	4
Mycosphaerella angelicae (Fr.) Petrak	On *Angelica sylvestris*			3	
M. ascophylli Cotton	Ubiquitous on *Ascophyllum* & *Pelvetia* so inevitably in 2,3,4				
M. aspidii (v. Höhn.) Holm & Holm	On *Dryopteris* & *Pteridium* (*M. aquilina* (Fr.) Schr.)	1			4
M. atomus (Desm.) Johans.	On *Fagus*	1			4
M. brassicicola (Fr.) Oud.	Anam. *Asteromella brassicae* (Chev.) Boer. & v. Kest.	1	2	3	4
M. buxicola (DC.) Tomilin (*M. limbalis* (Pers.) v. Arx)	On *Buxus*	1			
M. capronii (Sacc.) Tomilin	On *Salix*	1			
M. carinthiaca Jaap	On *Trifolium pratense*	1	2		
M. clymenia Sacc.	On *Lonicera periclymenum*	1			
M. conglomerata (Wallr.) Lindau	On *Alnus glutinosa*	1			
M. crataegi (Fuck.) Johans	On *Crataegus*				4
M. dianthi (Burt) Jørstad	On *Dianthus*, anam. *Cladosporium echinulatum* (Burt) de Vries	1		3	
M. fagi (Auersw.) Lindau	On *Fagus*	1			
M. fragariae (Tul.) Lindau	On *Fragaria × ananassa*, anam. *Ramularia brunnea* Peck	1	2	3	4
M. hedericola Lindau	On *Hedera helix*, anam. *Septoria hederae* Desm.	1	2		4
M. heraclei (Fr.) Petrak	On *Heracleum sphondylium*	1			4
M. iridis (Desm.) Schroet.	On *Iris pseudacorus*, anam. *Cladosporium gracile* Wallr.				4
M. isariophora (Desm.) Johans.	On *Stellaria*, anam. *Septoria stellariae* Desm.	1	2		4
M. juncaginacearum Schroet.	On *Triglochin*, anam. *Asteroma juncaginacearum* Rab.				4
M. killiani Petrak (*Cymadothea trifolii* Wolf)	Anam. *Polythrincium trifolii* Kunze	1	2	3	4
M. latebrosa (Cooke) Schroet.	On *Acer*, anam. *Septoria aceris* (Lib.) B. & Br.	1		3	
M. ligustri (Rob.) Lindau	On *Ligustrum*, anam. *Septoria ligustri* Kickx	1			4
M. macrospora (Kleb.) Jørstad	Anam. *Cladosporium iridis* (Fautr. & Roum) de Vries	1	2		4
M. maculiformis (Pers.) Schroet.		1	2	3	4
M. oedema (Fr.) Schroet.	On *Ulmus*	1			4
M. podagrariae (Fr.) Petrak	On *Aegopodium*, anam. *Phloeospora aegopodii* Grove	1			4
M. pteridis (Desm.) Schroet.	On *Pteridum aquilinum*		2		4
M. punctiformis (Pers.) Starb.		1	2	3	4
M. recutita (Fr.) Johans. (*M. chlouna* (Cooke) Lindau)	On *Phalaris* etc.	1			
M. ribis (Fuck.) Kleb.	On *Ribes nigrum*, anam. *Septoria ribis* Desm.	1			
M. sentina (Fr.) Schroet.	On *Pyrus communis*, anam. *Septoria pyricola* Desm.	1			
M. superflua (Auersw.) Petrak	On *Urtica dioica*, anam. *Ramularia urticae* Ces.	1	2	3	

M. tassiana (de Not.) Johans.   Anam. *Cladosporium herbarum* (Pers.) Link 1   2   3   4

M. ulmi Kleb.   On *Ulmus*, anam. *Phleospora ulmi* (Fr.)   1
    Wallr.

M. vaccinii (Cooke) Schroet.   On *Vaccinium*   1

Myriangium Mont. & Berk.
    Entomogenous parasite, *M. duriaei* Mont. & Berk. on scale insects on *Fraxinus*, Isle of Wight westwards.

Mytilidion Duby
    Saprophytes on coniferae, *M. decipiens* (Karst.) Sacc. in Hampshire, *M. gemmigenum* Fuck. in East Anglia, *M. mytillinellum* (Fr.) Zogg in Berkshire, *M. rhenanum* Fuck. in the west and *M. acicola* Wint. in the north.

Neopeckia Sacc.
    Saprophyte.

Neopeckia fulcita (Bucknall) Sacc.   On *Prunus*   1

Neotestudina Segretain & Destombes
    Saprophyte.

Neotestudina cunninghamiae (Hawkes. & Booth) v. Arx & Müller   1

Nodulosphaeria Rabh.
    Saprophytes, *N. jaceae* (Holm) Holm & *N. modesta* (Desm.) Munk in the north, *N. fruticum* (Rob.) Holm in E. Anglia.

Nodulosphaeria cirsii (Karst.) Holm   2

N. derasa (B. & Br.) Holm   1

N. erythrospora (Riess) Holm   4

N. ulnispora (Cooke) Holm   1   3

Ohleria Fuck.
    Lignicolous saprophytes, *O. obducens* Wint. on *Ulmus* in the west.

Ohleria rugulosa Fuck.   1

Omphalospora Theiss. & Syd.
    Saprophyte, *O. himantia* (Pers.) v. Höhn. in the north.

Ophiobolus Riess
    Saprophytes, see also *Nodulosphaeria*, additional species in the north and west.

Ophiobolus acuminatus (Sow.) Duby   1   3   4

O. typhae Feltg.   1

Ophiosphaerella Speg.
    Graminicolous saprophytes, *O. erikssonii* Walker in the north.

Ophiosphaerella herpotricha (Fr.) Walker   1

Otthia Nits.
    Saprophytes.

Otthia spiraeae (Fuck.) Fuck.   Anam. *Diplodia crataegi* West.   1   3   4
    (*O. pruni* Fuck.)

Paraphaeosphaeria O. Eriksson
Saprophytes.

Paraphaeosphaeria michotii (West.) O. Eriksson	Anam. *Coniothyrium scirpi* Trail	4
P. rusci (Wallr.) O. Eriksoon	On *Ruscus aculeatus*	1 2 3 4

Passeriniella Berl.
Saprophyte.

Passeriniella discors (Sacc. & Ellis) Apinis & Chesters	On *Spartina* & debris	2

Phaeocryptopus Naumov
Facultative parasite, *P. gaeumannii* (Rohde) Petrak on *Pseudotsuga* in the north and west.

Phaeosphaeria Miyake
Saprophytes or facultative parasites, segregates from *Leptosphaeria* with anamorphs often in *Phaeoseptoria* or *Septoria*; *P. epicalamia* (Riess) Holm on *Luzula*, *P. graminis* (Fuck.) Holm, *P. juncina* (Auersw.) Holm and *P. juncicola* (Rehm) Holm in the north and west, *P. nardi* (Fr.) Holm and *P. sowerbyi* (Fuck.) Holm in East Anglia.

Phaeosphaeria avenaria (Weber) O. Erikss.	Anam. *Septoria avenae* Frank, *S. tritici* Rob.	2
P. eustoma (Fuck.) Holm (*Leptosphaeria epicarecta* (Cke.) Sacc.)		1
P. fuckelii (Niessl) Holm	Anam. *Phaeoseptoria phalaridis* Sprague	1
P. herpotrichoides (de Not.) Holm		1 2 3
P. luctuosa (Niessl) Otani & Mikawa		1
P. microscopica (Karst.) O. Erikss.	Anam. *Phaeoseptoria airae* (Grove) Sprague	1 2 3
P. nigrans (Rob.) Holm		1
P. nodorum (Müller) Hedjaroude	Anam. *Septoria nodorum* (Berk.) Berk.	2 3 4
P. silvatica (Pass.) Hedjaroude		1
P. typharum (Desm.) Holm	Anam. *Scolecosporiella typhae* (Oud.) Petrak	1
P. vagans (Niessl) O. Erikss.		1 3

Platychora Petrak
Facultative parasite.

Platychora (*Systremma*) *ulmi* (Schleich.) Petrak	Anam. *Piggotia ulmi* (Grev.) Keissler	1 2 3 4

Platystomum Trev.
Saprophytes = *Lophidium* Sacc.

Platystomum compressum (Pers.) Trev.	2
P. compressum var *pseudomacrostomum* (Sacc.) Chesters & Bell	2

Pleomassaria Speg.
Lignicolous saprophytes, *P. holoschista* (B. & Br.) Sacc. on *Alnus* in East Anglia

Pleomassaria siparia (B. & Br.) Sacc.	On *Betula*, anam. *Prosthemium betulinum* Kunze	1	4

Pleospora Rabh.
Saprophytes, *P. abscondita* Sacc. & Roum., *P. halophila* Webster, *P. gaudefroyi* Pat., and *P. triglochinicola* Webster in East Anglia, *P. palustris* Berl. and *P. papaveracea* (de Not.) Sacc. in the Midlands, others in the north and west. See also *Phaeosphaeria vagans*

The Fungi of Southeast England

Pleospora bjoerlingii Byford	On *Beta*, anam. *Phoma betae* Frank	1	2		
P. calvescens (Fr.) Tul.	Anam. *Microdiplodia henningsii* Starb.	1			
P. herbarum (Pers.) Rabh.	Ubiquitous, anam. *Stemphylium botryosum* Wallr.	1	2	3	4
P. penicillus (Schm.) Fuck. (*P. media* Niessl, *P. phaeocomoides* (Sacc.) Wint.)		1			
P. pellita (Fr.) Rabh.		1		3	
P. scirpicola (DC.) Karst. (*P. scirpi* (Rabh.) Ces. & de Not.)		1			4
P. scrophulariae (Desm.) v. Höhn. (*P. infectoria* Fuck., *P. vulgaris* Niessl)		1			

Podonectria Petch
Entomogenous parasites, *P. tenuispora* Dennis in the north.

Poikiloderma Füisting
Lignicolous saprophyte.

Poikiloderma bufonium (B. & Br.) Fuisting	On *Quercus*	1

Polycoccum Sauter
Hyperparasites, sundry species on lichens in the north and west.

Preussia Fuck.
Saprophytes, *P. fleischakii* (Auersw.) Cain and *P. funiculata* (Preuss) Fuck. in the north and west.

Preussia vulgaris (Corda) Cain	1

Preussiella Lodha
Saprophytes, *P. dispersa* (Clum.) Lodha in Hertfordshire, *P. multispora* (Saito & Minoura) Lodha in the east midlands.

Protoventuria Berl. & Sacc.
Parasites of Ericaceae, *P. arxii* (Müller) Barr on *Rhododendron* in the midlands, *P. engleriana* (Henn.) Siv., *P. straussii* (Sacc. & Roum.) Siv. & *P. tetraspora* B. Erikss. in the north.

Pyrenidium Nyl.
Hyperparasite, *P. actinellum* Nyl. on *Peltigera* etc. in the north and west.

Pyrenophora Fr.
Facultative parasites of Gramineae with anamorphs in *Drechslera*, *P. typhaecola* (Cooke) Müller in E. Anglia, *P. semeniperda* (Brittleb. & Adams) Shoemaker on *Molinia* in the north.

Pyrenophora avenae Ito & Kuribay. (*Drechslera avenae* (Eidam) Scharif)	On *Avena*	1	
P. bromi (Died.) Drechsler		1	
P. dictyoides Paul & Parberry	On *Lolium*, anam. *Drechslera dictyoides* (Drechs.) Shoem.	1	
P. graminea Ito & Kuribay.	On *Hordeum*, anam. *Drechslera graminea* (Rab.) Shoemaker	1	
P. lolii Dovaston	Anam. *Drechslera siccans* (Drechs.) Shoemaker		4
P. teres Drechsler	On *Hordeum* etc., anam. *Drechslera teres* (Sacc.) Shoemaker	2	
P. triticirepentis Drechsler	Anam. *D. triticirepentis* (Died.) Shoemaker	1	

Rebentischia Karst.
Lignicolous saprophyte, *R. unicaudata* (B. & Br.) Sacc. on *Clematis vitalba* in the west.

Rhizodiscina Hafellner
  Saprophyte.

Rhizodiscina lignyota (Fr.)   On decorticated wood   1   4
  Hafellner

Rhopographus Nits.
  Saprophyte.

Rhopographus filicinum (Fr.) Nits.   On rhachides of *Pteridium aquilinum*   1   2   3   4

Rosenschoeldia Speg.
  Parasite, = *Naumovia* Dobrozr.

Rosenschoeldia (Naumovia) abundans   On *Prunella vulgaris*   1
  (Dobrozr.) Petrak

Saccothecium Fr.
  ?Saprophyte.

Saccothecium sepincola (Fr.) Fr.   On *Rosa* etc.   1   4
  (*Sphaerulina intermixta* (B. & Br.) Sacc.)

Schizothyrioma v. Höhn.
  Parasite.

Schizothyrioma ptarmicae (Desm.) v. Höhn. On leaves of *Achillea ptarmica*, anam.   1
                                       Leptothyrium ptarmicae Sacc.

Schizothyrium Desm.
  Saprophytes = *Microthyriella* v. Höhn., *S. speireum* (Fr.) Holm & Holm in East Anglia.

Schizothyrium pomi (Fr. ex Mont.) v. Arx   Anam. *Zygophiala jamaicensis* Mason;   2   4
                                       ubiquitous

Scirrhia Fuck.
  Saprophytes or facultative parasites, *S. agrostidis* (Fuck.) Wint. and *S. pini* Funk & Parker
  (*Dothistroma septospora*) (Dorog.) Morelet in the west and north.

Scirrhia aspidiorum (Lib.) Bub.   On *Athyrium* and *Pteridium*   1
S. rimosa (A. & S.) Fuck.   On *Phragmites*   2

Scirrhiachora Theiss. & Syd.
  Saprophyte, *S. groveana* (Sacc.) Theiss & Syd. in the west midlands.

Sclerodothis v. Höhn.
  ?Parasite, *S. aggregata* (Lasch) v. Höhn. on *Euphrasia* in the north.

Sphaerulina Sacc.
  Saprophytes and facultative parasites, others on lichens in the north.

Sphaerulina myriadea DC. Sacc.   On *Fagus* and *Quercus*   1
S. oraemaris Linder   On submerged timber   4
S. rehmiana Jaap   On *Rosa*, anam. *Septoria rosae* Desm   1

Splanchnonema Corda
  Lignicolous saprophytes = *Pteridiospora* Penz. & Sacc., *S. (Massaria) foedans* (Fr.) O.K. on
  Ulmus and *S. scoriadea* (Fr.) Barr on *Betula* in the north and west.

Splanchnonema ampullaceum (Pers.) Shoemaker (*Massariella curreyi* (Tul.) Sacc.)   2
S. argus (B.& Br.) Kuntze   On *Betula*, anam. *Myxocyclus polycistis* 1   4
                                           B. & Br.) Sacc.
S. pupula (Fr.) Kuntze   On *Acer*, anam. *Stilbospora ovata* Pers. 1   4

Sporormia de Not.
   Saprophytes.

Sporormia fimetaria de Not.   On dung   2

Sporormiella Ell. & Ev.
   Saprophytes, predominantly on dung, the ostiolate equivalent of *Preussia*, another ten British species to be recorded, with coprophiles locality can hardly be significant.

Sporormiella australis (Speg.) Ahmed & Cain   1
S. bipartis (Cain) Ahmed & Cain   2
S. intermedia (Auersw.) Ahmed & Cain   1   2   3
S. lageniformis (Fuck.) Ahmed & Cain (*Sporormia ambigua* Niessl)   2   3
S. leporina (Niessl) Ahmed & Cain (?*Sporormia longipes* Massee & Salmon) ?1   3
S. megalospora (Auersw.) Ahmed & Cain   1
S. minima (Auersw.) Ahmed & Cain   1   2
S. nigropurpurea Ell. & Ev.   1
S. ovina (Desm.) Ahmed & Cain   1
S. pulchella (Hansen) Ahmed & Cain   1

Stomiopeltis Theiss.
   Saprophytes, *S. cupressicola* Ellis in E. Anglia, *S. dryadis* (Rehm) Holm & *S. juniperina* (Grove) Holm & Holm in the north.

Stomiopeltis betulae Ellis   1
S. pinastri (Fuck.) v. Arx   Anam. *Sirothyriella pinastri* (Desm.) Minter 1

Strigopodia Bat.
   Saprophyte, *S. resinae* (Sacc. & Bres.) Hughes as *Helminthosporium resinaceum* Cooke in north.

Sydowia Bres.
   Facultative parasite.

Sydowia polyspora (Bref. & v. Tavel)   Only as the reputed anamorph   1
   Müller   *Sclerophoma pythiophila* (Corda) v. Höhn.

Systrema Theiss. & Syd.
   See *Platychora*, an old record of *S. frangulae* (Fuck.) Theiss. & Syd. was probably an error.

Taphrophylla Scheuer
   Saprophytes, *T. cornicapreoli* Scheuer in the west, *T. argyllensis*·Scheuer et al in the north.

Teichospora Fuck.
   Lignicolous saprophytes.

Teichospora obducens(Schum.) Fuck.   On *Fraxinus excelsior*   1

Thaxteriella Petrak
   Lignicolous saprophyte.

Thaxteriella pezicula (Berk. & Curt.)   As anam. *Helicoma muelleri* Corda   2
   Petrak

Thyridaria Sacc.
Lignicolous saprophytes, *T. minima* (Ell. & Ev.) Wehmeyer in the west.

Thyridaria rubronotata (B. & Br.) Sacc.	On *Acer, Crataegus, Hedera, Juglans, Ulmus,* anam. *Cyclothyrium juglandis* (Schum.) Sutton	1	4

Trematosphaeria Fuck.
Saprophytes, *T. alpestris* Toth on *Thymus, T. hydrela* (Rehm) Sacc., *T. paradoxa* Wint. in the north, *T. clarkii* Siv. on *Deschampsia* and *T. melina* (B. & Br.) Sacc. in the west.

Trematosphaeria heterospora (de Not.) Wint.	On *Iris germanica*	1		
T. pertusa (Pers.) Fuck. (*T. anglica* (Sacc.) Sacc.)	On *Carpinus, Quercus, Salix* etc.	1	2	3

Trichodelitschia Munk
Coprophilous saprophyte.

Trichodelitschia bisporula (Crouan) Lundq.		1	2	3

Trichothyrina (Petrak) Petrak
Saprophytes, *T. norfolciana* Ellis on *Carex* and *T. salicis* Ellis in East Anglia, *T. ammophilae* Ellis in the west, *T. pinophylla* (v. Höhn.) Petrak in the north.

Trichothyrina alpestris (Sacc.) Petrak	On *Phalaris* etc.	1
T. cupularum Ellis	On *Fagus*	1
T. fimbriata Ellis	On *Chamaecyparis* and *Cupressus*	1
T. nigroannulata (Webster) Ellis		1
T. parasitica (Fabre) v. Arx	On *Diatrype*	1

Tubeufia Penz. & Sacc.
Saprophyte, *T. paludosa* (Crouan) Rossman in E. Anglia, *T. trichella* (Bomm., Rouss. & Sacc.) Scheuer on *Ammophila* in the west, *T. hebridensis* Scheuer and *T. parvula* Dennis in the north.

Tubeufia cerea (B. & Curt.) Booth	Anam. *Helicosporium virescens* (Pers.) Siv.	1	2	4
T. helicoma (Phill. & Plowr.) Piroz.	Anam. *H. pannosum* (B. & Curt.) Moore	1		

Venturia Sacc.
Saprophytes and facultative parasites, *V. crataegi* Aderh. in E. Anglia, *V. geranii* (Fr.) Wint., *V. minuta* Barr, *V. nitida* (Petr.) Müller, *V. palustris* Sacc., Bomm. & Rouss in the north and west.

Venturia carpophila E. E. Fisher	On *Prunus persica,* anam. *Cladosporium carpophilum* Thum.				4
V. cerasi Aderh.	On *Prunus cerasus,* anam. *Fusicladium cerasi* (Rabh.) Sacc.				4
V. chlorospora (Ces.) Karst.	On *Salix* spp.	1			4
V. ditricha (Fr.) Karst.	On *Betula,* anam. *Fusicladium betulae* (Rob.) Aderh.	1			
V. fraxini Aderh.	On *Fraxinus,* anam. *Spilocaea fraxini* (Aderh.) Siv.	1			
V. inaequalis (Cooke) Wint.	On *Malus* and *Sorbus,* anam. *Spilocaea pomi* Fr.	1	2	3	4
V. integra Cooke	On *Corylus*	1			
V. maculiformis (Desm.) Wint.	On *Epilobium* spp.	1	2		
V. pirina Aderh.	On *Pyrus communis,* anam. *Fusicladium pyrorum* (Lib.) Fuck.	1	2	3	4
V. rumicis (Desm.) Wint.	On *Rumex* spp.	1	2	3	4

V. saliciperda Nüesch	On *Salix*, anam. *Pollaccia saliciperda* (All. & Tub.) v. Arx		4
V. tremulae Aderh.	On *Populus tremula*, anam. *Pollaccia radiosa* (Lib.) Bald. & Cif.	1	

Wentiomyces Koord.
?Parasites, *W. peltigericola* Haksw. and *W. sibirica* (Petr.) Müller in the north.

Wettsteinina v. Höhn.
Saprophytes, *W. dryadis* (Rostrup) Petrak and *W. niesslii* Müller in the north.

Xenomeris Syd.
?Parasites, *S. alpina* Petrak on *Vaccinium vitis-idaea* in the north.

Zopfia Rabh.
Saprophyte.

Zopfia rhizophila Rabh.	On roots of *Asparagus*	1	4

PYRENULALES

Aglaospora de Not.
Lignicolous saprophyte.

Aglaospora profusa (Fr.) de Not.	On *Robinia*	1

Massaria de Not.
Lignicolous saprophytes.

Massaria gigaspora Fuck.	On *Cornus* (*M. corni* (Mont.) Sacc.)	4
M. inquinans (Tode) de Not.	On *Acer pseudoplatanus*	4

PHYCOMYCETES

MUCORALES

Absidia v. Tiegh.
Saprophytes, largely in soil.

Absidia cylindrospora Hagem		1	4
A. glauca Hagem		2	4
A. orchidis (Vuill.) Hagem (*A. coerulea* Bainier)			
A. ramosa (Lindt.) Lendner (*A. corymbifera* (Cohn) Sacc. & Trott.)		1	
A. repens v. Tiegh.		1	

Acaulospora Gerdemann & Trappe
Endomycorrhizal, *A. nicolsonii* Walker et al. with Gramineae.

Actinomucor Schostakowitsch
Terrestrial saprophyte, *A. elegans* (Eid.) Benj. & Hesselt.

Azygozygum Chesters
Facultative parasite, *A. chlamydosporum* Chesters on roots of *Antirrhinum*.

Backusella Hesseltine & Ellis
Saprophytes.

Backusella lamprospora (Lendner) Benny & Benj.	1

Chaetocladium Fres.
Saprophytes and hyperparasites.

Chaetocladium brefeldii v. Tiegh. & Le Monnier	On other Mucorales	1	2	4
C. jonesii (B. & Br.) Fres.	On dung and in soil	1		

Chaetostylum v. Tiegh. & Le Monnier
Saprophyte.

Chaetostylum fresenii v. Tiegh. & Le Monnier	4

Circinella v. Tiegh. & Le Monnier
Saprophytes in dung and soil.

C. umbellata v. Tiegh. & Le Monnier	1

Coemansia v. Tiegh. & Le Monnier
Saprophytes or hyperparasites.

Coemansia erecta Bainier	1

Cunninghamella Matr.
Terrestrial saprophytes.

Cunninghamella elegans Lendner	4

Dicranophora Schroeter
Hyperparasite, *D. fulva* Schroet. on Boletales.

Dimargaris v. Tiegh.
Hyperparasites.

Dimargaris verticillata Benj.	On other Mucorales	1	

Dispira v. Tiegh.
Saprophytes or hyperparasites.

Dispira cornuta v. Tiegh.(*D. circinata* Elliott)			4

Endogone Link
Endomycorrhizal.

Endogone lactiflua B. & Br.			4

Glomus Tul.
Endomycorrhizal.

Glomus microcarpus Tul.		1	3

Gongronella Ribaldi
Terrestrial saprophyte, *G. butleri* (Lendn.) Peyronel & Dal Vasco.

Haplosporangium Thaxter
Parasites.

Haplosporangium bisporale Thaxter			4
H. decipiens Thaxter	1		

Helicostylum Corda
Saprophytes, *H. elegans* Corda and *H. venustulum* Lythgoe, cf. also *Chaetostylum*.

Kickxella Coemans
Coprophyllous saprophytes.

Kickxella alabastrina Coemans	1		4

Martensella Coemans
Saprophyte, *M. pectinata* Coemans in Hertfordshire.

Mortierella Coemans
Saprophytes, especially in soil.

Mortierella alpina Peyronel	1		4
M. atrogrisea v. Beyma	1		
M. bainieri Costantin	1		
M. elongata Linnemann		2	
M. gamsii Milko	1		
M. hyalina (Herz) Gams (*M. hygrophila* Linnemann)	1		
M. isabellina Oud.	1		
M. marburgensis Linnemann	1		
M. minutissima v. Tiegh.	1	2	4
M. parvispora Linnemann	1		
M. pulchella Linnemann	1		
M. ramanniana (Möller) Linnemann	1	3	

	1	2	3	4
M. turficola Ling Young		2		
M. verrucosa Linnemann	1			
M. vesiculosa Mehrotra, Baijal & Mehrotra				4
M. zychae Linnemann	1			

Mucor L.
Saprophytes, especially in dung and soil.

	1	2	3	4
Mucor circinelloides v. Tiegh.	1			
M. griseocyaneus Hagem		2		
M. hiemalis Wehmer	1			4
M. mucedo L.	1	2	3	4
M. piriformis Fischer	1			
M. plasmaticus v. Tiegh.	1			4
M. pusillus Lindt		2	3	
M. racemosus Fres.	1			4
M. racemosus f. *sphaerosporus* (Hagem) Schipper	1			
M. saturninus Hagem	1			
M. spinosus v. Tiegh. (*M. plumbeus* Bon.)				4
M. strictus Hagem				4

Mycotypha Fenner
Saprophytes, *M. dichotoma* Wolf in the east midlands.

Parasitella Bainier
Hyperparasite.

	1	2	3	4
Parasitella parasitica (Bainier) Syd. (*P. simplex* Bainier)				4

Phycomyces Kunze
Saprophytes, especially on dung.

	1	2	3	4
Phycomyces nitens Kunze	1	2		4
P. perrotianus Morini	1			

Pilaira v. Tiegh.
Coprophyllous saprophytes.

	1	2	3	4
Pilaira anomala (Ces.) Schroet.	1	2		4

Pilobolus Tode
Coprophyllous saprophytes.

	1	2	3	4
Pilobolus crystallinus Tode	1	2	3	4
P. kleinii v. Tiegh.	1			
P. longipes v. Tiegh.	1		3	
P. umbonatus Buller			3	

Piptocephalus de Bary
Hyperparasites of other Mucorales, *P. xenophila* Dobbs & English in Berkshire.

	1	2	3	4
Piptocephalus arrhiza v. Tiegh. & Le Monnier	1	2		4
P. cylindrospora Bainier	1			
P. fimbriata Richardson & Leadbeater	1			
P. freseniana de Bary	1	2		4
P. lepidula (Marchal) Benj.				4
P. microcephala v. Tiegh.	1	2		4

Rhizopus Ehrenb.
Saprophytes or facultative parasites of ripe fruit.

Rhizopus arrhizus Fischer	1		4
R. microsporus v. Tiegh.	?1		
R. stolonifer (Ehrenb.) Lind		2	4

Rhopalomyces Corda
Saprophytes, *R. elegans* Corda and *R. magnus* Berlese in the north.

Spinellus v. Tiegh.
Hyperparasites, especially of *Mycena* carpophores.

Spinellus chalybeus (Dozy & Wolkenboer) Vuill.	1		4
S. fusiger (Link) v. Tiegh.	1	2	4

Syncephalastrum Schroet.
Saprophyte.

Syncephalastrum racemosum Cohn	?1

Syncephalis v. Tiegh. & Le Monnier
Hyperparasites on other Mucorales.

Syncephalis intermedia v. Tiegh.	1	
S. nodosa v. Tiegh.		4
S. sphaerica v. Tiegh.	1	

Syzygites Ehrenb.
Hyperparasite of agaric carpophores.

Syzygites megalocarpus Ehrenb.	1	2	4

Thamnidium Link
Saprophyte.

Thamnidium elegans Link	1	4

Thamnostylum v. Arx & Upadhyay
Saprophyte.

Thamnostylum piriforme (Bainier) v. Arx & Upadhyay	On dung	1

Zygorhynchus Vuill.
Saprophytes.

Zygorhynchus heterogamus (Vuill.) Vuill.	1	
Z. moelleri Vuill.		4

ENTOMOPHTHORALES

Ancylistes Pfitzer
Parasites of Desmids, two species in the north.

Ballocephala Drechsler
Parasite, *B. verrucospora* Richardson on tardigrades in the north.

Basidiobolus Eidam
Saprophyte on dung of amphibia and reptiles, *B. ranarum* Eidam, perhaps coextensive with the "hosts" but localised records are hard to trace.

Completoria Lohde
Parasite of fern prothalli, *C. complens* Lohde.

Conidiobolus Bref.
Saprophytes or hyperparasites.

Conidiobolus rhysosporus Drechsler	1			

Entomophthora Fres.
Parasites of arthropoda, including *Empusa* Cohn, *Zoophthora* Batko.

Entomopthora americana (Thaxter) Sacc.	1			4
E. aphidis Hoffman	1	2	3	
E. conica Nowakowski	1			
E. culicis A. Braun		2		
E. dipterigena (Thaxter) Sacc.	1	2		
E. muscae (Cohn) Fres.	1	2	3	4
E. planchoniana Cornu		2		4
E. rhizospora Thaxter	1			
E. sphaerosperma Fres.		2		4
E. tenthredinis Fres.		2		
E. thaxteriana Petch				4

Tarichium Cohn
Parasites of arthropods.

Tarichium megaspermum Cohn				4

SAPROLEGNIALES

Achlya Nees
Saprophytes.

Achlya americana Humphrey	1
A. apiculata de Bary	1
A. colorata Pringsheim	1
A. flagellata Coker	1
A. glomerata Coker	1
A. hypogyna Coker & Pemberton	1
A. megasperma Humphrey	1
A. oblongata de Bary	1
A. polyandra Hildebrand	1
A. racemosa Hildebrand	1
A. radiosa Maurizio	1
A. spinosa de Bary	1
A. stellata de Bary	1
A. treleaseana (Humphrey) Kauffman	1

Althornea Gareth Jones & Alderm.
Endophyte in shell of *Ostrea*, *A. crouchii* G. Jones & Alderm. in Essex waters.

Aphanomyces de Bary
   Facultative root parasites of phanerogams, especially in wet soils.

Aphanomyces cochlioides Drechsler	"Sussex"
A. euteiches Drechsler	"Sussex"
A. laevis de Bary	1
A. stellatus de Bary	1

Aphanomycopsis Scherffel
   Parasites of freshwater algae, *A. bacillacearum* Scherffel and *A. desmidiella* Canter in the north.

Aplanopsis Höhnk
   Terrestrial saprophytes.

Aplanopsis spinosa Dick	4
A. terrestris Höhnk	4

Brevilegnia Coker & Couch
   Saprophytes, *B. diclina* Harvey in the west.

Brevilegnia linearis Coker & Braxton	4

Brevilegniella Dick
   Saprophyte.

Brevilegniella keratinophila Dick	In soil	4

Calyptralegnia Coker
   Saprophytes, *C. achlyoides* Coker & Couch.

Dictyuchus Leitgeb
   Saprophytes.

Dictyuchus monosporus Leitgeb		1	3
D. sterile Couch	Doubtfully distinct	1	

Ectrogella Zopf
   Parasites of diatoms, *E. bacillacearum* Zopf in Middlesex, *E. monostoma* Scherffel elsewhere.

Eurychasma Magnus
   Parasite of marine algae, *E. dicksonii* (Wright) Magnus in the north.

Eurychasmidium Sparrow
   Parasites of marine Rhodophyceae, *E. tumefaciens* (Magnus) Sparrow in *Ceramium* in the north.

Isoachlya Kauffman
   Terrestrial saprophytes.

Isoachlya anisospora (de Bary) Coker	1
I. toruloides Kauffman & Coker	1

Leptolegnia de Bary
   Saprophyte, *L. caudata* de Bary in Dorset.

Leptolegniella Honeycutt
Saprophyte, *L. keratinophyllum* Honeycutt in East Anglia, *L. marina* (Atkins) Dick on marine animals.

Nematophthora Kerry & Crump
Parasite, *N. gynophila* Kerry & Crump in cysts of *Heterodera avenae* in Hertfordshire etc.

Pythiopsis de Bary
Saprophytes in wet soils, also *P. humphreyana* Coker.

Pythiopsis cymosa de Bary	1

Saprolegnia Nees
Saprophytes or facultative parasites of fish.

Saprolegnia delica Coker	1	
S. diclina Humphrey	1	
S. eccentrica (Coker) Seymour	1	
S. ferax (Gruithuysem) Thuret	1	4
S. latvica Apinis	1	
S. litoralis Coker	1	
S. monoica Pringsheim	1	
S. paradoxa Maurisio	1	

Scoliolegnia Dick
Saprophytes, *S. blelhamensis* Dick and *S. subexcentrica* Dick in the north.

Scoliolegnia asterophora (de Bary) Dick	1

Sommerstorffia Arnaudow
Parasitic on *Cladophora*

Sommerstorffia spinosa Arnaudow	1

Thraustochytrium Sparrow
Parasites of marine algae, three species in Spithead, opposite 2.

Thraustotheca Humphrey
Saprophyte in wet soil.

Thraustotheca clavata (de Bary) Humphreys	1

LEPTOMITALES

Apodachlya Pringsheim
Saprophytic water moulds, *A. brachynema* in East Anglia.

A. pirifera Zopf	3

Araiospora Thaxter
Saprophytes, often on sodden twigs, *A. pulchra* Thaxter.

Leptomitus Agardh
Saprophytic water moulds.

Leptomitus lacteus (Roth) Agardh	1	3	4

Mindeniella Kanouse
Watermoulds on rotting rosaceous fruit, *M. spinospora* Kanouse.

Rhipidium Cornu
Watermoulds on sodden fruit and twigs, also *R. parthenosporum* Kanouse in the west.

Rhipidium americanum Thaxter	1
R. interruptum Cornu	1

Sapromyces Fritsch
Saprophytic water mould.

Sapromyces elongatus (Cornu) Coker	1

ZOOPAGALES

Acaulopage Drechsler
Saprophytes and facultative parasites of amoebae etc.

Acaulopage baculispora Drechsler	1	
A. rhicnospora Drechsler	1	
A. tetraceros Drechsler	1	4

Cochlonema Drechsler
Saprophytes and facultative parasites of amoebae etc.

Cochlonema dolichosporum Drechsler	1
C. linearis F. R. Jones	1

Stylopage Drechsler
Saprophytes and facultative parasites of nematodes.

Stylopage lepte Drechsler		2
S. rhabdospora Drechsler	1	
S. rhynchospora Drechsler	1	

Zoopage Drechsler
Parasites of amoebae etc.

Zoopage pachyblasta Drechsler	1
Z. thamnospira Drechsler	1
Z. virgispora Drechsler	1

AMOEBIDIALES

Parasites or endocommensals of freshwater arthropods.

Paramoebidium Leger & Dubosq
Endocommensal, *P. chattonii* Dubosq, Leger & Tuzet in hindgut of *Simulium* larvae in Dorset.

HARPELLALES

Presumably coextensive with their hosts but data not available, *Harpella melosirae* Leger & Dubosq, *Smittium* sp. and *Stipella* sp. in Dorset.

ASELLARIALES

Endocommensals, three genera in isopods, collembola and diptera respectively, presumably coextensive with the hosts but local data not available.

ECCRINALES

Thirteen genera of endocommensals to which the observation above also applies.

LAGENIDIALES

Lagenidium Schenk
Parasites, the following records are questionable and confirmation of these or additional species is desirable.

Lagenidium nodosum (Dangeard) Ingold	1
L. syncytiorum Kleb.	1

Lagenocystis Copeland
Parasite, *L. radicicola* (Vanterp. & Led.) Copeland on *Hordeum* roots.

Myzocytium Schenk
Parasites of freshwater algae in the midlands and north and on eggs of rotifers.

Myzocytium vermicolum (Zopf) Fischer	1

Olpidiopsis Cornu
Hyperparasites of other phycomycetes and algae, several species in E. Anglia, the midlands and west.

Protascus Dangeard
Parasite of nematodes.

Protascus subuliferus Dangeard	1

Rozellopsis Karling
Hyperparasites of other phycomycetes.

Rozellopsis septigena Karling	4
R. waterhouseii Karling	1

PERONOSPORALES

Albugo Pers.
Obligate parasites with anamorphs in *Cystopus*.

Albugo candida (Pers.) Kuntze — On many genera of Cruciferae, including *Alyssum, Cakile, Cochlearia, Diplotaxis, Nasturtium, Rorippa, Senebiera, Sinapis* elsewhere.

Host	1	2	3	4
On *Arabis caucasica* and *A. hirsuta*	1	2		4
On *Armoracia rusticana*	1			
On *Aubretia deltoides*	1	2		
On *Brassica oleracea* and *Brassica* spp.	1		3	4
On *Capsella bursapastoris*	1	2	3	4
On *Cardamine hirsuta*		2	3	
On *Lunaria annua*	1	2		4
On *Sisymbrium officinale*	1			

A. caryophyllacearum (Wallr.) Cif. & Biga — On *Spergularia* spp.

Host	1	2	3	4
On *Spergularia* spp.		2	3	4

A. tragopogi (Pers.) Schroet., — On *Crepis* elsewhere,

Host	1	2	3	4
On *Gerbera*		2		
On *Senecio vulgaris*	1			
On *Tragopogon*		2	3	4

A. tragopogi var *cirsii* Ciferri & Biga — On *Cirsium* spp.

Host	1	2	3	4
On *Cirsium* spp.				4

Bremia Regel
Obligate parasite of Compositae.

Bremia lactucae Regel — Races elsewhere on *Chrysanthemum, Cirsium, Crepis, Hypochaeris* and *Leontodon*

Host	1	2	3	4
On *Centauria cyanus* and *C. nigra*			3	
On *Cynara scolymus*			3	
On *Gaillardia grandiflora*	1			4
On *Helichrysum*		2		
On *Lactuca* spp.	1			4
On *Lapsana communis*	1			
On *Senecio* spp.	1			4
On *Sonchus* spp.	1	2		4

Peronospora Corda
Obligate parasites, also *P. hariotii* Gaum. in Hampshire, *P. scleranthi* Rab. in Hertfordshire, *P. potentillae-reptantis* Gaum. in Berkshire, *P. chlorae* de Bary, *P. conglomerata* Fuck., *P. digitalidis* Gaum., *P. jaapiana* Magn., *P. linariae* Fuck., *P. melandrii* Gaum., *P. oerteliana* Kuhn, *P. polygoni* (Thum.) Fischer, *P. rubi* Rab., *P. valerianae* Trail in East Anglia and a few others in the north and west.

Peronospora aestivalis Syd.
P. affinis Rossm.
P. agrestis Gaum.

Host	1	2	3	4
On *Medicago sativa*				4
On *Fumaria* spp.				4
On *Veronica agrestis, V. arvensis, V. chamaedrys, V. filiformis, V. persica* and *V. polita*	1	2		4

P. alsinearum Casp.
P. alta Fuck.
P. anemones Tramier
P. antirrhini Schroet.
P. aparines (de Bary) Gaum.
P. arabidopsis Gaum. (*P. parasitica* pp.)
P. arborescens (Berk.) Casp.

Host	1	2	3	4
On *Stellaria media*	1	2		
On *Plantago major*	1	2	3	4
On *Anemone coronaria*	"Sussex"			
On *Antirrhinum majus*	1		3	4
On *Galium aparine*	1	2		4
On *Arabidopsis thaliana*	1			
On *Meconopsis* spp. cult.	1	2		4
On *Papaver alpinum, P. argemone, P. rhoeas*	1	2		
On *Papaver somniferum*	1	2		

P. arenariae (Berk.) Tul.	On *Moehringia trinervia*		2		
P. arvensis Gaum.	On *Veronica hederifolia*	1			
P. brassicae Gaum. (*P. parasitica* pp.)	On *Brassica napus, B. oleracea*, B. rapa	1	2	3	4
	On *Raphanus raphanistrum* and *R. sativus*	1		3	4
	On *Sinapis* spp.	1			
P. calotheca de Bary	On *Galium odoratum*	1	2		
P. chenopodii-polyspermi Gaum. (*P. farinosa* Fr. pp.)	On *Chenopodium polyspermum*	1	2		
P. cheiranthi Gaum. (*P. parasitica* pp.)	On *Cheiranthus cheiri*	1	2		4
P. chenopodii Schlecht. (*P. farinosa* Fr. pp.)	On *Chenopodium* album	1			4
P. conferta (Ung.) Ung.	On *Cerastium vulgatum*	1	2		
P. coronopi Gaum. (*P. parasitica* pp.)	On *Coronopus didymus*	1			
P. cytisi Rostrup	On *Laburnum anagyroides*	1			4
P. debaryi Salmon & Ware	On *Urtica urens*	1			4
P. dentariae Rab. (*P. parasitica* pp.)	On *Cardamine pratensis*	1			
P. destructor (Berk.) Casp.	On *Allium cepa* (other species elsewhere)	1		3	4
P. dipsaci Tul.	On *Dipsacus fullonum*				4
P. effusa (Grev.) Tul. (*P. farinosa* (Fr.) Fr. f. sp. *spinaciae* Byford)	On *Spinacia*	1 "Sussex" 4			
P. erodii Fuck.	On *Erodium cicutarium*	1			
P. ervi A. Gust.	On *Vicia hirsuta*	1			
P. ficariae Tul.	On *Ranunculus ficaria*	1	2		
P. fragariae Roze & Cornu	On *Fragaria* sp.				4
P. fulva Syd. (*P. viciae* pp.)	On *Lathyrus pratensis*	1			
P. galii Fuck.	On *Galium mollugo*		2		
	On *Galium saxatile*				4
P. galligena Blumer (*P. parasitica* pp.)	On *Alyssum saxatile*	1 "Sussex"			
P. grisea (Ung.) Ung.	On *Hebe* sp.	"Sussex"			4
	On *Veronica beccabunga*	1	2		
P. jacksonii Shaw	On *Mimulus* sp.	1			
P. lamii Braun	On *Lamium maculatum*	1			
	On *Lamium purpureum*	1			4
P. lathyri-versicoloris Sav. & Rayss (*P. viciae* pp.)	On *Lathyrus nissolia*	1			
P. lepidii (McAlp.) Wilson (*P. parasitica* pp.)	On *Cardaria* (*Lepidium*) *draba*				4
P. leptosperma de Bary	On *Matricaria matricarioides*		2		
	On *Tripleurospermum inodorum*	1			
P. lotorum Syd.	On *Lotus corniculatus*	1	2		
	On *Lotus uliginosus*				4
P. manshurica (Naoum.) Syd.	On *Glycine max*	"Sussex"			
P. matthiolae Gaum. (*P. parasitica* pp.)	On *Matthiola*	1 "Sussex"			
P. minor (Casp.) Gaum.	On *Atriplex prostrata*		2		
P. myosotidis de Bary	On *Myosotis*	1	2		
P. niessleana Berl.	On *Alliaria petiolata*	1	2		
P. obovata Bon. (*P. lepigoni* Fuck.)	On *Spergularia media* & *S. rubra*	1	2		
P. parasitica (Pers.) Fr.	On *Capsella bursapastoris*	1	2		4
P. pisi Syd. (*P. viciae* pp.)	On *Pisum sativum*	1			4
P. pulveracea Fuck.	On *Helleborus foetidus*		2		
P. ranunculi Gaum. (on *R. acris* elsewhere)	On *Ranunculus bulbosus*		2		
	On *Ranunculus flammula*		2		
	On *Ranunculus repens*	1	2		4
P. radii de Bary (on *Tripleurospermum* elsewhere)	On *Chrysantheum* cult.		2		
P. romanica Sav. & Rayss.	On *Medicago lupulina*		2	3	
P. sanguisorbae Gaum.	On *Poterium sanguisorba*			3	4
P. schachtii Fuck. (*P. farinosa* f. sp. *betae* Byford)	On *Beta vulgaris*				4
	On *Beta vulgaris* ssp. *maritima*		2		4

Species	Host	1	2	3	4
P. sisymbrii-officinalis Gaum. (*P. parasitica*)	On *Sisymbrium officinale* and *S. irio*	1			
P. sordida B. & Br.	On *Scrophularia nodosa*	1	2		
P. sparsa Berk.	On *Rosa* sp.	1			4
P. symphyti Gaum.	On *Symphytum* spp.	1		3	
P. tabacina Adam	On *Nicotiana* spp.	1	2	3	4
P. thlaspeos-arvensis Gaum. (*P. parasitica* pp.)	On *Thlaspi arvense*	1			
P. trifolii-pratensis A. Gust.	On *Trifolium pratense*		2		
P. trifoliorum de Bary	On *Trifolium medium*	1			
P. valerianellae Fuck.	On *Valerianella carinata*	1			
P. verbasci Gaum.	On *Verbascum nigrum*	1	2		
	On *Verbascum thapsus*	1	2		
P. viciae (Berk.) De Bary	On *Vicia angustifolia*	1	2		
	On *Vicia faba*			3	4
	On *Vicia sativa*	1			
	On *Vicia sepium*		2		
P. violacea Berk.	(On *Knautia arvensis* elsewhere) On *Scabiosa columbaria*		2		
P. violae de Bary	On *Viola* spp.	1			4

Phytophthora de Bary

Saprophytes and facultative parasites, *P. megasperma* Drechsl. in Hampshire, *P. verrucosa* Foist. in Hertfordshire.

Species		1	2	3	4
Phytophthora cactorum (Lebert & Cohn) Schroet.		1		3	4
P. cinnamomi Rands		1	2		4
P. cambivora (Petri) Buism.				3	
P. citricola Sawada		1			4
P. cryptogea Pethybridge & Lafferty		1	2		4
P. cryptogea var. *richardiae* (Buism.) Ashby				3	4
P. fragariae Hickman				3	4
P. infestans (Mont.) de Bary	Ubiquitous since 1845	1	2	3	4
P. nicotianae v. Breda de Haan		1			
P. parasitica Dastur		1			4
P. porri Foister		1			
P. primulae Tomlinson				3	
P. syringae (Kleb.) Kleb.		1		3	4

Plasmopara Schroet.

Obligate parasites, also *P. epilobii* (Otth) Sacc. & Syd. on *Epilobium parviflorum* in Hampshire and *P. ribicola* Schroet. on *Ribes* spp. in East Anglia.

Plasmopara crustosa (Fr.) Jørstad

Collective name for races on Umbelliferae, of which the following have received individual names:

Species	Host	1	2	3	4
P. angelicae (Casp.) Trott.	On *Angelica sylvestris*	1			
P. chaerophylli (Casp.) Trott.	On *Anthriscus sylvestris*	1			
P. conii (Casp.) Trott.	On *Conium maculatum*	1			
P. pastinacae Sav. & Sav.	On *Pastinaca sativa*	1			4
P. petroselini Sav. & Sav.	On *Petroselinum sativum*	1			
P. podagrariae (Otth) Nannf.	On *Aegopodium podagraria*	1	2	3	4
P. saniculae Sav. & Sav.	On *Sanicula europaea*				4
P. smyrnii Sav. & Beck.	On *Smyrnium olusatrum*				4
P. densa (Rab.) Schroet. (elsewhere on *Euphrasia* & *Odontites*)	On *Rhinanthus minor*		2		
P. pusilla (de Bary) Schroet.	On *Geranium pratense*	1			4

		1	2	3	4
P. pygmaea (Ung.) Schroet. (*P. curta* (Berk.) Skalicky)	On *Aconitum* sp.	1			
	On *Anemone nemorosa*	1	2	3	4
P. viticola (Berk. & Curt.) Berlese & de Toni	On *Vitis coignetiae*	1			
	On *Vitis vinifera*	1			4

Pseudoperonospora Rostov

Obligate parasites, *P. cubensis* (Berk. & Curt.) Rostov on *Cucumis* in Essex.

		1	2	3	4
Pseudoperonospora humili (Fuck.) Wilson	On *Humulus lupulus*	1	2	3	4
P. urticae (Lib.) Salmon & Wand	On *Urtica dioica* and very rarely on *U. urens*	1	2		4

Records of *Basidiophora entospora* Roze & Cornu and *Sclerospora graminicola* (Sacc.) Schroet. appear to be based on casual introductions which did not persist. Both were in Surrey.

Pythiogeton v. Minden

Saprophyte.

	1
Pythiogeton uniforme A. Lund	1

Pythium Pringsheim

Saprophytes or weak facultative parasites associated with seedling damping off or root rots.

		1	2	3	4
Pythium acanthicum Drechsler		"London"			
P. arrhenomanes Drechsler					4
P. carolinianum Matthews		1			
P. catenulatum Matthews			2		
P. debaryanum Hesse	Confused with *P. ultimum,* confirmation desirable	1			
P. dissimile Vaartaja		1			
P. dissotochum Drechsler		1			
P. intermedium de Bary		1			4
P. mamillatum Meurs		1			
P. megalacanthum de Bary		"Sussex"			
P. perigynosum Sparrow		1			
P. spinosum Sawada		1			
P. splendens Braun				3	
P. tardicrescens Vanterpool					4
P. torulosum Coker & Patterson		1			
P. ultimum Trow			2	3	4
P. undulatum Peterson		1			
P. vexans de Bary		1			
P. volutum Vanterpool & Truscott					4

Zoophagus Sommerst.

Parasite of rotifers in fresh water.

	1
Zoophagus insidians Sommerstorff	1

MONOBLEPHARIDALES

Gonapodya Fischer
Saprophytes.

Gonapodya polymorpha Thaxter	1	3
G. siliquaeformis (Reinsch) Thaxter	1	3

Monoblepharis Cornu
Saprophytic water moulds, at least 5 other species in ponds elsewhere.

Monoblepharis ovigera Lagerheim	1	
M. polymorpha Cornu		3
M. sphaerica Cornu		3

BLASTOCLADIALES

Blastocladia Reinsch
Saprophytes on sodden twigs and fruit.

Blastocladia glomerata Sparrow	1	
B. gracilis Kanouse	1	
B. pringsheimii Reinsch	1	3
B. ramosa Thaxter	1	
B. truncata Sparrow	1	

Blastocladiella Matthews
Saprophytes, *B. brittannica* Willoughby in the north.

Catenaria Sorokin

Saprophytes or facultative parasites of nematodes and rotifers in fresh water, *C. anguillulae* Sorokin & *C. auxiliaris* (Köhn) Tribe in East Anglia.

Physoderma Wallr.
Parasites, including *Urophylctis; P. comari* (Berk. & White) Lagerh. & *P. schroeteri* Kruger in E. Anglia, *P. pulposum* Wallr. in Essex, *P. vagans* Schroet. in Hampshire, *P. gerhardti* Schroet. & *P. maculare* Wallr. in the midlands.

Physoderma alfalfae (Pat. & Lagerh.) Karling	On *Medicago*		4
P. heliocharidis (Fuck.) Schroet.			2
P. menthae Schroet.		1	
P. menyanthis (de Bary) de Bary			2

HARPOCHYTRIALES

Harpochytrium Lagerh.
Saprophytes.

Harpochytrium hedinii Wille	1

HYPHOCHYTRIALES

Parasites or hyperparasites, especially of freshwater algae and watermoulds.

Anisolpidium Karling
Parasites of marine algae, *A. sphacellarum* (Kny) Karling, widespread.

Canteriomyces Sparrow
Algal parasite, *C. stigeoclonii* (De Willd.) Sparrow on *Stigeoclonium* in the midlands.

Hyphochytrium Zopf
Parasites on algae, Helotales etc., *H. catenoides* Karling on Characeae etc. said to be in England.

Rhizidiomyces Zopf
Hyperparasite, *R. apophysatus* Zopf on watermoulds in the north.

Rhizidiomyces apophysatus Zopf   On bird droppings   1

Rhizidiomycopsis Sparrow
Hyperparasite, *R. japonicus* (Kob. & Ookubo) Sparrow on *Phytophthora erythroseptica* in the west midlands.

CHYTRIDIALES

Amphicypellus Ingold
Parasite of planktonic algae, *A. elegans* Ingold in the north and west midlands.

Asterophlyctis Petersen
Saprophyte on insect exuviae, *A. sarcoptoides* Petersen in the north.

Catenochytridium Berdan
Saprophyte, *C. carolinianum* Berdan in the north.

Catenomyces Hanson
Saprophyte.

Catenomyces persicinus Hanson   1

Chytridium Braun
Parasites of algae or hyperparasites, *C. cocconeidis* Canter and *C. lagenula* Braun in Middlesex, *C. olla* Braun, *C. sphaerocarpum* Dangeard in E. Anglia, several others in the north.

Chytridium chaetophyllum Scherffel	On pollen of *Typha* etc.	1	4
C. confervae (Wille) Minden	On *Tribonema bombycina*		4
C. inflatum Sparrow	On *Oedogonium*	1	
C. parasiticum Willoughby	On *Rhizidium, Rhizophlyctis, Entophylctis* etc.	1	
C. schenkii (Dangeard) Scherffel	On *Oedogonium*	1	
C. suburceolatum Willoughby	On *Rhizophylctis*	1	
C. versatile Scherffel	On diatoms	1	4
C. xylophilum Cornu	On sodden twigs	1	

Chytriomyces Karling
Saprophytes on insect exuviae, *C. poculatus* Willoughby et al in Buckinghamshire, four other species in the north.

Chytriomyces haylinus Karling		1	4

Cladochytrium Nowakowski
Saprophytes, *C. aurantiacum* Richards on Gramineae in the west, *C. hyalinum* Berdan and *C. tenue* Nowak. in the midlands.

Cladochytrium replicatum Karling	Ubiquitous in decaying submerged plant tissue	1

Cystochytrium I. Cook
Root parasite, perhaps referable to Hyphochytridiales.

Cystochytrium radicale I. Cook	On *Veronica*, needs clarification	4

Dangeardia Schröder
Parasites of freshwater algae.

Dangeardia mammillata Schröder	1

Diplophlyctis Schroet.
Saprophytes on Characeae, *D. laevis* Sparrow in the west.

Diplophlyctis intestina (Schenk) Schroet.	No definite locality but "it is difficult to find old plants of *Nitella* and *Chara* in which *D. intestina* is not present in abundance" so Massee's old record is acceptable	1

Endochytrium Sparrow
Saprophyte on Characeae, *E. digitatum* Karling in the north.

Endocoenobium Ingold
Parasite on *Eudorina elegans*, *E. eudorinae* Ingold in the midlands.

Endodesmium Canter
Parasite of desmids, *E. formosum* Canter in the north.

Entophlyctis Fischer
Parasites of algae, three species in the east midlands.

Entophylyctis confervae-glomeratae (Cienkowski) Sparrow	On *Oedogonium*	1

Karlingiomyces Sparrow
Saprophytes, *K. dubius* (Karling) Sparrow in Middlesex, *K. marilandicus* (Karling) Sparrow in the north.

Micromyces Dangeard
Parasites, including *Micromycopsis* Scherffel, six other species in the north.

Micromyces oedogonii (Roberts) Sparrow	On *Oedogonium*	1

Nephrochytrium Karling
Saprophytes, three species in the north.

Nowakowskiella Schroet.
Saprophytes, *N. elegans* (Nowak.) Schroet. in the east Midlands, *N. delica* Whiffen in the north.

Nowakowskiella hemisphaerospora Shanor	1
N. profusa Karling	1

Obelidium Nowakowski
Saprophytes on insect exuviae, *O. megarhizum* Willoughby & *O. mucronatum* Nowak. in the north.

Olpidium (Braun) Rab.
Parasites, several additional species reported from the east midlands and north.

Olpidium brassicae (Wor.) Dang.	In root hairs etc.	1
(*Asterocystis radicis* de Wild.)		
O. radicale Schwartz & Cooke	On *Veronica beccabunga*	4

Phlyctidium (Braun) Rab.
Parasites or saprophytes, *P. laterale* (Braun) Sorokin teste Sparrow, unlocalised, on *Ulothrix*; two other dubious species.

Phlyctochytrium Schroet.
Saprophytes or parasites, predominantly of algae, *P. proliferum* Ingold on *Chlamydomonas* in Berkshire, seven other species in the east midlands and north.

Phlyctorhiza Hanson
Keratinophilous saprophytes, *P. peltata* Sparrow and *P. variabilis* Karling in the east midlands.

Pleotrachelus Zopf
Parasites, *P. wildemani* Petersen in rhizoids of *Funaria* in the east midlands.

Podochytrium Pfitzer
Parasites of diatoms.

Podochytrium clavatum Pfitzer	On *Fragilaria*	1

Polyphagus Nowakowski
Parasites of *Euglena*, *P. laevis* (Nowak.) Bartsch. in the east midlands and north.

Polyphagus euglenae Nowakowski	Massee's unlocalised record is assumed to refer here	1

Pseudopileum Canter
Parasite, *P. unum* Canter on *Mallomonas* in the north.

Rhizidium Braun
Saprophytes or parasites, *R. variabile* Canter on *Spirogyra* in Middlesex, *R. mycophilum* Braun in the east midlands and *R. windermerense* Canter in the north.

Rhizidium richmondense Willoughby	In soil	1

Rhizoclosmatium Petersen
Saprophyte, *R. globosum* Petersen on insect exuviae in the east midlands and north.

Rhizophlyctis Fischer
Saprophytes or parasites.

Rhizophlyctis ingoldii Sparrow	On chitin		3	4
R. (Karlingia) rosea (de Bary & Wor.) Fischer	In soil and bird droppings	1		

Rhizophydium Schenk
Saprophytes and parasites, *R. gibbosum* (Zopf) Fischer, *R. graminis* Ledingham, *R. granulosporum* Scherffel, *R. stipitatum* Sparrow in E. Anglia, *R. elyensis* Sparrow, *R. fulgens* Canter, *R. laterale* (Braun) Rab., *R. pollinis-pini* (Braun) Zopf, *R. goniosporum* Scherffel, *R. subangulosum* (Braun) Rab. and *R. vermicola* Sparrow in the east midlands, about 16 others in the north.

Rhizophydium carpophylum (Zopf) Fischer	On eggs of *Dictyuchus*	1		
R. contractophilum Canter	On *Eudorina*	1		
R. fusus (Zopf) Fischer	On *Melosira*	1		
R. hyperparasiticum Karling	On phycomycete hyphae		3	
R. keratinophyllum Karling			3	
R. megarhizum Sparrow	On *Oscillatoria*			4
R. simplex (Dang.) Fischer	On *Chlorococcum*	1		
R. sphaerotheca Zopf	On dead pollen grains			4

Rhizosiphon Scherffel
Parasites of freshwater algae, *R. crassum* on *Anabaena* in the midlands, *R. akinetum* Canter in Cheshire.

Rozella Cornu
Hyperparasites, includes *Pleolpidium* Fischer; *R. septigena* Cornu in E. Anglia, *R. polyphagi* (Sparrow) Sparrow in the east midlands.

Rozella blastocladiae (Minden) Sparrow		1	
R. irregularis (Butler) Sparrow	On *Pythium*	1	

Scherffeliomyces Sparrow
Parasite on *Euglena*, *S. parasitans* (Sparrow) Sparrow in the east midlands.

Septolpidium Sparrow
Diatom parasite.

Septolpidium lineare Sparrow	On *Synedra*	1	

Septosperma Whiffen
Hyperparasite, *S. anomala* (Couch) Whiffen in the north.

Siphonaria Petersen
Saprophyte on insect exuviae, *S. variabilis* Petersen in the north.

Synchytrium de Bary & Woronin
Obligate parasites, *S. succisae* de Bary & Wor. in Hampshire, *S. aureum* Schroet. in E. Anglia.

Synchytrium anemones (DC.) Wor.	On *Anemone nemorosa*	1		
S. endobioticum (Schilb.) Perc.	On *Solanum tuberosum*, introduced c. 1896, virtually extinct through prohibition of cultivation of susceptible varieties of the host	1		4
S. mercurialis Fuck.	On *Mercurialis perennis* Fuck.	1		3
S. taraxaci de Bary & Wor.	On *Taraxacum officinale*	1	2	

Zygorhizidium Löwenthal
　Parasitic on algae, *Z. melosirae* Canter & *Z. parvum* Canter in midlands, 6 others in the north.

Zygorhizidium willei Lowenthal	On *Mougeotia, Spirogyra* etc.	4

PLASMODIOPHORALES

Ligniera Maire & Tison
　Obligate parasites especially of root hairs.

Ligniera junci (Schwartz) Maire & Tison (*L. graminis* (Schwartz) Winge, *L. bellidis* Schwartz *L. menthae* Schwartz)	In *Juncus, Lolium, Poa, Spartina* etc.	4

Plasmodiophora Woronin
　Obligate parasites.

Plasmodiophora brassicae Wor.	In hypertrophied roots of Cruciferae, especially *Brassica* spp. but also *Gramineae, Fragaria, Papaver, Rumex* etc.	1	4

Polymyxa Ledingham
　Root parasite of Gramineae, *P. graminis* Ledingham on *Hordeum* in East Anglia.

Sorosphaera Schroeter
　Obligate parasites.

Sorosphaera radicalis Cook & Schwartz	In root hairs of *Glyceria, Poa, Catabrosa* etc.	3	4
S. veronica (Schroet.) Schroet.	In tumours on *Veronica chamaedrys* and allies		4

Spongospora Brunchorst
　Obligate parasites.

Spongospora subterranea (Wallr.) Lagerheim	Powdery scab of *Solanum tuberosum* and root gall of *Lycopersicon esculentum*	"Sussex"
S. subterraneam f. sp. *nasturtii* Tomlinson	Crook root of *Nasturtium officinale*	4

Tetramyxa Goebel
　Obligate parasites, forming galls on *Ruppia* & *Zanichellia* spp. in Hampshire.

Genus incertae sedis:
Frankiella Maire & Tison
　Apparently an endosymbiont of *Alnus* and invariably present in galls on the roots of the host.

Frankiella alni (Wor.) Maire & Tison (*Plasmodiophora alni*(Wor.) Moeller)	On *Alnus*	1

LABYRINTHULALES

Labyrinthula Cienk.
  Parasites.

Labyrinthula macrocystis Cienk. Involved in Wasting Disease of Zostera 4

DEUTEROMYCOTINA

SPHAEROPSIDALES-MELANCONIALES (COELOMYCETES)

Actinonema Fr.
  Facultative parasites.

Actinonema aquilegiae (Roum. & Pat.)	Leafspot of *Aquilegia*	1 2 3

Actinothyrium Kunze

Actinothyrium graminis Kunze	On *Molinia* and other grasses and carices 1	4

Amarenographium O.Eriksson

Amarenographium metableticum (Trail) O.Eriksson	On *Ammophila arenaria*	2

Amerosporium Speg.
  Saprophyte, *A. polynematoides* Speg. On Gramineae &c. in Hertfordshire.

Ampelomyces Ces.
  Hyperparasites of Erysiphales.

Ampelomyces quisqualis Ces. (*Cicinnobolus cesatii* de Bary)	Ubiquitous	1	4

Aposphaeria Sacc.
  Lignicolous saprophytes, in part anamorphs of *Melanomma,* which see

Aposphaeria freticola Speg.	On *Fagus*	1

Apostrasseria NagRaj
  Anamorphs of *Phacidium,* *A.*(*Ceuthospora*) *lunata* (Shear) NagRaj. On Ericaceae in England.

Ascochyta Lib.
  Saprophytes and facultative parasites, see also *Didymella* anamorphs; *A. caricis-arenariae* Melnik in Middlesex, *A. allii-cepae* Punith et al., *A. asteris* (Bres.) Gloyer, *A. doronici* Allesch., *A. equiseti* Grove, *A. ligustri* Sacc. & Speg., *A. pseudacori* Sm. & Rams., *A. rhodesii* Punith., *A. stellariae* Fautr., *A. tenerrima* Sacc. & Roum., *A. viburni* and *A. wisconsinensis* Davis in E. Anglia, *A. bohemica* Kab. & Bub. and *A. scabiosae* Rab. in Berkshire, *A. straminea* Punith. and *A. teretiuscula* Sacc. & Roum. in Hampshire, *A. cinerariae* Tassi in Bedfordshire.

Ascochyta ari Died. (*A. pellucida* Bub.)	On *Arum maculatum*		4
A. avenae (Petrak) Sprague	On *Avena, Hordeum, Lolium, Triticum*		4
A. boltshauseri Sacc.	On *Phaseolus*		4
A. brassicae Thüm.	On *Brassica* spp.		4
A. calami (Bres.) Punith. (*Stagonospora calami* Bres.)	On *Acorus calamus*		4
A. carpinea Sacc.	On *Carpinus betulus*		3
A. chenopodii Rostrup	On *Atriplex* and *Chenopodium*	1	
A. clematidina Thüm.	On *Clematis* (cult.)		3
A. cynarae Died.	On *Cynara scolymus*	1	
A. cytisi Lib.	On *Laburnum anagyroides*	1	

A. dahliicola (Brun.) Petrak	Leafspot of *Dahlia* (*Phyllosticta dahliicola* Brun)			4
A. dianthi Berk.	On *Dianthus barbatus*			"Sussex"
A. fabae Speg.	On *Vicia faba*			"Sussex"
A. garryae Sacc.	On *Garrya elliptica*			3
A. glaucii (Cke. & Mass.) Died.	On *Glaucium fulvum* & *G. luteum*	1	2	3
A. gracilispora Punith.	On *Bromus erectus* & *Dactylis glomerata*	1		
A. hordei Hara var *europaea* Punith.	On *Hordeum, Secale* and *Triticum*		2	
A. humuli Kab. & Bub.	On *Humulus lupulus*			4
A. lathyri Trail	On *Lathyrus sylvestris*		3	
A. leptospora (Trail) Hara	On *Agropyron, Agrostis, Avena, Festuca, Holcus, Phragmites, Poa* &c. Most old records as *A. graminicola* Sacc. belong here	1		
A. malvicola Sacc.	On *Hibiscus syriacus*			4
A. mercurialis (Desm.) Bres.	On *Mercurialis perennis*	1		
A. metulispora B. & Br.	On *Fraxinus excelsior*	1		
A. nymphaeae Pass.	On *Nuphar japonica*	1		
A. phaseolorum Sacc.	On *Phaseolus*	1		4
A. philadelphi Sacc. & Speg.	On *Philadelphus coronarius*	1		
A. phleina Sprague	On *Bromus, Festuca, Phleum, Poa* and other grasses			4
A. pisi Lib.	Leaf and pod spot on *Pisum sativum*	1		
A. plantaginis Sacc. & Speg.	On *Plantago major*	1		
A. psammae Oud.	On *Ammophila arenaria*		2	4
A. scabiosae Rab.	On *Scabiosa*			4
A. sparganii Ellis	On *Sparganium ramosum* (but see *typhoidearum*)	1		
A. teucrii Lasch	On *Teucrium scorodonia*			4
A. typhoidearum (Desm.) v. Höhn.	On *Sparganium* & *Typha*	1		
A. violae Sacc. & Speg.	On *Viola* spp.	1		
A. vulgaris Kab. & Bub. v. *symphoricarpi* (West.) Grove	On *Symphoricarpus racemosus*	1		

Ascochytula (Potebnia) Died.
Saprophytes.

Ascochytula deformis Grove	On *Sambucus nigra*	1

Asteroma DC.
Facultative parasites but the following records largely relate to sterile material and are of little value.

Asteroma aceris Rob.	On *Acer campestre*		4
A. betulae Rob.	On *Betula*	1	
A. corni Desm.	On *Cornus sanguinea*		4
A. delicatulum Desm.	On *Colutea arborescens*	1	
A. robergei (Desm.) Sacc.	On *Conium maculatum*	1	
A. solidaginis Cooke	On *Solidago elliptica*	1	

Bachmanniomyces Hawksw.
Hyperparasite, *B. uncialicola* (Zopf) Hawksw. on *Cladonia* in the north.

Blennoria Fr.
Saprophyte

Blennoria buxi Fr.	On leaves of *Buxus sempervirens*	1

Camarographium Bub.
Saprophytes, *C. stephensii* (B. & Br.) Bub. on *Pteridium* in the north & west. See also *Amarenographium*.

Camarosporium Schulz.
Lignicolous saprophytes, at present largely "host-species" in need of revision; *C. obiones* Jaap & *C. rosae* Grove in East Anglia, *C. karstenii* Sacc. & Syd. on *Malus* in Berkshire.

Camarosporium ambiens (Cooke) Grove	On *Acer* and *Fagus*	1	
C. berberidis Cooke	On *Berberis vulgaris*	1	
C. caprifolii Brun.	On *Lonicera periclymenum*	1	
C. cistinum Cooke	On *Cistus* (alien)	1	
C. crataegi Oud.	On *Crataegus*	1	
C. eleagni Potebnia	On *Elaeagnus* (alien)	1	
C. ephedrae Cooke & Massee	On *Ephedra andina* (alien)	1	
C. ilicis Oud.	On *Ilex aquifolium*	1	
C. lauri Grove	On *Laurus nobilis*	1	
C. limoniae Cooke	On *Aegle sepiaria* (*Citrus trifoliata*) (alien)	1	
C. mori Sacc.	On *Morus alba*	1	
C. oreades (Dur. & Mont.) Sacc.	On *Quercus*		4
C. pini (West.) Sacc.	On *Juniperus, Pinus radiata, P. sylvestris*	1	4
C. propinquum Sacc.	On *Quercus*	1	
C. quercus (Lib.) Sacc.	On *Quercus coccinea*	1	4
C. robiniae (West.) Sacc.	On *Robinia pseudacacia*	1	
C. rubicolum (Sacc.) Sacc.	On *Rubus fruticosus*	1	
C. salicinum Grove	On *Salix*	1	
C. spartii Trail	On *Sarothamnus scoparius*		4
C. spiraeae Cooke	On *Spiraea opulifolia* &c. (alien)	1	
C. staphyleae Cooke	On *Staphylea pinnata* (alien)	1	
C. syringae Cooke & Massee	On *Syringa vulgaris*	1	
C. tamaricis (Cooke) Grove	On *Tamarix gallica*	1	
C. tiliae (Lév.) Sacc. & Penz.	On *Tilia*	1	
C. vitalbae Roum.	On *Clematis vitalba*	1	

Catinula Lév.
Saprophyte.

Catinula aurea Lév. (*Lemalis aurea* (Lév.) Fr.)	On debris of *Pinus sylvestris*	1

Cenangiomyces Dyko & Sutton
Saprophyte.

Cenangiomyces luteus Dyko & Sutton	On needles of *Pinus*	1

Ceuthospora Grev. non Fr.
Saprophytes mainly on fallen leaves, many "host-species" seem inseparable from *C. lauri*.

Ceuthospora feurichii Bub.	On *Vinca minor*				4
C. lauri Grev.	(*C. phacidioides* Grev. on *Ilex*, *C. rhododendri* Grove) on *Prunus laurocerasus*)	1	2	3	4
C. pinastri (Fr.) v. Höhn.	On *Pinus*	1			4

Chaetomella Fuck.
Lignicolous or terrestrial saprophytes, *C. acutiseda* Sutton & Sarbhoy on *Quercus* in the midlands.

Chaetospermum Sacc.
Saprophyte.

Chaetospermum chaetosporum (Pat.) Sm. & Rams.	Plurivorous	1	

Cheirospora Moug. & Fr.
Saprophyte, pulvinate on dead twigs.

Cheirospora botryospora (Mont.) B. & Br.	Especially on *Hedera*, also *Cornus*, *Fagus*	1	3

Coleophoma v. Höhn.
Saprophytes, especially on the more coriaceous fallen leaves.

Coleophoma cylindrospora (Desm.) v. Höhn.	Plurivorous	1	2
C. empetri (Rostr.) Petrak	Plurivorous	1	

Colletotrichum Corda
Saprophytes and facultative parasites, in part anamorphs of *Glomerella*, which see

Colletotrichum acutatum Simmonds		1			4
C. coccodes (Wallr.) Hughes (*C. atramentarium* (B. & Br.) Taub.)	On Solanaceae	1	2	3	4
C. circinans (Berk.) Vogl.	On *Allium*	1			4
"*Gloeosporium*" *colubrinum* Sacc.	On *Sanseviera* (alien)	1			
C. crassipes (Speg.) v. Arx	On *Agave, Cymbidium* &c. (alien)			3	4
C. dematium (Pers.) Grove	Plurivorous	1	2	3	4
C. fuscum Laub.	On *Digitalis*				4
C. gossypii Southw.	On *Dieffenbachia* (alien)	1			
C. holci Grove	On native Gramineae	1			
C. lindemuthianum (Sacc. & Magn.) Briosi & Cavara	On *Phaseolus*	1			
C. malvarum (Braun & Casp.) Southw.	On *Lavatera* & *Malva*			"Sussex"	
C. musae (B. & C.) v. Arx	On *Musa* under glass and imported fruit (alien)	1			
C. sasaecola Hino & Katumoto	On "bamboo" (alien)			3	
C. trichellum (Fr.) Duke	On *Hedera helix*	1	2		
C. sp. cf *C. capsici* (Syd.) Butl. & Bisby	On *Impatiens*	1			
C. sp. cf *C. omnivorum* Halst.	On *Aspidistra lurida*			3	

Coniella v. Höhn.
Saprophytes or facultative parasites.

Coniella eucalypti (Bat. & Peres) Sutton		1	
C. fragariae (Oud.) Sutton (*C. pulchella* v. Höhn.)		1	4

Coniothyrium Corda
Saprophytes and facultative parasites, including and confused with *Microsphaeropsis* v. Höhn. see also *Melanomma coniothyrium* of which *C. fuckelii* Sacc. is the reputed anamorph.

Coniothyrium aucubae Sacc.	On *Aucuba japonica* (alien)	1		
C. buddleiae (Cooke) Grove	On *Buddleja globosa* (alien)	1		
C. cassiicola Cooke	On *Cassia marylandica* (alien)	1		
C. concentricum (Desm.) Sacc.	On *Yucca* spp. (alien)	1		4
C. conoideum Sacc.	On *Chamaenerion, Epilobium, Urtica*		3	4
C. ephedrinum Grove	On *Ephedra* (alien)	1		
C. equiseti Lamb. & Fautr.	On *Equisetum telmateia*	1		
C. hellebori Cooke & Massee	On *Helleborus niger*	1	3	
C. ilicis Sm. & Rams.	On *Ilex aquifolium*	2		

C. melanconieum Sacc.	On *Ribes uvacrispa*	1		
C. minitans Campbell	On sclerotia of *Sclerotinia sclerotiorum* & *S. trifoliorum*		2	
C. olivaceum Bon.	On *Atropa, Colutea, Cydonia, Elaeagnus, Ginkgo, Mahonia, Ulex* etc.	1	3	4
C. pyrinum (Sacc.) Sheldon	On *Malus, Prunus, Pyrus*			4
C. sarothamni (Thüm.) Sacc.	On *Sarothamnus scoparius*	1		
C. subolivaceum Sacc.	On *Lupinus*		2	
C. tamaricis Oud.	On *Tamarix gallica*	1		

Additional "species" in adjacent counties are not listed because critical revision is likely to reduce many of these and the above to synonymy under *C. fuckelii* or *C. olivaceum.*

Cornutispora Piroz.
Hyperparasites of lichens, *C. lichenicola* Hawksw. & Sutton on *Parmelia* in the west.

Coryneum Nees
Anamorphs of *Pseudovalsa,* which see

Cryptocline Petrak
Facultative parasites, see also *Europolella* & *Trochila*; *C. cyclamenis* (Sib.) v. Arx in Herts.

Cryptocline effusa Petrak	On *Abies*		4
C. taxicola (All.) Petrak	On *Taxus baccata*		4

Cryptosporiopsis Bub. & Kab.
Anamorphs of *Pezicula,* which see

Cryptosporium Kunze
Anamorphs of *Ophiovalsa, Melanconis.*

Cyclothyrium Petrak
Anamorphs of *Thyridaria,* which see

Cylindrosporium Grev.
The type is the anamorph of *Pyrenopeziza brassicae*; the following may not be congeneric:

Cylindrosporium montenegrinum Petrak	On *Trollius*	4

Cytoplacosphaeria Petrak
?Parasite, *C. rimosa* (Oud.) Petrak on *Phragmites* in Suffolk, perhaps *Scirrhia* anamorph.

Cytospora Ehrenb.
Anamorphs of *Valsa,* which see

Dendrophoma Sacc.
Lignicolous saprophytes.

Dendrophoma pulvispyrius Sacc.	On *Betula, Tilia* etc.	2

Dichomera Cooke
Saprophytes or ?facultative parasites.

Dichomera mutabilis Sacc.	On *Platanus*	1
D. ribicola Grove	On *Ribes sanguineum*	1
D. saubinetii (Mont.) Cooke	On *Acer, Corylus, Carpinus, Frangula*	1

Dilophospora Desm.
Anamorph of *Lidophia*, which see

Diplodia Fr.
Lignicolous saprophytes or ?weak facultative parasites, including *Botryodiplodia* Sacc. and anamorphs of *Botryosphaeria*, *Cucurbitaria*, *Otthia*; most names are mere "hostspecies", *D. opuntiae* Sacc. on *Phyllocactus* in Middlesex.

Species	Host				
Diplodia aesculi Lév.	On *Aesculus hippocastanum*	1			
D. amorphae Sacc.	On *Amorpha fruticosa*	1			
D. betae Potebnia	On *Beta vulgaris*				4
D. buxi Fr.	On *Buxus sempervirens*	1			
D. buxi var *minor* Grove	On *Buxus sempervirens*	1			
D. celtidis Roum.	On *Celtis occidentalis*	1			
D. cistina Cooke	On *Cistus laurifolius*	1			
D. clematidis Sacc.	On *Clematis*				4
D. coryphae Cooke	On *Corypha*	1			
D. elaeagni Pass.	On *Elaeagnus angustifolius*	1	2		
D. (Botryodiplodia) fraxini Fr.	On *Fraxinus excelsior*	1		3	4
D. griffoni Sacc. & Trav.	On *Malus* (? cf. *Botryosphaeria obtusa*)		2		
D. henriquesii Thüm.	On *Smyrnium olusatrum*			3	
D. herbarum Lév.	Plurivorous on the larger herbaceous stems .				4
D. ilicicola Desm.	On *Ilex aquifolium*	1		3	
D. inquinans West.	On *Fraxinus*, doubtfully distinct from *D. fraxini*	1			
D. lantanae Fuck.	On *Viburnum lantana*	1			4
D. laurina Sacc.	On *Laurus nobilis*	1	2		
D. ligustri West.	On *Ligustrum*	1			4
D. lonicerae Fuck.	On *Lonicera caprifolium*	1			
D. magnoliae West.	On *Magnolia grandiflora*	1			
D. mamilana Fr.	On *Cornus sanguinea*	1			
D. melaena Lév.	On *Ulmus*	1			4
D. mori West.	On *Morus* spp.	1			
D. paulowniae Cooke	On *Paulownia imperialis*	1			
D. rhois Sacc.	On *Rhus glabra*	1			
D. rosarum Fr.	On *Rosa* spp.	1			
D. rubi Fr.	On *Rubus fruticosus*	1		3	4
D. rudis Desm. & Kickx	On *Laburnum anagyroides*				4
D. saccardiana Speg. var *anglica*Grove	On *Sarothamnus scoparius*		2		
D. salicina Lév.	On *Salix* spp.	1			
D. sambucina Sacc.	On *Sambucus nigra*	1			4
D. sarmentorum Fr.	On *Menispermum canadense* (?*Otthia spiraeae*)	1			
D. sarothamni Cooke & Harkn.	On *Sarothamnus scoparius*				4
D. siliquastri West.	On *Cercis canadensis*	1			
D. subtecta Fr. (*D. atrata* (Desm.) Sacc.)	On *Acer* spp.	1			
D. tecta B. & Br.	On *Prunus laurocerasus* & *P. lusitanica*	1			
D. tiliae Fuck.	On *Tilia*	1	2		
D. ulicis Sacc. & Speg.	On *Ulex*			3	
D. viticola Desm.	On *Vitis vinifera*	1			

Diplodina West.

Anamorphs of *Cryptodiaporthe* but also currently referred here are: *D. teretiuscula* Died. in Middlesex and:

Species	Host			
Diplodina aloysiae Grove	On *Aloysia citriodora*	1		
D. cirsii Grove	On *Cirsium*			3
D. dahliae Hollos	On *Dahlia*	1		

D. eurhododendri Voss	On *Rhododendron*	1	3	4
D. malvae Togn.	On *Althaea* & *Sidalcea*	1	"Sussex"	
D. millefolii All.	On *Achillea millefolium*	1		
D. passerinii All.	On *Antirrhinum majus*	1		
D. phlogis Fautr.	On *Phlox paniculata*	1		
D. semi-immersa Karst. & Har.	On *Fagus*	1		

Discogloeum Petrak
Facultative parasite.

Discogloeum veronicae (Lib.) Petrak   On *Hebe* & *Veronica* spp.   1

Discosia Lib.
Saprophyte.

Discosia artocreas (Tode) Fr.   Plurivorous on fallen leaves, *Vaccinium*   1
etc.

Discosporium v. Höhn.
Anamorph of *Cryptodiaporthe populea* which see

Discula Sacc.
Anamorph of *Apiognomonia errabunda*, also:

Discula junci Sm. & Rams.   On *Juncus*   4

Dothistroma Hulbary
Anamorph of *Scirrhia pini* which see

Elachopeltis Sacc.
Anamorphs of Microthyriaceae.

Eleutheromyces Fuck.
Saprophytes on decaying basidiomycete carpophores, *E. subulatus* (Tode) Fuck. in E.
Anglia.

Entomosporium Lév.
Anamorph of *Diplocarpon* which see

Epicladonia Hawksw.
Lichenicolous hyperparasite.

Epicladonia sandstedii (Zopf) Hawksw.   On podetia of *Cladonia conista*   2

Eriospora B. & Br.
Saprophyte, *E. leucostoma* B. & Br. on *Carex, Juncus, Typha* in the west.

Fujimyces Minter & Cain
Saprophyte.

Fujimyces oodes (Elliott) Minter & Cain   On *Juniperus* & *Pinus*   1

Gloedes Colby
Saprophyte.

Gloeodes pomigena (Schw.) Colby   On fruit of *Malus* & *Prunus*   4

Hapalosphaeria Syd.
Parasite, *H. deformana* (Syd.) Syd. On anthers of *Rubus* in East Anglia.

Haplosporella Speg.
Saprophytes.

Haplosporella francisci D. Sacc.	On *Frangula alnus*	2

Hendersonia auct. non Berk.
Saprophytes, largely "hostspecies" requiring redisposition.

Hendersonia coronillae Cooke	On *Coronilla emerus*	1		
H. culmiseda Sacc.	On Gramineae generally	1		
H. epicalamia Cooke	On *Phragmites*	1	2	4
H. fiedleri West. (*H. corni* Fuck.)	On *Cornus sanguinea*	1		
H. fiedleri var *symphoricarpi* Cooke	On *Symphoricarpos rivularis*	1		4
H. grossulariae Oud.	On *Ribes uvacrispa*	1		
H. maculans (Corda) Lév.	On *Camellia*	1		
H. phormii Naito	On *Phormium tenax*			4
H. sarmentorum West.	On *Vitis vinifera*	1		
H. vagum Fuck.	On *Salix*	1		

Heteropatella Fuck.
Saprophytes or feebly parasitic anamorphs of *Heterosphaeria* which see, also:

Heteropatella antirrhini Buddin & Wakef. (*Pseudodiscosia antirrhini* (Wakef.)) Buddin & Wakef.		1	
H. valtellinensis (Trav.) Woll. (*H. dianthi* Budd. & Wakef.)	On *Dianthus*	3	4

Hyalopycnis v. Höhn.
Saprophyte, *H. blepharistoma* (*Berk.*) Seeler in E. Anglia.

Hymenopsis Sacc.
Saprophytes.

Hymenopsis typhae (Peck) Sutton. (*Cryptomela typhae* (Peck) Died.)	On *Typha latifolia*	1

Hysterodiscula Petrak
Parasite. *Hysterodiscula empetri* (White) Petrak (*Rhystima empetri* (White)) on *Empetrum* in the north.

Kabatia Bub.
Faculative parasites, anamorphs of *Guignardia*.

Kabatia periclymeni (Desm.) Morelat	On *Lonicera periclymenum*	1

Kabatiella Bub.
Faculative parasites.

Kabatiella berberidis (Cooke) v. Arx	On *Berberis asiatica*	1	
K. caulivora (Kirschn.) Karak.	On *Trifolium pratense*		4
K. microsticta Bub.	On *Convallaria, Lilium umbellatum, Polygonatum* spp.	1 "Sussex"	

Lamproconium (Grove) Grove
Saprophytes.

Lamproconium desmazierii (B. & Br.) Grove	On *Tilia* spp.	1	3

Lemalis Fr. See *Catinula*

Leptostroma Fr.
Saprophytes, anamorphs of *Hypoderma* which see, also:

Leptostroma juncacearum Sacc.	On *Juncus*	1

Leptothyrina v. Höhn.
?Saprophyte.

Leptothyrina rubi (Duby) v. Höhn.	On *Rubus fruticosus*	4

Libertiella Speg. & Roum.
Hyperparasite, *L. malmedyensis* Speg. & Roum. on *Peltigera* in the west midlands.

Lichenoconium Petrak & Syd.
Lichenicolous hyperparasites, *L. usneae* (Anzi) Hawks., *L. pyxidatae* (Oud.) Petr. & Syd., and *L. xanthoriae* Christ. in the north and west.

Lichenoconium erodens Christ. & Hawks.	On *Cladonia, Evernia, Hypogymnia, Parmelia, Pertusaria*	3
L. lecanorae (Jaap) Hawks.	On *Lecanora conyzaeoides*	1

Lichenodiplis Dyko & Hawks.
Lichenicolous hyperparasite, *L. lecanorae* (Vouaux) Dyk. & Hawks on *Pertusaria* and other crustaceous lichens in the north and west.

Marssonina Magnus
Facultative parasites, anamorphs of *Diplocarpon, Drepanopeziza, Gnomonia* which see, also:

Marssonia clematidis All.	On *Clematis vitalba*	2
Marssonina daphnes (Desm.) Magn.	On *Daphne mezereum*	1
M. sambuci (Rostr.) Magn.	On *Sambucus nigra*	1

Massariothea Syd.
?Saprophytes, mainly on tropical Gramineae but *M. scotica* Sutton & Rizwi on *Quercus* in the north and west.

Melanconium Link
Anamorphs of *Melanconis* which see, also:

Melanconium hederae Preuss	On *Hedera helix*	1	3

Microdiplodia All.
Saprophytes: *M. mamma* All. see *Diplodia ligustri*, *M. subtecta* All. see *Diplodia subtecta*

Microdiplodia palmarum (Corda) Died.	On *Chamaerops humilis* (alien)	1
M. perpusilla (Desm.) Grove	On *Foeniculum vulgare*	4

Microdiscula v. Höhn.
Saprophyte.

Microdiscula phragmitis (West.) v. Höhn.	On *Phragmites*	1

Micropera Lév.
Anamorphs of *Dermea* which see

Monochaetia (Sacc.) All.
Saprophytes.

Monochaetia karstenii (Sacc. & Syd.) Sutton On leaves of *Camellia* etc.　　　　2　3

Monostichella v. Höhn.
?Saprophytes, *M. salicis* (West.) v. Arx in the north and west.

Monostrichella robergei (Desm.) v. Höhn.　On leaves of *Carpinus*　　　　3

Myxocyclus Riess
Saprophytes.

Myxocyclus cenangioides (Ell. & Roth) Petrak (*Camarosporium abietis* Wils. & Anders.)　　3
M. polycistis (B. & Br.) Sacc.　　　　On *Betula*　　　　3

Myxosporium Link
Anamorphs of *Cryptodiaporthe, Diaporthe, Pezicula,* see also *Cryptosporiopsis.*

Naemaspora Sacc.
Saprophytes, see *Roscoepoundia,* but also:

Naemaspora strobi All.　　　　On *Pinus sylvestris*　　　　1

Neoalpakesa Punith.
Saprophyte, *N. poae* Punith. on *Poa glauca* in the north.

Neohendersonia Petrak
Lignicolous saprophyte.

Neohendersonia kickxii (West.) Sutton &　On *Fagus*　　　　1
Pollack (*N. piriformis* (Otth.) Petr.)

Neophoma Petrak & Syd.
Saprophyte, *N. graminella* (Sacc.) Petr. & Syd. (*Phoma stagonosporoides* Trail) abundant
on several genera of Gramineae in the north and west.

Neottiospora Desm.
Saprophyte.

Neottiospora caricina (Desm.) v. Höhn.　On *Carex pendula* and other carices, *Iris*　　3
etc.

Neottiosporina Subram.
Saprophyte.

Neottiosporina australiensis Sutton &　On *Phragmites*　　　　1
Alcorn

Oncospora Kalchbr. & Cooke
Saprophyte.

Oncospora pinastri Died.　　　　On *Pinus sylvestris* bark　　　　1

Oramasia Urries see *Vermiculariopsella*

Patellina (Speg.) Speg.
Saprophytes, *P. caesia* Elliott & Stansfield and *P. diaphana* Elliott in the midlands.

Peltasterinostroma Punith.
Saprophyte, *P. rubi* Punith. on *Rubus* in the north.

Pestalotia de Not.
Saprophyte, *P. pezizoides* de Not. on *Vitis vinifera* not confirmed as British.

Pestalotiopsis Steyaert
Saprophytes, *P. monochaetoides* (Doyer) Stey. in Hampshire, *P. neglecta* (Thüm.) Stey. in Essex, *P. disseminata* (Thüm.) Stey. in E. Anglia, *P. gracilis* (Kleb.) Stey. in Middlesex, *P. clavispora* (Atk.) Stey. & *P. foedans* (Sacc. & Ellis) Stey. in the west.

Pestalotiopsis decolorata (Speg.) Stey.	On *Myrtus communis*	1			
P. funerea (Desm.) Stey. (*P. conigena* Lév.)	On *Chamaecyperis, Cupressus, Juniperus, Pinus, Rhododendron*	1	2		4
P. funerioides Stey.	On *Cupressus*			3	
P. guepini (Desm.) Stey.	On *Camellis japonica, Rhododendron*	1			
P. guepini var *macrotricha* (Kleb.) Stey.	On *Juniperus*				4
P. montellica (Sacc. & Vogl.) Kobayasi	On *Quercus ilex*	1			
P. sydowiana (Bres.) Sutton	On *Calluna, Erica, Rhododendron, Prunus laurocerasus*	1	2	3	4
P. versicolor (Speg.) Stey.	On *Acer, Fuchsia, Rhododendron*	1	2		4

Pestalozziella Sacc. & Ell.
Facultative parasite, *P. subsessilis* on *Geranium* in the north.

Phacidiopycnis Potebnia
Facultative parasites, anamorph of *Potebniamyces* also:

Phacidiopycnis tuberivora (Gussow & Foster) Sutton	On *Humulus* & *Solanum tuberosum*	3

Phaeoseptoria Speg.
Anamorphs of *Phaeosphaeria*.

Phialophorophoma Linder
Saprophyte, *P. litoralis* Linder on driftwood & *Spartina* in the Isle of Wight and E. Anglia.

Phleospora Wallr.
Facultative parasites, anamorphs of *Mycosphaerella*, which see, also:

Phleospora pseudoplatani (Rob.) Bub. (*P. aceris* Sacc.)	On *Acer*	1	4
P. robiniae (Lib.) v. Höhn.	On *Robinia pseudacacia*	1	3

Phoma Sacc.
Saprophytes and facultative parasites, anamorphs of *Didymella, Leptosphaeria* etc. but the catalogue contains many imperfectly studied "hostspecies" likely eventually to be relegated to synonymy; elsewhere *P. ammophilae* Dur. & Mont., *P. arundinacea* Berk., *P. minutula* Sacc. (on *Lonicera*), *P. putaminum* Speg., *P. telephii* (Vestergr.) v. Kest. (on *Sedum*) in East Anglia. *P. allostoma* Died. on *Taxus* in the west, *P. dennisii* Boerema on *Solidago* & *P. macrocapsa* Trail on *Mercurialis* in the north.

Phoma aculeorum Sacc.	On prickles of *Rosa*	1		
P. amorphae Sacc.	On *Amorpha*	1		
P. barbari Cooke	On *Lyceum barbarum*	1		
P. chrysanthemicola Hollos	On *Chrysanthemum*		2	
P. complanata (Tode) Desm	On many Umbelliferae	1		4

Species	Host / notes	1	2	3	4
P. crataegi Sacc.	On *Crataegus*	1			
P. destructiva Plowr.	On *Lycopersicon* fruit	1			
P. domestica Sacc.	On *Jasminum officinale*	1			
P. epicoccina Punith., Tulloch & Leach	On *Beta, Malus* etc.				4
P. eupyrena Sacc.	In soil, on dead roots, tubers etc.	1	2		4
P. exigua Desm. (*P. tuberosa* Melh., Ros. & Sch.; *P. herbarum* auct. non West.)		1	2	3	4
P. exigua var *sambuci-nigrae* (Sacc.) Boer. & How.		1			
P. fimeti Brun.	On wood				4
P. galactis Cooke	On *Galax aphylla*	1			
P. gentianae Kühn	On *Gentiana acaulis*	1			
P. glomerata (Corda) Woll. & Hochst.	Plurivorous	1	2		4
P. grossulariae Schulz. & Sacc.	On *Ribes uva-crispa*				4
P. herbarum West. (*P. urticae* Schulz. & Sacc.)	*Plurivorous*	1	2		4
P. hoehnelii v. Kest.	Anamorph of *Leptosphaeria doliolum* which see, ubiquitous				
P. (Phyllostictina) hysterella Sacc.	On *Taxus baccata*	1	2	3	
P. jacquiniana Cooke & Massee	On *Thalictrum jacquinianum*	1			
P. lantanoides (Peck) Griff. & Boer.	On *Viburnum tinus*	1			
P. lavandulae Gaboto	On *Lavandula vera*	1			
P. leguminum West	On *Laburnum anagyroides*	1			
P. leveillei Boer. & Bollen	On *Ceanothus, Fragaria, Rhododendron* etc.		2		4
P. libertiana Speg. & Roum.	On *Larix* and other conifers	1			
P. lignicola Rennerfelt	On wood pulp	1			4
P. macrostoma Mont. (*P. pigmentivora* Massee)	Plurivorous	1			4
P. macrostoma var *incolorata* (Horne) Boer. & Dor.		1			
P. mahoniae Thüm.	On leaves of *Mahonia aquifolium*	1			4
P. medicaginis Malbr. & Roum.	On *Medicago sativa*	1			
P. medicaginis var *pinodella* (Jones) Boerema	On *Lathyrus, Pisum, Trifolium*	1		3	
P. mororum Sacc.	On *Morus alba*	1			
P. nebulosa (Pers.) Berk.	*On Scrophularia*	1			4
P. nelumbii Cooke & Massee	On *Nelumbium speciosum*	1			
P. onagracearum Cooke	On *Oenothera biennis*	1			
P. paulowniae Thüm.	On *Paulownia imperialis* needs confirmation	1			
P. phlogis Roum.	On *Phlox paniculata*	1			
P. polemonii Cooke	On *Polemonium caeruleum*	1			
P. pomorum Thüm.	On *Rhododendron* etc.	1			
P. rhodorae Cooke	On *Rhododendron*	1		3	
P. selaginellae Cooke & Massee	On *Selaginella willldenovii*	1			
P. syringae (Fr.) Sacc.	On *Syringa vulgaris*	"Sussex"			4
P. tamaricina Thüm.	On *Tamarix*	1			
P. typharum Sacc.	On *Typha latifolia*	1			
P. valerianellae Boer. & de Jong	On *Valerianella olitoria*		2		
P. verbascicola Cooke	On *Verbascum nigrum*	1			
P. viburni (Roum.) Boer. & Griffin	On *Viburnum tinus*		2		4
P. violacea (Bertel) Eveleigh	Plurivorous	1			4
P. vulgaris Sacc.	On *Clematis vitalba*	1			

Phomopsis (Sacc.) Sacc.
Anamorphs of *Diaporthe*, which see, also:

Species	Host / notes	1	2	3	4
Phomopsis cinerascens (Sacc.) Trav.	On *Ficus carica*	1	2		
P. conorum (Sacc.) Died.	On *Picea excelsa*	1	2		
P. diachenii Sacc. (*Cyphellopycnis pastinacae* Tehon & Stout)	On *Pastinaca sativa*				4
P. iridis (Cooke) Hawks. & Punith.	On *Iris*	1			
P. gardeniae Buddin & Wakef.	On *Gardenia*	1	2		
P. sclerotioides v. Kesteren	On *Cucumis sativa*		2		

Phragmotrichum Kunze

Lignicolous saprophytes, *P. rivoclarinum* (Peyr.) Sutton & Piroz (*Septotrullula bacilligera* v. Höhn v. *cambrica* Grove) on *Alnus* in the north and west.

Phyllosticta Pers.

Saprophytes and facultative parasites, anamorphs of *Guignardia*, *Mycosphaerella* etc. but often loosely employed for species of *Phoma* on leaves; *P. alismatis* Sacc. & Speg. and *P. draconis* Berk. in Essex, *P. forsythae* Sacc., *P. helianthemi* Roum., *P. rhamni* West., *P. tiliae* Sacc. & Speg., *P. tormentillae* Sacc. & *P. ulmariae* Thüm. in Hampshire, *P. spartinae* Brun. in East Anglia, others in the north and west.

Phyllosticta aceris Sacc.	On *Acer campestre*	1	2		4
P. aizoon Cooke	On *Sedum aizoon*	1			
P. ajugae Sacc. & Speg.	On *Ajuga reptans*		2		
P. antirrhini Syd.	On *Antirrhinum majus*	1			
P. apii Halst.	On *Apium graveolens*	"Sussex"			
P. arbuti Sacc.	On *Arbutus unedo*	1			4
P. argyrea Speg.	On *Elaeagnus pungens*	1			4
P. aucubicola Sacc. & Speg.	On *Aucuba japonica*	1			
P. berberidis Rab.	On *Berberis vulgaris*				4
P. berolinensis Henn.	On *Rhododendron ponticum*	1			
P. bolleana Sacc.	On *Euonymus japonicus*	1		3	
P. buxina Sacc.	On *Buxus sempervirens*	1			
P. camelliae West	On *Camellia japonica*			3	
P. cirsii Desm.	On *Cirsium arvense*	1	2		
P. cornicola (DC.) Rab.	On *Cornus alba* & *C. sanguinea*		2		4
P. cunninghami All.	On *Rhododendron*			3	
P. destructiva Desm.	On *Malva sylvestris*	1			4
P. digitalis Bell	On *Digitalis purpurea*	1			4
P. dulcamarae Sacc.	On *Solanum dulcamara*	1			
P. epimedii Sacc.	On *Epimedium alpinum*	1			
P. erysimi West.	On *Alliaria petiolata*	1			
P. fraxinicola Sacc.	On *Fraxinus excelsior*		2		
P. garryae Cooke & Harkn.	On *Garrya elliptica*	1			
P. glechomae Sacc.	On *Glechoma hederacea*	1			
P. grossulariae Sacc.	On *Ribes uva-crispa*				4
P. hedericola Dur. & Mont.	On *Hedera helix*	1	2		
P. humuli Sacc. & Speg.	On *Humulus lupulus*				4
P. hydrophila Speg.	On *Nymphaea alba*	1			
P. ilicicola (Cooke & Ell.) Ell. & Ev.	On *Ilex aquifolium*	1			
P. impatientis Fautr.	On *Impatiens parviflora*	1			
P. japonica Thüm.	On *Mahonia japonica*		2		
P. lappae Sacc.	On *Arctium*	1			
P. lauri West.	On *Laurus nobilis*				4
P. ligustri Sacc. (?*Mycosphaerella ligustri*)	On *Ligustrum vulgare*	1			
P. limbalis Pers.	On *Buxus sempervirens*	1			
P. lonicerae West.	On *Lonicera pericymenum*	1			
P. lutetiana Sacc.	On *Circaea lutetiana*	1			
P. magnoliae Sacc.	On *Magnolia grandiflora*	1			
P. mahoniana All.	On *Mahonia aquifolium*	1	2		
P. masdevalliae Henn.	On *Masdevallia corniculata*	1			
P. medicaginis Sacc.	On *Medicago sativa*	1			
P. mercurialis Desm. dubious teste Grove	On *Mercurialis perennis*				4
P. paulowniae Sacc.	On *Paulownia imperialis*	1			
P. penstemonis Cooke	On *Penstemon grandiflorus*	1			
P. phillyreae Sacc.	On *Phillyrea*	1			
P. plantaginis Sacc.	On *Plantago major*	1			4
P. podophylli Wint.	On *Podophyllum peltatum*	1			
P. populina Sacc.	On *Populus robusta*	"Sussex"			
P. primulicola Desm.	On *Primula vulgaris* and cultivars	1			4

P. rhamnicola Desm.	On *Rhamnus catharticus*		4
P. rhododendri West.	On *Rhododendron*	1	
P. saccardoi Thüm.	On *Rhododendron*	1	
P. sagittifoliae Brun.	On *Sagittaria sagittifolia*	2	
P. sanguinea (Desm.) Sacc.	On *Cotoneaster frigida*	1	
P. sorbicola (Rab.) All.	On *Sorbus aucuparia*	1	
P. straminella Bres.	On *Rumex acetosa*	1	
P. syringae West.	On *Syringa vulgaris*	1	4
P. tinea Sacc.	On *Viburnun tinus*	1	4
P. trollii Trail	On *Trollius*	1	
P. vincaemajoris All.	On *Vinca major*	1	

Piggotia B. & Br.
Anamorph of *Platychora* but also:

Piggotia (Labrella) coryli (Desm.) Sutton	On *Corylus avellana*	4

Pilidium Kunze
Saprophytes.

Pilidium acerinum Kunze (*Leptothyrium acerinum* Kunze) Cda, *L. medium* Cooke)		1 2 3 4
P. concavum (Desm.) v. Höhn. (*L. macrothecium* Fuck., *L. protuberans* Sacc.)		4

Plasia Sherwood
Anamorphs of *Durella* which see.

Plectophomella Moesz
Parasite, *P. concentrica* Redfern & Sutton on *Ulmus* in the north.

Pleurophoma v. Höhn.
Saprophyte, *P. pleurospora* (Sacc.) v. Höhn. on wood of *Salix* in the north.

Polymorphum Chev.
Anamorph of *Ascodichaena* which see.

Prosthemium Kunze
Anamorphs of *Pleomassaria* which see.

Psammina Rouss. & Sacc.
Facultative parasites.

Psammina bommerae Rouss. & Sacc.	On *Ammophila arenaria*		2 4
P. stipitata Hawks.	On *Lecanora conizaeioides*	1	

Pseudocenangium Karst.
Saprophyte, *P. succineum* (Spree) Dyko & Sutton on needles of *Pinus sylvestris* in the north.

Pseudodiplodia (Karst.) Sacc.
Lignicolous saprophytes, *P. corticis* Grove on *Acer* in the midlands.

Pseudolachnea Ranojevic
Saprophytes.

Pseudolachnea hispidula (Schrad.) Sutton	On wood and woody herbaceous stems	1 2 3

Pseudopatellina v. Höhn.
  Saprophyte.

Pseudopatellina conigena v. Höhn.   On cones of *Pinus sylvestris,* also *Ulex*   2

Pseudorobillarda Morelet
  Saprophytes on Gramineae, *P. agrostis* (Sprague) Nag Raj et al. in the west, *P. phragmitis* (Cunnell) Nag Raj et al. in Middlesex.

Pseudoseptoria Speg.
  Facultative parasites of Gramineae

Pseudoseptoria donacis (Pass.) Sutton   On *Agropyron, Bromus, Hordeum, Triticum*   1
                                            etc.
P. stomaticola (Bauml.) Sutton   On *Dactylis, Phleum* etc.   1

Pycnidiella v. Höhn.
  Anamorphs of *Sarea* which see

Pycnofusarium Punith.
  Saprophyte, *P. rusci* Haws. & Punith. On *Ruscus aculeatus* in the west.

Pycnothyrium Diet.
  Facultative parasites, *P. gentianicola* (Baum.) Grove & *P. junci* Grove in the north and west.

Pyrenochaeta de Not.
  Saprophytes, *P. terrestris* (Hansen) Gorenz et al. in the north; see also *Phoma leveillei.*

Pyrenochaeta fallax Bres.   On *Urtica dioica*   1
P. ilicis Wilson   On *Ilex aquifolium*   1
P. lycopersici Schneid. & Gerlach On *Lycopersicon esculentum*   2
P. nobilis de Not.   On *Laurus nobilis*   1
P. phlogis Massee   On *Phlox*

Readeriella Syd. & Syd.
  ?Saprophyte, *R. mirabilis* Syd. on leaves of *Eucalyptus* spp. in the north and west.

Rhabdospora (Sacc.) Sacc.
  See in the main under *Septoria* or *Septocyta* but also:

Rhabdospora acantophila Mass.   On cupules of *Castanea sativa*   4
R. cervariae Syd.   On *Peucedanum officinale*   4
R. lupini Buchwald   On *Lupinus*   4

Rhizosphaera Mangin & Hariot
  Parasites on needles of conifers, *R. kalkoffii* Bub. & *R. pini* (Cda) Maubl. in north and west.

Rhodesia Grove
  Saprophyte, *R. subtecta* (Rob.) Grove on *Ammophila arenaria* in E. Anglia and generally.

Rhodesiopsis Sutton & Campbell
  ?Saprophyte, *R. gelatinosa* Sutton & Campbell on *Phormium tenax* in the west.

Rhynchosporium Heinsen
  Facultative parasites, *R. alismatis* (Oud.) Davis in East Anglia.

Rhynchosporium orthosporum Caldwell	On *Dactylis glomerata* & *Lolium*	1	2
R. secalis (Oud.) Davis	On *Hordeum, Phleum, Secale* etc.	1	2

Roscoepoundia O.K.
  Lignicolous saprophytes (*Naemaspora* Sacc. non Pers.), anamorphs of *Diatrype* etc.

Roscoepoundia croceola (Sacc.) O.K.	On *Acer, Castanea, Tilia*	1

Schizothyrella Thüm.
  Saprophytes.

Schizothyrella quercina (Lib.) Thüm.	On *Quercus*	1

Sclerophoma v. Höhn.
  Facultative parasite.

Sclerophoma pithiophila (Corda) v. Höhn.	On *Pinus sylvestris*	1	4

Sclerophomella v. Höhn. see *Phoma*.

Scolecosporium Lib.
  Saprophyte, probable anamorph of *Asteromassaria* which see

Seimatosporium Corda
  Anamorphs of *Discostroma*, including species of *Coryneopsis* Grove, and *Amphichaeta* McAlp., also:

Coryneopsis tamaricis Grove	On *Tamarix*	4

Seiridium Nees
  Saprophytes or facultative parasites, *S. intermedium* (Sacc.) Sutton on *Ulmus* in the west.

Seiridium cardinale (Wagener) Sutton & Gibson	On *Cupressus macrocarpa*	1	3	4

Selenophoma Maire
  Saprophytes or facultative parasites, *S. asterina* (B. & Br.) Sutton in the west, *S. moravica* Petrak on *Centaurea* in the north, *S. juncea* (Mont.) v. Arx anamorph of *Guignardia cytisi*.

Septocyta Petrak
  Facultative parasite.

Septocyta ruborum (Lib.) Petrak (*Rhabdospora ramealis* (Rob.) Sacc.)	On *Rubus*	3	4

Septogloeum Sacc.
  Facultative parasites.

Septogloeum carthusianum (Sacc.) Sacc.	On *Euonymus europaeus*	1

Septoria Sacc.
  Saprophytes or facultative parasites, anamorphs of *Mycosphaerella* or *Sphaerulina*, also in Berkshire *S. socia* Pass., in Hampshire *S. cerastii* Rob., *S. elaeagni* (Chev.) Desm., *S. holci*

Pass., *S. hyperici* Rab., *S. posoniensis* Bauml., *S. senecionis-silvatici* Syd., and *S. sisymbrii* Niessl, in Hertfordshire *S. helenii* Ell. & Ev., *S. lactucae* Pass. & *S. leontodontis* Sm. & Rams., in E. Anglia *S. arundinacea* Sacc., *S. astericola* Ell. & Ev., *S. brissaceana* Sacc. & Let., *S. caricicola* Sacc., *S. caricis* Pass., *S. dianthi* Desm., *S. ebuli* Desm., *S. fulvescens* Sacc., *S. matricariae* Syd., *S. menyanthes* Desm., *S. scleranthi* Desm., *S. sedi* West., *S. sii* Rab. & *S. tanaceti* Niessl, in the midlands *S. lycopersici* Speg., *S. scillae* West. & *S. villarsiae* Desm., over 40 others reported from the north and west.

Septoria acetosae Oud.	On *Rumex acetosa*	1			
S. anemones Desm.	On *Anemone nemorosa*	1	2		4
S. antirrhini Rob.	On *Antirrhinum majus*	1			4
S. apii Chester (*S. apiigraveolentis* Dorog.)	On *Apium graveolens*	1	2		4
S. aristolochiae Sacc.	On *Aristolochia clematitis*	1			
S. armeriae All.	On *Armeria* cult.	"London"			
S. armoraciae Sacc. (*Ascochyta armoraciae* Fuck.)	On *Armoracia rusticana*				
S. astragali Rob.	On *Astragalus glyciphyllus*				4
S. avellanae B. & Br.	On *Corylus avellana*	1	2		
S. azaleae Vogl.	Usually on imported pot plants. On *Rhododendron* sp.	1		3	4
S. badhami B. & Br.	On *Vitis vinifera*	1			
S. bellidis Desm. & Rob.	On *Bellis perennis*	1	2		4
S. berberidis Niessl	On *Berberis vulgaris*				4
S. betae West.	On *Beta vulgaris* ssp. *maritima*		2		
S. castaneicola Desm.	On *Castanea sativa*				4
S. centaureae Sacc.	On *Centaurea nigra*	1			
S. chelidonii Desm.	On *Chelidonium majus*	1			
S. chenopodii West.	On *Atriplex hastata*	1	2		
S. chrysanthemella Sacc. (?*S. chrsanthemi* All.)	On *Chrysanthemum morifolium*	1	2		4
S. cisti Urries	On *Cistus* sp.	1			
S. clematidis Rob.	On *Clematis vitalba*				4
S. convolvuli Desm.	On *Convolvulus arvensis* &c.		2		
S. cornicola Desm.	On *Cornus sanguinea*	1	2		4
S. cytisi Desm.	On *Laburnum anagyroides*	1			
S. digitalis Pass.	On *Digitalis purpurea*				4
S. divaricata Ell. & Ev. (*S. drummondii* Ell. & Ev.)	On *Phlox* spp.	1			
S. doronici Pass.	On *Doronicum pardalianches*	1			
S. epilobii West.	On *Epilobium montanum*	1			4
S. euonymi Rab.	On *Euonymus japonicus*				4
S. exotica Speg.	On *Hebe* spp.				4
S. ficariae Des.	On *Ranunculus ficaria*	1	2	3	
S. fragariae Desm.	On *Fragaria vesca*	1			4
S. fraxini Desm.	On *Fraxinus excelsior*	1	2		
S. gei Rob.	On *Geum urbanum*	1			4
S. geranii Rob.	On *Geranium lucidum, G. dissectum*	"London"			
S. heterochroa Desm.	On *Malva sylvestris*	1			
S. hippocastani B. & Br.	On *Aesculus hippocastanum*	1			
S. humuli West.	On *Humulus lupulus*				4
S. hydrocotyles Desm.	On *Hydrocotyle vulgaris*	1			
S. lamii Sacc.	On *Lamium album* & *L. purpureum*				4
S. lavandulae Desm.	On *Lavandula vera*				4
S. leucanthemi Sacc. & Speg.	On *Chrysanthemum leucanthemum* & *C. maximum*	1	2		4
S. ligustri Kickx	On *Ligustrum vulgare*	1			
S. lonicerae All.	On *Lonicera*	1			
S. lycopi Pass.	On *Lycopus europaeus*	1			
S. lysimachiae West.	On *Lysimachia nemorum, L. nummularia, L. vulgaris* &c.				4

	Host	1	2	3	4
S. mougeotii Sacc. & Roum.	On *Hieracium*	1			
S. obesa Syd.	On *Chrysanthemum morifolium*	1	2		4
S. oenanthes Ell. & Ev.	On *Oenanthe crocata*		2	3	
S. oenotherae West.	On *Oenothera biennis*		2	3	
S. paeoniae West. var. *berolinensis* All.	On *Paeonia* sp. cult.	1	2		
S. petroselini Desm.	On *Petroselinum sativum*	1	2		
S. polemonii Thüm.	On *Polemonium*	1			
S. polygonorum Desm.	On *Polygonum lapathifolium* & *P. persicaria*				4
S. populi Desm.	On *Populus nigra* & *P.* × *canadensis*	1			4
S. primulae Bucknall	All material is dubious according to Grove, reported in 1 & 4				
S. quercicola Sacc.	On *Quercus pedunculata*				4
S. rubi West.	On *Rubus fruticosus*	1	2	3	4
S. saponariae Savi & Becc.	On *Saponaria officinalis*	1		3	
S. scabiosicola Desm.	On *Knautia arvensis* & *Succisa pratensis*	1			4
S. scutellariae Thüm.	On *Scutellaria galericulata* and *S. minor*		2		
S. selenophomoides Cash & Watson	On *Scaphopetalum*	1			
S. slaptonensis Hawks.	On *Ulex europaeus*	1			
S. sorbi Lasch	On *Sorbus aucuparia*	1			
S. stachydis Rob.	On *Stachys sylvatica*	1			
S. tormentillae Rob.	On *Potentilla erecta*	1			
S. unedonis Rob.	On *Arbutus unedo*	"London"			
S. urticae Rob.	On *Urtica dioica*	1			
S. verbenae Rob.	On *Verbena officinalis*	1			
S. veronicae Rob.	On *Veronica chamaedrys* &c.	1			
S. violae West.	On *Viola* spp.	1	2		
S. virgaureae Desm.	On *Solidago virgaurea*				4
S. wistariae Brum.	On *Wistaria sinensis*	1			

Septoriella Oud.
Saprophytes, *S. junci* (Desm.) Sutton & *S. phragmitis* Oud. in East Anglia.

Sirexcipula Bub.
Saprophytes.

Sirexcipula kabatiana Bub.	On *Allium* & *Funkia*	1

Sirococcus Preuss
Saprophyte.

Sirococcus (*Discella*) *strobilinus* Preuss (*Phoma conigena* Karst.)	On *Picea,Pinus* &c.	1 2 3

Sirodothis Clem.
Anamorphs of *Tympanis*, which see.

Sirothyriella v. Höhn.
Anamorphs of *Stomiopeltis*, which see.

Sirozythiella v. Höhn.
Saprophyte.

Sirozythiella sydowiana (Sacc.) v. Höhn.	On *Phragmites*	1

Sphaceloma de Bary
Anamorphs of *Elsinoe*, which see

Sphaerellopsis Cooke
Hyperparasites = *Darluca* Cast., reputed anamorph of *Eudarluca*.

Sphaerellopsis filum (Biv.Bern.) Sutton   Ubiquitous on uredosori of Uredinales   1   2   3   4

Sphaeropsis Sacc.
Facultative parasites.

Sphaeropsis sapinea (Fr.) Dyko & Sutton   On conifers   1   2   4
(*Diplodia pinea* (Desm.) Kickx)

Sphaerothyrium Bub.
Saprophytes.

Sphaerothyrium filicinum Bub.   On *Pteridium aquilinum*   1

Sporonema Desm.
Anamorphs of *Leptotrochila*, which see.

Stagonospora
Predominantly saprophytes on monocotyledons, a few facultative parasites, *S. anglica* Cunn., *S. cylindrica* Cunn., *S. macropycnidia* Cunn. & *S. paludosa* (Sacc. & Speg.) Sacc. in Middlesex, *S. bromi* Sm. & Rams. in Hampshire, *S. arenaria* Sacc., *S. gigaspora* (Niessl) Sacc., *S. heliocharidis* Trail, *S. spartinicola* Sprague & *S. trimera* Sacc. in E. Anglia, *S. innumerosa* (Desm.) Sacc. & *S. schoeni* Keissl. in Dorset, *S. suaedae* Syd. in the E. Midlands, several others in the north and west. *S. compta* (Sacc.) Died. on *Trifolium repens* recorded in Berks.

Stagonospora aquatica Sacc. var. *luzulicola* Sacc. & Scalia	On *Luzula*				4
S. atriplicis (West.) Lind.	On *Atriplex* and *Chenopodium*	"London"			
S. calystegiae (West.) Grove	On *Calystegia sepium, C. soldanella* and *Convolvulus*	2			
S. caricinella Brun	On *Carex acutiformis, C. arenaria, C. atrata, C. nigra*			3	
S. caricis Sacc.	On many species of *Carex, Cladium mariscus* &c.	1	2		
S. curtisii (Berk.) Sacc.	On *Amaryllis, Narcissus*			3	4
S. elegans (Berk.) Sacc. & Trav.	On *Phragmites*	1			
S. hysterioides Sacc.	On *Phragmites*				4
S. maritima Syd.	On *Scirpus maritimus*				4
S. subseriata (Desm.) Sacc.	On *Deschampsia, Festuca, Molinia* &c.	1			
S. typhoidearum (Desm.) Sacc.	On *Typha*	1			
S. vitensis Unamuno	On *Carex disticha, C. flacca, C. hirta, Juncus* &c.	1			

Stegonsporium Corda
Lignicolous saprophytes.

Stegonsporium pyriformis (Hoffm.) Corda   Especially on *Acer*   1   3   4

Stilbospora Pers.
Lignicolous saprophytes.

Stilbospora cistina (Cooke) Sutton   On *Cistus laurifolius* &c.   1
S. meridionalis (Sacc.) Dyko & Sutton   On *Tamarix gallica*   2

Strasseria Bres. & Sacc.
Saprophytes = *Cytotriplospora* Bayless-Elliott & Chance.

Strasseria geniculata (B. & Br.) v. Höhn.   On *Abies, Lycopersicon, Malus, Picea, Pinus*   1

Thyriostroma Died.
  Saprophyte.

Thyriostroma spireae Died.   On *Filipendula ulmaria*   3

Tiarospora Sacc. & March.
  Saprophyte.

Tiarospora perforans (Rob.) v. Höhn.   On *Ammophila arenaria*   2

Tiarosporella v. Höhn.
  Saprophyte, *T. schizochlamys* (Ferd. & Winge) v. Höhn. on *Trichophoron caespitosum* in E.
  Anglia.

Trullula Ces.
  Saprophytes.

Trullula melanochlora (Desm.) v. Höhn.   On *Abies, Pteridium*   1
  (*T. olivascens* (Sacc.) Sacc.)

Truncatella Stey.
  Saprophytes, segregated from *Pestalotia*.

Truncatella angustata (Pers.) Hughes	On dead herbaceous material & twigs	1	4
T. hartigii (Tub.) Stey.	On *Picea, Thuja plicata*	1	4
T. laurocerasi (West.) Stey.			4

Vouauxiella Petrak & Syd.
  Hyperparasites of lichens, *V. lichenicola* (Linds.) Petr. & Syd. on *Lecanora* apothecia in
  the midlands, *V. uniseptata* Hawks. on *Parmelia* in the north, *V. verrucosa* (Vouaux) Petr. &
  Syd. on *Lecanora* in Hampshire.

Vouauxiomyces Dyko & Hawks.
  Anamorphs of *Abrothallus*

Wojnowicia Sacc.
  Saprophyte.

Wojnowicia hirta Sacc.   On *Triticum*   4

HYPHALES

Acremoniella Sacc.
  Saprophytes.

Acremoniella atra (Corda) Sacc.   Plurivorous   1   3   4

Acremonium Link
  Predominantly saprophytes = *Cephalosphorium* Corda spp., *A. inflatum* (Dickinson)
  Gams in the east midlands, see also *Aphanocladium*.

Acremonium alternatum Link	1	4
A. balanoides (Drechsl.) Subramanian	1	

A. fimicolum Massee & Salmon	1	
A. furcatum Moreau	1	
A. kiliense Grutz	1	
A. strictum Gams		2

Acroconidiella Lindquist & Alippi
Facultative foliar parasite.

Acroconidiella tropaeoli (Bond) Lindq. & Alippi	On *Tropaeolum*	1

Acrodictyopsis Kirk
Saprophyte, *A. lauri* Kirk on *Laurus nobilis* in Essex.

Acrodontium de Hoog
Saprophytes, including the anamorph of *Ascocorticium anomalum*.

Acrodontium echinulatum Kirk	On *Ulex europaeus*	1	
A. hydnicola (Peck) de Hoog	On *Aesculus, Crataegus, Rhododendron*	1	4

Acrogenospora Ellis
Lignicolous saprophytes, anamorphs of *Farlowiella*.

Acrogenospora sphaerocephala (B. & Br.) Ellis	1	2

Acrophialophora Edwards
Saprophyte, *A. fusispora* (Saksena) Ellis in the midlands.

Acrospeira B. & Br.
Falcultative parasite, see also *Monodictys*.

Acrospeira mirabilis B. & Br.	On fruit of *Castanea sativa*, especially imports	1

Acrostalagmus Corda
= *Verticillum* which see, but apparently not transferred:

Acrostalagmus zeosporus Dreschler	1

Actinocladium Ehrenb.
Saprophyte, plurivorous on bark and wood of angiosperms.

Actinocladium rhodosporum Ehrenb. (*Triposporium cambrense* Hughes)	4

Actinospora Ingold
Aquatic saprophytes, including *Miladina* anamorph.

Aegerita Fr.
Anamorph of *Bulbillomyces*.

Akanthomyces Leb.
Entomogenous parasite.

Akanthomyces aculeata Leb.	On Lepidoptera	4

Alatospora Ingold
Aquatic saprophytes, others in the north and west.

Alatospora acuminata Ingold ... 2 ... 4

Alternaria Nees

Saprophytes or facultative parasites, *A. godetiae* Neerg. in Essex, *A. linicola* Groves & Skolko & *A. maritima* Sutherl. in Hampshire, *A. dauci* (Kuhn) Groves & Skolko, *A. dennisii* Ell., *A. petroselini* (Neerg.) Simmons, *A. porri* (Ellis) Cif. & *A. ramulosa* (Sacc.) Joly in E. Anglia, also anamorphs of *Pleospora* spp.

Alternaria alternata (Fr.) Keissler (*A. tenuis* Nees)	Plurivorous	1	2	3	4
A. anagallidis Raabe	On *Anagallis arvensis*	1			
A. brassicae (Berk.) Sacc.	On *Brassica* and *Raphanus*	1			4
A. brassicicola (Schw.) Wiltshire	On *Brassica* and *Raphanus*	1			4
A. cheiranthi (Lib.) Bolle	On *Cheiranthus*	1		3	4
A. cinerariae Hori & Enjoji (*A. senecionis* Neerg.)	On *Senecio cruentus*	"Sussex"			4
A. dianthi Stevens & Hall	On *Dianthus* spp.	1 "Sussex"			
A. radicina Meier, Dreschler & Eddy	Black rot of *Daucus carota*	1			4
A. ramulosa (Sacc.) Joly	On *Anthriscus* & *Petroselinum*	London			
A. raphani Groves & Skolko (*A. matthiolae* Neerg.)	On *Matthiola* & *Raphanus*				4
A. solani Sorauer	On *Lycopersicon esculentum* & *Solanum tuberosum*		2		4
A. sonchi J.J. Davis	On *Sonchus* spp.	1			4
A. tenuissima (Kunze) Wiltshire	Plurivorous	1			4
A. zinniae Pape	On *Zinnia*	1			

Alysidium Kunze

Lignicolous saprophytes.

Alysidium dubium (Pers.) Ellis	1	2		4
A. resinae (Fr.) Ellis (*Torula ramosa* Fuck.)	1	2		

Amallospora Penzig

Saprophyte, *A. dacrydion* Penzig reported in the north.

Amblyosporium Fres.

Hyperparasite.

Amblyosporium spongiosum (Pers.) Hughes (*A. botrytis* Fres.)	On *Lactarius*	1	2	3

Ampulliferina Sutton

Saprophyte, *A. lauri* Kirk in the north.

Anavirga Sutton

Saprophyte.

Anavirga laxa Sutton	On *Fagus* litter, cupules etc.	1	3

Anguillospora Ingold

Aquatic saprophytes, several others in the north and west.

Anguillospora crassa Ingold		4
A. longissima (Sacc. & Syd.) Ingold	2	4

Anungitea Sutton

Saprophytes.

Anungitea continua Matsushima	On *Pinus sylvestris* needles in litter	1	
A. fragilis Sutton	On *Fagus* leaf litter and *Laurus nobilis*	1 2	4
A. heterospora Kirk	On stems of *Rosa* & *Rubus*	1	
A. rhabdospora Kirk	On *Fagus* leaf litter	1	

Aphanocladium Gams
Hyperparasite on myxomycete sporangia.

Aphanocladium album (Preuss) Gams (*Acremonium album* Preuss)　　　　　1 2

Arachnophora Hennebert
Saprophyte.

Arachnophora fagicola Hennebert　　　On *Fagus* cupules in litter　　　　1

Arthrinium Kunze
Saprophytes, in part anamorphs of *Apiospora* and *Pseudoguignardia*, *A. caricicola* Kunze in East Anglia.

Arthrinium curvatum Kunze v. *minor* Ellis	On *Carex*	1	
A. phaeospermum (Corda) Ellis (*Papularia*	On Gramineae	1 2 3	4
sphaerosperma (Pers.) v. Höhn.)			
A. puccinioidea (DC.) Kunze	On Carices	1	
A. sporophlaeum Kunze	On Carices	3	

Arthrobotrys Corda
Saprophytes and parasites of nematodes in soil and litter, see also *Dactylariopsis*.
Arthrobotrys cladodes Dreschler	1
A. cladodes var. *macroides* Dreschler	1
A. conoides Dreschler	1
A. dactyloides Dreschler	1
A. oligospora Fres.	1
A. robusta Duddington	1
A. scaphoides (Peach) Schenck et al.	1
A. superba Corda	1

Arthrobotryum Ces.
Lignicolous saprophyte, *A. stilboideum* Ces. On *Quercus* in the north and west.

Arthrographis Cochet
Saprophytes.

Arthrographis sulphurea (Grev.) Stalpers & v. Oorschot (*A. kalrae* (Tewari & Macpherson) Sigler & Carmichael)　　　　　1

Arthrosporium Sacc.
Saprophyte.

Arthrosporium elatum Massee　　　On grass　　　　1

Articulospora Ingold
Aquatic saprophytes, others in the north and west.

Articulospora tetracladia Ingold　　　　　2

Aspergillus Mich.
Saprophytes, anamorphs of *Emericella* and *Eurotium*.

Aspergillus candidus Link　　　　　"London"

229

	1	2	3	4
A. *carneus* (v. Tiegh.) Blochwitz	1			
A. *clavatus* Desm.	1			
A. *flavus* Link				4
A. *niger* v. Tiegh.	1			4
A. *penicilloides* Speg.				4
A. *sejunctus* Bain. & Sart. (A. *ruber* (Spieckermann & Bremer) Thom & Church)	1			
A. *terreus* Thom	?1			
A. *thomii* Smith	"London"			
A. *versicolor* (Vuill.) Tiraboschi	"London"			

Asteromyces Moreau
Saprophyte.

		1	2	3	4
Asteromyces cruciatus Moreau	On *Ammophila arenaria*				4

Aureobasidium Viala & Boyer
Saprophyte, often a constituent of "Sooty moulds".

	1	2	3	4
Aureobasidium pullulans (de Bary) Arnaud (*Pullularia pullulans* (deBy.) Berkhout)	1	2	3	4

Bactridium Kunze
Saprophyte.

		1	2	3	4
Bactridium flavum Kunze	On decaying decorticated wood	1	2		

Bactrodesmiella Ellis
Saprophyte.

		1	2	3	4
Bactrodesmiella masonii (Hughes) Ellis	On *Fagus* cupules in litter	1			

Bactrodesmium Cooke
Lignicolous saprophytes, *B. betulicola* Ellis & *B. longisporum* Ellis in East Anglia.

	1	2	3	4
Bactrodesmium abruptum (B. & Br.) Mason & Hughes	1	2		
B. *arnaudii* Hughes	1			
B. *atrum* Ellis	1			
B. *betulicola* Ellis	1			
B. *esheri* Kirk	1			
B. *ovatum* (Oud.) Ellis	1	2		4
B. *pallidum* Ellis	1			
B. *spilomeum* (B. & Br.) Mason & Hughes	1			4
B. *submoniliforme* Holubova-Jechova	1			

Balanium Wallr.
Lignicolous saprophyte.

		1	2	3	4
Balanium stygium Wallr.	On *Sambucus nigra*	1			

Beauveria Vuill.
Insectivorous parasites and animal mycoses.

		1	2	3	4
Beauveria alba (Limber) Saccas	On a cat	1			
B. *bassiana* (Bals.-Criv.) Vuill.		1			
B.*brongniartii* (Sacc.) Petch		1			
B. *densa* (Link) Picard		1	2		

Belemnospora Kirk
Foliicolous saprophytes.

Belemnospora epiphylla Kirk	On *Rhododendron* etc.	1	
B. pinicola Kirk	On *Pinus sylvestris*	1	

Beltrania Penzig
Foliicolous saprophytes.

Beltrania querna Harkn.	On *Quercus ilex* & *Laurus nobilis*	1	4
B. rhombica Penzig	On *Quercus ilex*		4

Beltraniella Subramanian
Foliicolous saprophytes.

Beltraniella pirozynskii Kirk	On *Laurus nobilis*	4

Beverwijkella Tubaki
Saprophyte, *B. pulmonaria* (Bev.) Tubaki in the west.

Bispora Corda
Predominantly lignicolous saprophytes but *B. christiansenii* Hawks. on *Caloplaca* in Essex.

Bispora antennata (Pers.) Mason (*B. monilioides* Cda.)	On *Corylus, Fagus, Quercus, Ulmus* etc.	1	2	3	4
B. betulina (Corda) Hughes	On *Fagus, Ilex, Populus, Quercus* etc.	1	2		

Blastomyces Costantin & Rolland
Pathogens.

Blastomyces luteus Costantin & Rolland	3

Blistum Sutton
Hyperparasites of myxomycetes, anamorphs of *Byssostilbe*, also:

Blistum ovalisporum (Smith) Sutton	1	4

Bloxamia B. & Br.
Lignicolous saprophyte, *B. sanctae-insulae* Minter & Coppins in the north.

Bloxamia truncata B. & Br.	On *Ulmus*	4

Bostrychonema Ces.
Foliicolous parasites, *B. alpestre* Ces. On *Polygonum viviparum* in the north.

Botryophialophora Linder
Lignicolous saprophyte, *B. marina* Linder on timber in the sea.

Botryosporium Corda
Saprophytes, *B. madrasense* Rajhukumar on marsh plants in East Anglia.

Botryosporium longibrachiatum (Oud.) Maire	On rotting vegetation, especially *Lycopersicon*	1	2
B. pulchrum Corda	On *Dahlia, Pteridium, Senecio, Urtica* etc.	1	4

Botryotrichum Sacc. & Marchal
Saprophytes, anamorphs of *Chaetomium*.

Botryotrichum piluliferum Sacc. & Marchal	On dung	1	4

Botrytis Mich.
Facultative parasites, anamorphs of *Botryotinia* which see, also unassigned:

Botrytis anthophila Bondarzew	On anthers of *Trifolium pratense*				4
B. allii Munn	Neck rot of *Allium cepa*	1			4
B. elliptica (Berk.) Cooke	On *Lilium candidum* etc.	1			4
B. croci Cooke	On *Crocus*	1			
B. fabae Sardina	Chocolate spot of *Vicia faba*				4
B. galanthina (B. & Br.) Sacc.	On *Galanthus* spp.	1			
B. gladiolorum Timmerm.	Leaf spot & corm rot of *Gladiolus* spp.				4
B. paeonia Oud.	On *Paeonia* cult.	1		3	4
B. tulipae Lindl	"Fire" of *Tulipa*	1	2		
B. sp. undescribed	On *Endymion nonscriptum*	1			

Brachydesmiella Arnaud
Lignicolous saprophyte.

Brachydesmiella biseptata Arnaud	On *Fagus* & *Fraxinus*	1

Brachysporiella Batista
Saprophyte, *B. laxa* (Hudson) Ellis on *Fagus* leaves in East Anglia.

Brachysporium Sacc.
Lignicolous saprophytes.

Brachysporium bloxami (Cooke) Sacc.	1	2	4
B. britannicum Hughes	1	2	4
B. dingleyae Hughes	1		
B. masonii Hughes	1		4
B. nigrum (Link) Hughes (*B. apicale* (B. & Br.) Sacc.)	1	2	4
B. obovatum (Berk.) Sacc.	1	2	4

Cacumisporium Preuss, see *Pleurothecium*.

Calcarisporium Preuss
Saprophytes or ?hyperparasites, *C. thermophile* Evans in soil in the midlands.

Calcarisporium arbuscula Preuss	In or on *Lactarius, Mycena, Russula*	1	4

Campsosporium Harkness
Lignicolous saprophytes.

Campsosporium antennatum Harkness	On *Laurus nobilis*	1
C. cambrense Hughes	On *Fagus* cupules, *Ilex, Laurus, Quercus* etc.	1
C. hyalinum Abdullah	On sodden *Fagus* cupules	1
C. pellucidum (Grove) Hughes	On *Fagus* cupules, *Quercus ilex, Laurus* & dead stems	1 2 4

Campylospora Ranzoni
Aquatic saprophyte.

Campylospora parvula Kuzuha		4

Candelabrella Rifai & Cooke
Predacious on microfauna, *C. musiformis* (Dreschler) Rifai & Cooke in the midlands.

Candelabrum v. Beverwijk
Saprophytes, *C. brocchiatum* Tubaki in the west.

Candelabrum spinulosum v. Bev.	Lignicolous and in leaf litter	1	2	4

Candida Berkhout
Yeasts, saprophytic or facultative parasites of man and other mammals.

Candida albicans (Robin) Berkhout	1
C. brumptii Lang. & Guerra	1
C. curvata (Diddens & Lodder) Lodder & Kreger v. Rij	1
C. guilliermondii (Cast.) Langeron & Guerra	1
C. intermedia (Cif. & Ashf.) Langeron & Guerra	1
C. krusei (Cast.) Berkhout	1
C. mycoderma (Reess) Lodder & Kreger v. Rij	1
C. parapsilosis (Ashf.) Langeron & Talice	1
C. pelliculosa Redselli	1
C. pulcherrima (Lindner) Windisch	1
C. rugosa (Anderson) Diddens & Dodder	1
C. solani Lodder & v. Rij.	1
C. tropicalis (Cast.) Berkhout	1
C. zeylanoides (Cast.) Langeron & Guerra	1

Listed for Surrey as having been reported from the Weybridge laboratory, with ultimate sources not disclosed.

Capnobotrys Hughes
Saprophytes, anamorphs of *Metacapnodium*, which see.

Casaresia Fragoso
Saphrophyte = *Ankistrocladium*, *C. sphagnorum* Fragoso in Berkshire.

Catenularia Grove
Saprophytes, anamorphs of *Chaetosphaeria*, which see.

Cenococcum Fr.
Terrestrial saprophyte.

Cenococcum geophilum Fr.	1	3

Cephalosporiopsis Peyronel
Saprophytes, *C. alpina* Peyronel in soil in the north.

Cephalosporium Corda, see *Acremonium* and *Verticillium*

Cephalotrichum Link, see *Doratomyces*

Ceratocladium Corda
Lignicolous saprophyte.

Ceratocladium microspermum Corda	On *Fagus*	1

Ceratosporella v. Höhn.
Lignicolous saprophyte.

Ceratosporella stipitata (Goidanich) Hughes	On *Carpinus* & *Castanea*	1

Ceratosporium Schwein.
Lignicolous saprophytes, *C. rilstonii* Hughes in the west.

Ceratosporium fuscescens Schw.	1	3	4

Cercospora Fres.
Facultative foliar parasites, *C. exitiosa* Syd. in Middlesex, *C. physospermi* Deighton in Buckinghamshire, *C. armoraciae* Sacc., *C. ferruginea* Fuck., *C. plantaginis* Sacc., *C. resedae* Fuck., *C. rhamni* Fuck., *C. violae* Sacc., *C. zebrina* Pass. in E. Anglia, *C. nasturtii* Pass. in the west midlands, a few others in the north and west. See also *Mycovellosiella* & *Pseudo-cercospora*.

Cercospora beticola Sacc.	On *Beta*			4
C. chenopodii Fres.	On *Atriplex* & *Chenopodium*		2	
C. crataegi Sacc. & Massal.	On *Crataegus*	1		
C. depazeoides (Desm.) Sacc.	On *Sambucus nigra*	1		4
C. fabae Fautr. (*C. zonata* Wint.)	On *Vicia faba*			4
C. handelii Bub.	On *Rhododendron*	1		4
C. loti Hollos	On *Lotus*			4
C. mercurialis Pass.	On *Mercurialis perennis*	1	2	4
C. myrticola Speg. (*C. myrti* Erikss.)	On *Myrtus communis*		2	
C. nigellae Hollos	On *Nigella*			4
C. odontoglossi Prill. & Del.	On *Odontoglossum crispum* etc.	1	3	

Cercosporella Sacc.
Facultative parasites, see also *Pseudocercosporella* and *Ramularia*

Cercosporella antirrhini Wakefield	On *Antirrhinum majus*	1	4
C. oxalidis Grove	On *Oxalis*	1	
C. pastinacae Karst.	On *Pastinaca*	1	
C. primulae All.	On *Primula*	2	

Cercosporidium Earle
Facultative foliar parasites.

Cercosporidium depressum (B. & Br.) Deighton	On *Angelica sylvestris*	1
C. graminis (Fuck.) Deighton	On *Agrostis, Alopecurus, Arrhenatherum, Glyceria, Phleum, Poa* etc.	1

Cercosporula Arnaud
Saprophytes, *C. corticola* Arnaud in the west midlands.

Cercosporula crassiuscula Arnaud	On *Rubus fruticosus*	1

Chaetendophragmia Matsushima
Saprophyte, *C. britannica* Kirk on *Arundinaria* in the west.

Chaetochalara Sutton & Pirozynski
Saprophytes, *C. bulbosa* Sutton & Piroz. On *Ilex* leaves in the west.

C. cladii Sutton & Piroz.	On *Juncus* (& *Cladium*)	1

Chaetopsina Rambelli
Saprophyte.

Chaetopsina fulva Rambelli	On *Buxus*	1

Chaetopsis Grev.
Lignicolous saprophyte.

Chaetopsis grisea (Ehrenb.) Sacc.	On *Fagus, Fraxinus, Tilia, Ulmus*	1	2

Chaetostroma Corda
A *nomen confusum,* two dubious old records rejected.

Chalara (Corda) Rab.
Saprophytes, *C. insignis* (S., R. & B.) Hughes, *C. kendrickii* NagRaj & *C. rhynchophialis* NagRaj & Kendrick in the north and west, *C. cladii* Ellis in E. Anglia.

Chalara affinis Sacc. & Berl.	Especially on cupules & leaf litter	1	
C. aotearoae NagRaj & Hughes	On *Rubus*	1	
C. aurea (Corda) Hughes	Lignicolous, on *Betula, Quercus* etc.	1	2
C. cylindrica Karst.	On *Castanea, Picea*	1	
C. cylindrosperma (Corda) Hughes	On *Fagus* cupules & leaf litter	1	
C. fungorum Sacc.	On *Fagus* cupules, *Pteridium,* *Moellerodiscus* apothecia etc.	1	
C. fusidioides (Corda) Rab.	On *Pinus* and *Laurus nobilis*	1	
C. heterospora Sacc.	Lignicolous	1	
C. hughesii NagRaj & Kendrick	On litter under *Laurus, Quercus ilex, Salix*	1	
C. inflatipes (Preuss) Sacc.	Corticolous on *Larix, Pinus, Quercus*	1	
C. longipes (Preuss) Cooke	On *Pinus*		2
C. microspora (Corda) Hughes	On *Pinus*		2
C. ovoidea NagRaj & Kendrick	On *Fagus* & *Pinus*	1	
C. parvispora NagRaj & Hughes	On *Pteridium aquilinum*	1	
C. pteridina Syd.	On *Pteridium aquilinum*	1	
C. spiralis NagRaj & Kendrick	On *Fagus*	1	
C. urceolata NagRaj & Kendrick	On *Eupatorium, Heracleum, Rumex* stems	1	

Chalaropsis Peyronel
Facultative parasite.

Chalaropsis thielavioides Peyronel	On *Daucus carota, Juglans regia* etc.	1	4

Cheiromycella v. Höhn.
Saprophyte, anamorph of *Hyaloscypha* (*C. microscopica* (Karst.) Hughes = *Torula gyrosa* C. & M.).

Chlamydomyces Bainier
Saprophyte.

Chlamydomyces palmarum (Cooke) Mason (*C. diffusum* Bainier)	1	"Sussex"

Chloridium Link
Saprophytes, anamorphs of *Chaetosphaeria, C. caudigenum* (v. Höhn.) Hughes in E. Anglia.

Chloridium botryoideum (Corda) Hughes	On *Fagus* bark	1	
C. (*Bisporomyces*) *chlamydosporis* (Van Beyma) Hughes	On *Fagus*		2
C. lignicola (Mangenot) Gams & Hol.-Jech.	On *Betula, Quercus*	1	
C. preussii Gams & Hol.-Jech.	On *Castanea, Quercus, Rubus*	1	
C. virescens (Pers.) Gams & Hol.-Jech. v. *virescens*	On *Quercus, Tilia*	1	

Chromelosporium Corda
Saprophytes, anamorphs of *Peziza* and *Plicaria*.

Chromelosporium tuberculatum (Pers.) Hennebert (*Hyphelia terrestria* Fr.)   1

Chrysonilia v. Arx
Anamorphs of *Neurospora*.

Chrysosporium Corda
Saprophytes, including anamorphs of Gymnoascales, Sordariales.

Chrysosporium luteum (Cost.) Carm.	On soil, straw	2	4
(*Myceliopthora lutea* Cost.)			
C. pannorum (Link) Hughes (*Sporotrichum carnis* Brooks & Hansford)		1	4

Circinotrichum Nees
Saprophytes.

Circinotrichum britannicum Kirk	On *Laurus nobilis* leaves	2

Cirrenalia Meyers & Moore
Lignicolous saprophytes.

Cirrenalia lignicola Kirk	On *Fagus* & *Quercus*	1

Cladobotryum Nees
Anamorphs of *Hypomyces*.

Cladorhinum Sacc. & Marchal
Saprophytes, anamorphs of *Apiosordaria*, which see.

Cladosporium Link
Saprophytes, anamorphs of *Mycosphaerella* & *Venturia*, see also *Heterosporium*. In East Anglia *C. aecidiicola* Thüm., *C. orchidis* Ellis & Ellis, *C. phlei* (Gregory) de Vries, *C. uredinicola* Speg.

Cladosporium britannicum Ellis		1	
C. chlorocephalum (Fres.) Mason & Ellis	On *Paeonia*	1	
C. cladosporioides (Fres.) de Vries (*Hormodendrum cladosporioides* (Fres.) Sacc.)		1	3
C. cucumerinum Ellis & Archer	On *Cucumis sativa*	"Sussex"	4
C. macrocarpum Preuss		1	
C. sphaerospermum Penzig		2	4

Clasterosporium Schwein
Facultative parasite.

Clasterosporium caricinum Schwein.	On *Carex riparia*	1

Clathrosphaerina van Beverwijk
Saprophyte.

Clathrosphaerina zalewskii van Beverwijk	4

Clavariopsis de Wild.
Aquatic saprophytes, *C. tenuis* de Wild. in the west.

Clavariopsis aquatica de Wild	2

Clavatospora Nilsson
Aquatic saprophytes, *C. flagellata* Gönczöl in the west.

Clavatospora longibrachiata (Ingold) Nilsson	1
C. stellata (Ingold & Cox) Nilsson	4

Clonostachys Corda
Saprophytes, *C. dichotoma* Bayliss Elliott is perhaps *Acrodontium hydnicola* teste de Hoog.

Clonostachys compactiuscula (Sacc.) Hawks	On fallen twigs, *Fagus, Salix* etc.	4
C. simmonsii Massee	On caterpillar frass	1

Codinaea Maire
Saprophytes, *C. setosa* Hughes & Kendrick on *Ilex* in East Anglia.

Codinaea britannica Ellis	1	
C. fertilis Hughes & Kendrick	1	2
C. simples Hughes & Kendrick	1	2

Conioscypha v. Höhn
Lignicolous saprophytes, *C. hoehnelii* Kirk.

Conioscyphe hoehnelii Kirk	1

Coniosporium Link
Lignicolous saprophytes, anamorphs of *Hysterium*, *C. olivaceum* Link in the west.

Coniosporium ilicinum Kirk	1

Coniothecium Corda
Nomen dubium, see *Trimmatostroma*.

Conoplea Pers.
Saprophytes, *C. fusca* Pers. in the west.

Cordana Preuss
Saprophytes, *C. boothii* Ellis in the west, *C. crassa* Tóth in East Anglia.

Cordana pauciseptata Preuss	On *Castanea, Fagus, Quercus*	1	2	3

Cordella Speg.
Saprophytes, *C. clarkii* Ellis on *Carex* in the west midlands.

Coremiella Bub. & Krieger
Saprophyte.

Coremiella cubispora (Berk. & Curt) Ellis	On *Chamaenerion, Eleocharis, Filipendula, Lapsana, Lythrum, Ranunculus, Rosa, Rubus* etc.	1

Corynespora Güssow
Saprophytes, *C. olivacea* (Wallr.) Ellis on *Tilia* in E. Anglia, *C. cambrensis* Ellis, *C. cespitosa* (Ell. & Barth.) Ellis, *C. proliferata* Loerakker in the north and west.

Corynespora biseptata Ellis	On wood	2	
C. cassiicola (Berk. & Curt.) Wei (*C. melonis* (Cooke) Lind)		1	"Sussex"
C. foveolata (Pat.) Hughes	On *Arundinaria*	1	
C. pruni (Berk. & Curt.) Ellis	On *Fagus, Symphoricarpos, Taxus*	1	
C. smithii (B. & Br.) Ellis	On *Carpinus, Fagus, Ilex*	1	4

Corynesporopsis Kirk
 Saprophytes.

Corynesporopsis quercicola (Borowska) Kirk	On *Populus* etc.	1
C. uniseptata Kirk	On *Laurus nobilis*	1

Costantinella Matruchot
 Saprophytes on soil and vegetable debris, *C. micheneri* (Berk. & Curt.) Hughes in Buckingham.

Costantinella terrestris (Link) Hughes (*C. tillettii* (Desm.) Mason & Hughes)   1 2

Cristulariella v. Höhn.
 Facultative parasite.

Cristulariella depraedens (Cooke) v. Höhn	On *Acer pseudoplatanus*	1 2 3 4

Cryptococcus Kützing
 Animal mycoses, listed under Surrey because reported from the Weybridge laboratory.

Cryptococcus albidus (Saito) Skinner	1
C. diffluens (Zach) Lodder & van Rij	1
C. laurentii (Kuff.) Skinner	1
C. neoformans (Sanf.) Vuill.	1

Cryptocoryneum Fuck.
 Lignicolous saprophytes.

Cryptocoryneum condensatum (Wallr.) Mason & Hughes	On angiosperm bark and wood	1 2 3 4
C. rilstonii Ellis	On *Fraxinus* etc.	1

Cryptostroma Gregory & Waller
 Facultative parasite.

Cryptostroma corticale (Ell. & Ev.) Gregory & Waller	On *Acer pseudoplatanus* since 1945	1

Culcitalna Meyers & Moore
 Marine saprophyte, *C. achraspora* Meyers & Moore on timber in the sea.

Culicidospora Petersen
 Aquatic saprophytes, *C. aquatica* Petersen & *C. gravida* Petersen in the west.

Curvularia Boedijn
 Saprophytes or weak facultative parasites, mainly in the tropics, *C. crepini* (West.) Boedijn on *Ophioglossum* in E. Anglia, *C. trifolii* (Kauffm.) Boedijn f.sp. *gladioli* Parm. & Luttrell in Hampshire, *C. lunata* (Wakk.) Boedijn v. *aeria* Ellis & *C. protuberata* Nelson & Hodges in the north.

Cylindrium Bon.
 Saprophytes, especially on leaf litter in woods.

Cylindrium flavovirens (Ditmar) Bon. (*Fusidum aeruginosum* Link)		1
C. griseum (Link) Bon.	On *Carpinus, Fagus, Quercus, Ulmus* etc.	1

Cylindrocarpon Woll.
Anamorphs of *Nectria* which see, also:

Cylindrocarpon lucidum	On *Acer* and *Malus*	1
C. olidum Woll. v. *crassum*	In soil and Cactus roots	4

Cylindrocladium Morgan
Saprophytes, anamorphs of *Calonectria*, which see.

Cylindrocladium parvum Anderson	On *Pinus*	4

Cylindrocolla Bon.
Anamorph of *Calloria*, which see

Cylindrodendrum Bon.
Saprophyte or entomogenous parasites, *C. suffultum* Petch on dipterous pupae in E. Anglia.

Cylindrodendrum album Bon.	On *Aesculus, Alnus, Betula, Craetaegus, Fagus, Tilia*	1

Cylindrotrichum Bon.
Lignicolous saprophytes, *C. clavatum* Gams in Hampshire, *C. ellisii* Morgan Jones in E. Anglia, *C. zignoellae* (v. Höhn.) Gams & Hol.-Jech. on *Filipendula* in the west.

Cylindrotrichum oligospermum (Corda) Bon.	On *Angelica, Eupatorium, Fagus, Festuca, Laurus nobilis, Filipendula, Gunnera* etc.	1	2

Cystodendron Bub.
Endophyte, *C. dryophilum* (Pass.) Bub. in *Erica* in the west (but the type host was *Quercus*!).

Dactylaria Sacc.
Saprophytes, for parasites formerly so called see *Arthrobotrys* or *Dactylariopsis*; *D. junci* Ellis in East Anglia, *D. lepida* Winter in the west.

Dactylaria (Diplorhinotrichum) candidula (v. Höhn.) Bhatt & Kendrick		1
D. chrysosperma (Sacc.) Bhatt & Kendrick	On Quercus	1
D. obtriangularia Matsushima	On *Laurus nobilis*	1
D. orchidis Cooke & Massee		1
D. purpurella (Sacc.) Sacc.	On *Quercus*	1
D. scolecospora Kirk	On *Ulex europaeus*	1

Dactylariopsis Mekhtieva
Parasites of nematodes, probably = *Arthrobotrys*.

Dactylariopsis asthenopaga (Drechsler) Mekhtieva	1
D. brochopaga (Drechlser) Mekhtieva	1

Dactyella Grove
Parasites of nematodes in soil or litter, some *Massarina* anamorphs & see *Monacrosporium*.

Dactyella arnaudii Yadav	4
D. candida (Nees) de Hoog & v. Oorschot (*Arthrobotrys candida* (Nees) Schenck et al.)	1
D. cionopaga Drechsler	1
D. heptameres Drechsler	1
D. lobata Duddington	1

Dactylellina Morelet
Saprophyte, *D. leptospora* (Drechsler) Morelet in the midlands.

Dactylium Nees
Anamorphs of *Hypomyces*, which see

Dactylosporium Harz
Saprophyte, *D. macropus* (Corda) Harz in East Anglia.

Deightoniella Hughes
Facultative parasite.

Deightoniella (Napicladium) arundinaceum On *Phragmites* (Corda) Hughes			4

Dematophora Hartig
Facultative parasite.

Dematophora necatrix Hartig	White rootrot of herbaceous crops and fruit trees	1	4

Dendrodochium Bon.
Anamorphs of *Nectria*, *D. citrinum* Grove in East Anglia.

Dendrospora Ingold
Aquatic saprophytes, several species in the north and west.

Dendrostilbella v. Höhn.
Anamorphs of *Claussenomyces*, which see

Dendryphiella Bub. & Ranojevic
Saprophytes, *D. infuscans* (Thüm.) Ellis in East Anglia, see also *Scolecobasidium*.

D. vinosa (Berk. & Curt.) Reisinger (*D. interseminata*) (Berk. & Rav.) Bub. & Ran.)		1	4

Dendryphion Wallr.
Saprophytes.

Dendryphion comosum Wallr. (*D. curtum* B. & Br., *D. griseum* B. & Br.)	1	2	3	4
D. nanum (Nees) Hughes (*D. laxum* B. & Br.)	1	2	3	4

Dendryphiopsis Hughes
Lignicolous saprophyte.

Dendryphiopsis atra (Corda) Hughes	On *Buxus*, *Fagus* etc.	1

Dichotomophthora Mehrlich & Fitzpatrick
Facultative parasite, *D. portulacae* Werlich & Fitzpatrick on *Portulaca oleracea*, casual alien.

Dicoccum Corda
Nomen ambiguum.

Dicranidion Harkness
Aquatic saprophyte.

Dicranidion fragile Harkness	4

Dictyochaeta Speg.
Saprophyte.

Dictyochaeta querna Kirk	On *Quercus* litter	1

Dictyopolyschema Ellis
Corticolous saprophyte, *D. pirozynskii* Ellis on *Picea* in Dorset.

Dictyosporium Corda
Saprophytes, mainly lignicolous, *D. foliicola* Kirk in the north, *D. oblongum* (Fuck.) Hughes in Dorset.

Dictyosporium elegans Corda	On stubble and bare wood			4
D. pelagica (Linder) Hughes				4
D. toruloides (Corda) Guegan (*D. boydii* Sm. & Rams.)	On cupules, stems and wood	1	2	4

Didymaria Corda
Facultative parasites on angiosperm leaves.

Didymaria (*Ramularia*) *didyma* (Unger) Pound (*D. ungeri* Cda.)	On *Ranunculus repens*	1	2	4
D. kriegeriana Bres.	On *Melandrium rubrum* (*Lychnis diurna*)	1		

Didymopsis Sacc. & March
Hyperparasite, *D. helvellae* (Corda) Sacc. & March. on *Helvella* spp. in E. Anglia.

Digitodesmium Kirk
Lignicolous saprophyte.

Digitodesmium elegans Kirk	On *Quercus*	1

Diheterospora Kamyschko
Saprophytes, *D. cylindrospora* Barron in soils.

Dimorphospora Tubaki
Aquatic saprophytes = *Fluminispora* Ingold.

Dimorphospora foliicola Tubaki (*Fluminispora ovalis* Ingold)	1	4

Diplocladiella Matsushima
Saprophyte.

Diplocladiella scalaroides Arnaud	On litter of *Laurus* & *Quercus ilex*	1	2	4

Diplocladium Bon.
Anamorphs of *Hypomyces*, which see, also:

Diplocladium minus Bon.	On wood	3

Diplococcium Grove
Anamorph of *Helminthosphaeria*, also lignicolous saprophytes, *D. clarkii* Ellis in E. Anglia, *D. lawrencei* Sutton in the north.

Diplococcium spicatum Grove	On *Fagus, Pinus, Prunus, Quercus, Sorbus* etc.	1	4

Diplorhinotrichum v. Höhn. = *Dactylaria*

Doratomyces Corda
Saprophytes, *Echinobotryum atrum* Corda is a state of *D. stemonitis.*

Doratomyces microsporus (Sacc.) Morton & Smith	1	2		
D. nanus (Ehrenberg) Morton & Smith	1			4
D. purpureofuscus (Fr.) Morton & Smith	1			4
D. putredinis (Corda) Morton & Smith (*Graphium cuneiferum* (B. & Br.) Mason & Ellis)	1			4
D. stemonitis (Pers.) Morton & Smith	1	2	3	4

Drechmeria Gams & Jansson
Parasite.

Drechmeria coniospora (Drechsler) Gams & Jansson	1

Drechslera Ito
Facultative parasites, anamorphs of *Pyrenophora*, which see; *D. avenaceae* (Curtis) Shoemaker on *Avena*, *D. festuca* Scharif on *Festuca*, *D. phlei* (Graham) Shoemaker on *Phleum* in Berkshire.

Drechslera andersenii Scharif	On *Lolium perenne*	1		
D. biseptata (Sacc. & Roum.) Richardson & Fraser	On *Dactylis, Triticum* etc.			4
D. dematioidea (Bub. & Wrobl.) Subramanian & Jain	On *Anthoxanthum odoratum*	1		
D. fugax (Wallr.) Shoemaker (*Helminthosporium stenacrum* Drechsler	On *Agrostis*	1		
D. iridis (Oud.) Ellis		1	"Sussex"	4
D. triseptata (Drechsler) Subramanian & Jain	On *Holcus*	1		

Duddingtonia R.C. Cooke
Parasite of nematodes.

Duddingtonia flagrans (Dudd.) R.C. Cooke (*Trichothecium flagrans* Duddington)	"London"

Dwayaangam Subramanian
Aquatic saprophyte.

Dwayaangam cornuta Subramanian	1

Ectostroma Fr.
Saprophyte.

Ectostroma iridis Ehrenberg	On *Iris pseudacorus*	4

Embellisia Simmons
Saprophytes or feebly parasitic on bulb scales.

Embellisia allii (Campanile) Simmons	On *Allium sativum*, presumably an import	"London"	
E. hyacinthi de Hoog & Muller	On *Freesia, Hyacinthus, Scilla*	1	4
E. phragmospora (von Emden) Simmons	In soil		4

Endoconidium Prill. & Del.
Anamorph of *Gloeotinia*, which see.

Endoconospora Gjaerum
Facultative parasite, *E. cerastii* Gjaerum in the north.

Endophragmia Duvernoy & Maire
Lignicolous saprophytes, see also *Ityorhoptrum* and *Kienocephala*; *E. nannfeldtii* Ellis, *E. prolifera* (Sacc., Rouss. & Bomm.) Ellis & *E. stemphyllioides* (Corda) Ellis in East Anglia, *E. boewi* Crane, *E. cesatii* (Mont.) Ellis & *E. microaquatica* (Tubaki) Matsushima in the west.

Endophragmia alternata Tubaki & Saito	1			
E. atra (B. & Br.) Ellis	1	2		
E. biseptata Ellis		2		4
E. dennisii Ellis (*Sporidesmium dennisii* (Ellis) Kirk)	1		3	
E. elliptica (B. & Br.) Ellis	1	2	3	4
E. hyalosperma (Corda) Morgan Jones & Cole	1			

Endophragmiella Sutton
Lignicolous saprophytes, *E. uniseptata* (Ellis) Hughes anamorph of *Phaeotrichosphaeria*, *E. cambrensis* Ellis & *E. fagicola* Kirk in E. Anglia, *E. resinae* Kirk in the west.

Endophragmiella biseptata (Peck) Hughes	1		
E. boothii (Ellis) Hughes	1		
E. corticola Kirk	1		
E. eboracensis Sutton	1		
E. ellisii Hughes		2	4
E. fallacia Kirk	1		
E. lauri Kirk		2	
E. lignicola Hughes	1		
E. ovoidea Kirk	1		
E. pallescens Sutton	1		
E. pinicola (Ellis) Hughes	1		
E. socia (Ellis) Hughes		2	
E. subolivacea (Ell. & Ev.) Hughes	1		
E. suttonii Kirk	1		
E. taxi (Ellis) Hughes	1	2	
E. valdiviana (Speg.) Hughes			4

Endostilbum Malençon
Saprophyte.

Endostilbum albidum (Berk.) Reid	On *Pinus*, *Prunus* etc.	1

Engyodontium de Hoog
Saprophyte.

Engyodontium rectidentatum (Matsushima) de Hoog & Samson	On cork	1

Epicoccum Link
Saprophyte.

Epicoccum purpurascens Ehrenberg	Ubiquitous	1	2	3	4

Excipularia Sacc.
Saprophyte.

Excipularia fusispora (B. & Br.) Sacc.	On *Clematis vitalba*	1	4

Exochalara Gams & Holubova-Jechova
Saprophyte.

Exochalara longissima (Grove) Gams & Hol.-Jech.	On *Pinus*, *Pteridium* etc.	1

243

Exophiala J.W. Carm.
   Saprophytes, *E. jeanselmei* (Langer.) McGinnis & Padhye in Berkshire.

Exophiala dermatidis (Kane) de Hoog	In oil	"London"
E. mansonii (Castellani) de Hoog		1

Exosporium Link
   Saprophyte.

Exosporium tiliae Link	On *Tilia*	1

Filosporella Nawawi
   Aquatic saprophyte, *F. annelidica* (Shearer & Crane) Crane & Shearer in the west.

Flabellospora Alasoadura
   Aquatic saprophyte, *F. acuminata* Descals in the west.

Flagellospora Ingold
   Aquatic saprophyte, *F. penicillioides* Ingold in Berkshire, *F. fusarioides* Iqbal in the west.

Flagellospora curvula Ingold		2	4

Fulvia Ciferri
   Facultative parasite.

Fulvia fulva (Cooke) Ciferri	Ubiquitous on *Lycopersicon esculentum* under glass since 1887	1	2	3	4

Fusariella Sacc.
   Saprophytes, *F. atrovirens* Sacc. on *Allium* in the midlands.

Fusariella hughesii Chabelska-Frydman	On *Foeniculum, Lupinus, Sambucus, Urtica* etc.	1

Fusarium Link
   Anamorphs of *Gibberella* & *Nectria,* which see, also:

Fusarium culmorum (W.G.Smith) Sacc.	On Gramineae, causing root rot in cereals			2	4
F. merismoides Corda	Saprophyte in soil and polluted water	1			
F. oxysporum Schlecht.	Ubiquitous saprophyte with many parasitic races			2	4
f.sp. *dianthi* (Pril. & Del.) Snyder & Hansen	On *Dianthus* & *Lychnis*			2	4
f.sp. *gladioli* (Massey) Snyder & Hansen	Corm rot and yellows in *Gladiolus*	1			
f.sp. *lycopersici* (Sacc.) Snyder & Hansen	On *Lycopersicon*			2	4
F. oxysporum var. *redolens* (Woll.) Gussow	Plurivorous vascular wilt				4
F. poae (Peck) Woll	Especially on Gramineae but also on *Dianthus*				4
F. solani (Mart.) Sacc. var. *coeruleum* (Sacc.) Booth	On *Solanum tuberosum*				4
F. sporotrichoides Sherb.	Infected grain induces toxic aleukia in man and mammals	1			

Fusicladium Bon.
   Anamorphs of *Venturia,* which see.

Fusoma Corda
Facultative parasite.

Fusoma biseptata Sacc.	On *Agrostis*	1

Gabarnaudia Samson & Gams. See *Paecilomyces*.

Genicularia Rifai & R.C. Cooke
Nematode parasites.

Genicularia cystosporis (Duddington) Rifai & R.C. Cooke	1
G. inaequalis (Massee & Salmon) Subramanian	1
G. psychrophila (Drechsler) Rifai	1

Geniculospora Nilsson
Aquatic saprophytes, *G. inflata* (Ingold) Nilsson in the west.

Geniculosporium Chesters & Greenhalgh
Anamorphs of *Hypoxylon*, which see.

Geomyces Traaen
Saprophytes in soil and vegetable debris, see also *Chrysosporium pannorum*.

Geosmithia Pitt
Saprophytes *G. argillacea* (Stolk et al.) Pitt & *G. cylindrospora* (G. Smith) Pitt in midlands.

Geosmithia namyslowskii (Zalewski) Pitt (*Penicillium namyslowskii* Zalewski)	1
G. putterilii (Thom) Pitt (*Penicillium pallidium* G. Smith)	1

Geotrichum Link
Saprophytes, also a few potential human parasites not listed, *G. lutescens* (Sacc.) Lind (*Oospora microsperma* (B. & Br.) Sacc. & Vogl.) on bark in the midlands & north.

Geotrichum candidum (Link) Sprengel	In sap flows etc.	1 2 3 4
(*Oospora lactis* (Fres.) Sacc.)		
G. roseum Grove	On dead grass	1

Gibellula Cavara
Parasites, anamorphs of *Torrubiella*, which see, also:

Gibellula aranearum (Schwein.) Sydow	On spiders	1

Gilmaniella Barron
Saprophytes.

Gilmaniella humicola Barron	In soil, *Lycopersicon* roots etc.	1 2

Gliocladium Corda
Saprophytes, anamorphs of *Hypomyces* & *Nectria*, also:

Gliocladium album (Preuss) Petch	On Myxomycetes	1	3
G. catenulatum Gilman & Abbott	On bark, bulbs, hay etc.	1	
G. deliquescens Sopp	In soil	1	4
G. macropodium Marchal	In rotting vegetation	1	

Gliomastix Guégen
Saprophytes.

Gliomastix cerealis (Karst.) Dickinson (*G. guttuliformis* Brown & Kendrick)		1

G. luzulae (Fuck.) Mason (*Fusidium viride* Grove)	1	
G. murorum (Corda) Hughes (*G. convoluta* (Harz) Mason)	1	
G. murorum var. *felina* (Marchal) Hughes	1	4
G. musicola (Spreg.) Dickinson	1	

Glomospora Henderson
Parasite, *G. empetri* Henderson on *Empetrum* in the north.

Gonatobotrys Corda
Saprophyte.

Gonatobotrys simplex Corda	On dead stems and bark	1	4

Gonatobotryum Sacc.
Saprophyte.

Gonatobotryum fuscum (Sacc.) Sacc.	On *Fagus* & *Quercus*	1

Gonatorrhodiella Thaxter
Hyperparasite, perhaps = the preceding, *G. parasitica* Thaxter on *Hypomyces*.

Gonytrichum Nees
Anamorph of *Melanopsamella*, which see.

Graphiothecium Fuck.
Saprophyte.

Graphiothecium parasiticum (Desm.) Sacc.	4

Graphium Corda
Saprophytes, *G. smaragdinum* (A. & S.) Sacc. in E. Anglia, see also *Doratomyces putredinis*.

Graphium calicioides (Fr.) Cooke & Massee	1	3
G. comatrichoides Massee & Salmon	1	
G. penicilloides Corda	1	
G. stercorarius Marchal	1	
G. stilboideum Corda	1	

Gymnodochium Massee & Salmon
Saprophyte.

Gymnodochium fimicola Massee & Salmon	1

Gyoerffyella Kol.
Aquatic saprophytes, several others in the north and west.

Gyoerffyella tricapillata (Ingold) Marvanova	4

Gyrothrix Corda
Saprophytes, *G. podosperma* (Corda) Rab. in East Anglia.

Gyrothrix verticillata Pirozynski	On *Daphne, Galium, Urtica*	1

Hadrotrichum Fuck.
Facultative parasites, reputed anamorphs of *Scirrhia* but?

Hadrotrichum anceps Sacc.	On *Arrhenatherum*, doubtfully distinct from the next	1	
H. phragmitis Fuck. (*H. virescens* Sacc.)	Especially on *Agrostis*	1	4

Hansfordia Hughes
Saprophytes or hyperparasites, *H. caricis* Kirk in Suffolk

Hansfordia pulvinata (Berk. & Curt.) Hughes 3

Haplariopsis Oud.
Saprophyte.

Haplariopsis fagicola Oud. On *Fagus* cupules and litter 1 4

Haplobasidion Eriksson
Facultative foliar parasites, *H. thalictri* Erikss. on *Aquilegia* & *Thalictrum* in the west.

Haplographium B. & Br.
Anamorph of *Dematioscypha*, which see.

Harpographium Sacc.
Saprophyte.

Harpographium fasciculatum Sacc. 1

Harposporium Lohde
Aquatic saprophytes, several others in the west.

Harposporium anguillulae Lohde 1

Helicodendron Peyronel
Saprophytes, some *Lambertella* anamorphs; several others in the north and west.

Helicodendron articulatum GlenBott 1
H. conglomeratum GlenBott 1
H. luteoalbum GlenBott 1 4
H. triglitziensis (Jaap) Linder 1 4

Helicoma Corda
Saprophytes, see also *Troposporella* & *Zalerion*, some anamorphs of *Thaxteriella* which see

Helicoma olivaceum (Karst.) Linder 4

Helicomyces Link
Saprophytes.

Helicomyces roseus Link 1

Helicoon Morgan
Saprophytes, *H. ellipticum* (Peck) Morgan, *H. richonis* (Boud.) Linder & *H. sessile* Morgan in the west.

Helicoon chlamydosporum Abdullah & Webster 2
H. fuscosporum Linder 1

Helicorhoidion Hughes
Saprophytes, *H. irregulare* Mulder on *Typha* in the midlands.

Helicosporium Nees
Anamorphs of *Tubeufia*, which see.

Heliscella Marvanova, see *Clavatospora*

Heliscus Sacc. & Therry
Saprophytes, anamorphs of *Nectria* which see, compare also *Clavatospora*.

Heliscus aquaticus Ingold	4
H. longibrachiatus Ingold	4

Helminthophora Bon.
Saprophyte, *H. sphaerocephala* (Berk.) de Hoog in Hampshire

Helminthosporium Link
Saprophytes, see also *Corynespora, Drechslera*

Helminthosporium solani Dur. & Mont. (*Spondylocladium atrovirens* (Harz) Harz	1		3	4
H. velutinum Link	1	2	3	4

Hemibeltrania Pirozynski
Saprophytes, *H. mitrata* Kirk in the north and west.

Hemibeltrania echinulata Kirk	1

Henicospora Kirk & Sutton
Saprophytes.

Henicospora minor Kirk & Sutton	1

Heteroconium Petrak
Perhaps hyperparasites, *H. chaetospira* (Grove) Ellis in the north and west.

Heteroconium tetracoilum (Corda) Ellis (*Septocylindrium pallidum* Grove)	1

Heterosporium Klotzsch
Facultative parasites, anamorphs of *Mycosphaerella*, also

Heterosporium allii Ellis & Martin	On *Allium schoenoprasum*		3	
H. allii var. *cepivorum* Nicolas & Aggery	On *Allium cepa*		3	
H. minutulum Cooke & Massee		1		
H. syringae Oud.	On *Syringa vulgaris*		2	
H. variabilis Cooke	On *Spinacia oleracea*			4

Hirsutella Pat.
Entomogenous parasites, *H. gregis* Minter et al., *H. kirchneri* (Rostr.) Minter et al., *H. necatrix* Minter et al. in Berkshire, *H. acridiorum* Petch, *H. brownorum* Minter & Brown, *H. citriformis* Speare, *H. saussurei* (Cooke) Speare & *H. subulata* Petch in E. Anglia, *H. nodulosa* Petch & *H. thompsoni* Fisher in east midlands, *H. dipterigena* Petch widespread in caves, others in the north and west.

Hirsutella eleutheratorum (Nees) Petch	On beetles	1

Histoplasma Darling
Agent of histoplasmosis in man and animals, *H. capsulatum* Darling.

Hobsonia Berk.
Mainly saprophytic, *H. mirabilis* (Peck) Linder in the north.

Hobsonia christiansenii Brady & Hawks.	On *Lecanora & Xanthoria*	3

Hormiactella Sacc.
Saprophytes.

Hormiactella asetosa Hol.-Jech.	On bark & cones of *Picea* & *Pinus*	1
H. fusca (Preuss) Sacc.	On *Pinus sylvestris*	1

Hormiactis Preuss
Hyperparasite.

Hormiactis alba Preuss	On *Agaricus bisporus*	1	2

Hormonema Lagerberg & Melin
Saprophyte, *H. prunorum* (Dennis & Buhagia) Hermanidis & Nijhof in Essex.

Humicola Traaen
Saprophytes, widespread in soils, see also *Thermomyces*; *H. insolens* Cooney & Emerson and *H. fuscoatra* Traaen in the midlands, *H. allopalonella* Meyers & Moore on timber in the sea, *Humicola grisea* Traaen in Berkshire.

Hyalodendron Diddena
Saprophyte, *H. lignicola* Diddens in the west, but see also *Erostrotheca*.

Hymenella Fr.
Saprophytes, *H. cerealis* Ell. & Ev. on *Triticum* in Hertfordshire.

Hymenostilbe Petch
Entomogenous parasites, anamorphs of *Cordyceps*, which see, also:

Hymenostilbe arachnophila (Ditmar) Petch On spiders		3

Hyphelia Fr.
See *Chromelosporium*.

Idriella Nelson & Wilheim
Saprophytes, *I. grisea* (Sutton et al.) v. Arx. Occasional on *Laurus nobilis*

Illosporium Mart.
Hyperparasites of lichens, anamorphs of *Nectriella*, which see.

Infundibura NagRaj & Kendrick
Saprophyte, *I. adhaerens* NagRaj & Kendrick on forest litter in the west.

Ingoldia Petersen see **Gyoerffyella**

Isaria Hill
See in the main *Paecilomyces* but also:

Isaria cretacea v. Beyma thoe Kingma	1	
I. muscigena Cooke & Muller		3

Isariopsis Fres.
See *Ramularia* but also *I. carnea* Oud. on *Lathyrus pratensis* in East Anglia.

Isthmolongispora Matsushima
Saprophytes, casual on *Eucalyptus* & *Laurus* elsewhere.

Isthmotricladia Matsushima
Aquatic saprophyte, *I. britannica* Descals in the west.

Ityorhoptrum Kirk
Saprophyte.

Ityorhoptrum verruculosum (Ellis) Kirk   On Fagus   1

Jaculispora Hudson & Ingold
Aquatic saprophyte, *J. submersa* Hudson & Ingold in the west.

Junctospora Minter & Hol.-Jech.
Saprophytes, *J. pulchra* Minter & Hol.-Jech. on *Pinus* debris in the north.

Keratinomyces Vanbreus
Saprophytes, anamorphs of *Arthroderma*

Kienocephala Kirk
Saprophyte.

Kienocephala catenulata (Ellis) Kirk   On *Fagus* cupules   1

Lateramulosa Matsushima
Aquatic saprophytes, *L. bi-inflata* Matsushima & *L. uni-inflata* Matsushima in the west.

Laterispora Uecker et al
Hyperparasites, *L. brevirama* Uecker et al. on sclerotia of *Sclerotinia* in soil, apparently widespread.

Leightoniomyces Hawks. & Sutton
Hyperparasites, *L. phillipsii* (Leighton & Berk.) Hawks. & Sutton on lichen thalli in the west.

Lemonniera de Wild.
Aquatic saprophytes, several others in the west.

Lemonniera aquatica de Wild.   4
L. brachyclada Ingold   4

Leptodontidium de Hoog
Saprophytes.

Leptodontidium elatius (Mangenot) de   On *Quercus*   1
Hoog (*Rhinocladiella elatior* Mang.)

Leptographium Lagerb. & Melin
Saprophytes associated with bark beetles in conifers, *L. lundbergii* Lagerberg & Melin, *L. microsporum* Davidson, *L. truncatum* (Wingfield & Marasas) Wingfield & *L. wingfieldii* Morelet in East Anglia.

Lichenopuccinia Hawks. & Haff.
Hyperparasite, *L. poeltii* Hawks. & Haff. On *Parmelia saxatilis* elsewhere.

Linodochium v. Höhn.
Anamorph of *Pseudohelotium pineti* which see.

Lobatopedis Kirk
   Saprophyte.

Lobatopedis foliicola Kirk.　　　　　On Fagus　　　　　　　　　　　　　1

Lunulospora Ingold
   Aquatic saprophyte, *L. curvula* Ingold in Berkshire.

Magdalaenaea Arnaud
   Aquatic saprophyte, *M. monogramma* Arnaud.

Malbranchea Sacc.
   Anamorphs of Gymnoascaceae, *M. graminicola* Lacey et al. & *M. sulfurea* (Miehe) Sigler
   & Carm. in hay and straw.

Malbranchea pulchella Sacc. & Penz.　　　On cardboard　　　　　　　　　　4

Mammaria Ces.
   Saprophyte.

Mammaria echinobotryoides Ces.　　　In soil & decaying timber　　　　　1　　　4

Margaritispora Ingold
   Aquatic saprophytes, *M. aquatica* Ingold & *M. monticola* Dyko in the west.

Mariannaea Arnaud
   Lignicolous saprophytes, *M. camptospora* Samson in the north and west.

Mariannaea elegans (Corda) Samson　　On *Abies, Betula, Fagus, Fraxinus, Larix,*
　　　　　　　　　　　　　　　　　　　　Pinus &c.　　　　　　　　　　1　2

Masoniella G. Smith see **Scopulariopsis**

Mastigosporium Riess
   Facultative parasites of Gramineae, *M. deschampsiae* Jørstad in the east midlands, *M.
   album* Riess on *Alopecurus geniculatus* & *A. pratensis* in Middlesex.

Mastigosporium cylindricum Sprague　　On *Phleum*　　　　　　　　　　　4
M. rubricosum (Dearn. & Barth.) Sprague　On *Dactylis glomerata*　　　　1　　3
　(var. *agrostidis* Bollard in the north & west)

Memnoniella v. Höhn.
   Saprophyte, *M. echinata* (Riv.) Galloway on damp paper &c. probably a casual alien.

Menispora Pers.
   Lignicolous saprophytes, *M. caesia* Preuss anamorph of *Chaetosphaeria pulviscula*, which
   see.

Menispora britannica Ellis　　　　On *Fagus* cupules & *Filipendula* stems　1
M. ciliata Corda　　　　　　　　On *Corylus, Fagus, Quercus, Rubus, Ulex*　1　2　3
M. glauca Pers.　　　　　　　　On *Acer, Betula, Fagus*　　　　　　1
M. tortuosa Corda　　　　　　　On *Betula, Fagus*　　　　　　　　1

Meria Vuill.
   Parasites, see also *Drechmeria*.

Meria laricis Vuill.　　　　　　needle-cast of *Larix decidua*　　　　　1

Metarhizium Sorokin
Parasite widespread on many arthropods.

Metarhizium anisopliae (Metsch.) Sorokin   London

Microdochium Syd.
Facultative parasites, *M. bolleyi* (Sprague) de Hoog & *M. dimerum* (Penz.) v. Arx in Berkshire.

Microdochium (Marssonina) panattonianum On *Lactuca sativa* (Berl.) Sutton et al.		1	4

Microsporum Gruby
Agents of animal mycoses, anamorphs of *Nannizzia*, which see.

Microsporum audounii Gruby	Unlocalised but presumably prevalent, or formerly so	
M. canis Bodin	On *Felis*	1
M. equinum (Del. & Bodin) Gueguen	On *Equus*	1

Microstroma Niessl
Parasites, *M. juglandis* (Bereng.) Sacc. on *Juglans regia* in the west.

M. album (Desm.) Sacc.	On *Quercus*	1	3

Microxyphium (Pers.) Hughes
Anamorph of *Dennisiella babingtonii*.

Milospium Hawks.
Hyperparasite.

Milospium graphideorum (Nyl.) Hawks.	On *Opegrapha atra, O. lyncea* &c.	2	3	4

Minimedusa Weresub
Saprophyte.

Minimedusa polyspora (Hotson) Weresub in soil in the midlands.

Mirandina Arnaud
Saprophyte.

Mirandina corticola Arnaud	1

Moeszia Bub.
Saprophyte, *M. cylindrioides* Bub. in *Quercus* litter in the west.

Monacrosporium Oud.
Parasites of nematodes, several additional species in the midlands and East Anglia.

Monacrosporium bembicoides (Drechsler) Subramanian	1
M. ellipsospora (Preuss) Grove	1

Monilia Bon.
Anamorphs of *Monilinia* which see, but compare also *Chrysonilia* & *Scopulariopsis*.

Monocillium Saksena
Anamorphs of *Niesslia*, which see.

Monodictys Hughes
Saprophytes, *M. antiqua* (Corda) Hughes on bark in the north and west.

Monodictys aspera (Corda) Hughes		1		
M. asperispora (Cooke & Massee) Ellis	On *Sambucus*	London		
M. castaneae (Wallr.) Hughes	On damp paper, *Lachnellula* apothecia &c.	1		
M. lepraria (Berk.) Ellis	On wood and *Lecanora* thalli	1		
M. levis (Wiltshire) Hughes	In soil, decaying roots &c.	1	2	
M. melanops (Ach.) Ellis	On bark of *Fraxinus* & *Malus*	"Sussex"		
M. paradoxa (Corda) Hughes	On bark of *Betula*	1		
M. putredinis (Wallr.) Hughes	On *Betula*, driftwood &c.	1		3

Monosporium Bon.
Anamorphs of *Allescheria*.

Monotosporella Hughes
Saprophyte, *M. tuberculata* Gönczöl in the west.

Myceliophthora Cost. see **Chrysosporium**

Mycocentrospora Deighton
Saprophytes and facultative parasites, four aquatic saprophytes in the west.

Mycocentrospora acerina (Hartig) Deighton	On *Apium, Daucus* &c.	1	2	4
M. cantuariensis (Salmon & Wormald) Deighton	On *Humulus lupulus*			4

Mycogone Link
Saprophytes and hyperparasites.

Mycogone cervina Ditmar			3	
M. perniciosa Magnus	On *Agaricus*		2	4
M. rosea (Link) Link	On *Agaricus*	1	2	3

Mycosylva Tulloch
Saprophyte, *M. clarkei* Tulloch on dung in the west midlands.

Mycotorula Will.
Saprophyte

M. verticillata Red. & Cif.	In cattle	?1

Mycovellosiella Rangel
Facultative parasites, *M. murina* (Ell. & Kell.) Deighton on Viola in E. Anglia.

Mycorellosiella (*Cercospora*) *ferruginea* (Fuck.) Deighton	On *Artemisia vulgaris*	1

Myrothecium Tode
Saprophytes or weak facultative parasites, *M. atroviride* (B. & Br.) Tulloch in the west, *M. carmichaelii* Grev., *M. cinctum* Sacc. & *M. leucotrichum* (Peck) Tulloch in E. Anglia.

Myrothecium inundatum Tode	On effete carpophores of *Russula* spp.	1	
M. roridum Tode	Widespread saprophyte sometimes parasitic on *Viola*		4
M. verrucaria (A. & S.) Fr.	In soil	1	2

Myxormia B. & Br. see **Myrothecium**

Nematoctonus Drechsler
Parasite of nematodes.

Nematoctonus tylosporus Drechsler   1

Nematogonum Desm.
Hyperparasite.

Nematogonum ferrugineum (Pers.) Hughes   On *Nectria coccinea*   1

Neta Shearer & Crane
Lignicolous saprophyte.

Neta patuxentica Shearer & Crane   On wet wood   1

Nigrospora Zimmermann
Saprophytes, mainly in the tropics.

Nigrospora sphaerica (Sacc.) Mason   On *Arundinaria* &c.   1

Nodulisporium Preuss
Lignicolous saprophytes, anamorphs of *Hypoxylon*, which see, also:

Nodulisporium (Isaria) umbrinum (Pers.) Deighton   1   2   4

Oedemium Link
Anamorph of *Chaetosphaerella*, which see.

Oedocephalum Preuss
Saprophytes, anamorphs of *Peziza*, *Plicaria* and allies, also:

Oedocephalum glomerulosum (Bull.) DC.   1   4
O. ochraceum Massee & Salmon   1
O. pallidum (B. & Br.) Cost.   1   4
O. preussii Sacc.   1
O. roseum Cooke   1   4

Oidiodendron Robak
Lignicolous saprophytes, *O. flavum* Szilvinyi & *O. fuscum* Robak in E. Anglia, *O. maius*
Barron in the north.

Oidiodendron griseum Robak   1   4
O. rhodogenum Robak   1
O. tenuissimum (Peck) Hughes   1

Oidiopsis Scalia
Anamorph of *Leveillula*.

Oidium Link
Anamorphs of *Erysiphaceae* but see also *Acladium* anamorphs of *Botryobasidium*

Oncopodiella Arnaud
Saprophyte but also *O. hyperparasitica* Hawks. on *Lasiosphaeria spermoides* in E. Anglia.

Oncopodiella trigonella (Sacc.) Rifai   On various woods   1   2
   (*O. tetraedrica* Arnaud)

Onychophora Arnaud
   Saprophyte, *O. coprophila* Gams et al. on rabbit dung in the west.

Ostracoderma Fr.
   Anamorphs of *Peziza*, see also *Chromelosporium*

Ostracoderma carnea (Ehrenberg) Hughes	1	3	4
O. ochraceum (Corda) Hughes	1	3	

Ovularia Sacc.
   Falcultative foliar parasites, better referred to *Ramularia*, *O. haplospora* (Speg.) Magnus in East Anglia.

Ovularia berberidis Cooke	On *Berberis asiatica*	1		
O. duplex Sacc.	On *Scrophularia*	1		
O. primulana Karst.	On *Primula vulgaris*	1	2	3
O. rubi Bub.	On *Rubus*	2		

Ovulariella Bub. & Kab.
   Facultative parasite.

Ovulariella nymphaearum (All.) Bub. & Kab.	1

Ovulariopsis Pat. & Har.
   Anamorph of *Phyllactinia*.

Pachnocybe Berk. see Auriculariales

Paecilomyces Bainier
   Saprophytes or entomogenous parasites, *P. inflatus* (Burnside) Carm. in Hampshire, *P. fumosoroseus* (Wize) Brown & Smith in East Anglia, several others elsewhere. See also *Mariannaea*.

Paecilomyces betae (Del.) Cornford					4
P. carneus (Duché & Heim) Brown & Smith			2		
P. (Isaria) farinosus (Dickson) Brown & Smith	On wood and insect pupae &c.	1	2	3	4
P. lilacinus (Thom) Samson			2		
P. marquandii (Massee) Hughes	In soil	1	2		
P. varioti Bainier		1			

Papulaspora Preuss
   Saprophytes & hyperparasites, *P. coprophila* (Zukal) Hotson in E. Anglia, *P. halima* Anastasiou.

Papulaspora byssina Hotson	Brown plaster mould of mushroom beds	2	3	4
P. candida Sacc.	On *Geoglossum* spp.	1	2	

Paradendryphiopsis Ellis
   Lignicolous saprophyte.

Paradendryphiopsis cambrensis Ellis	On *Quercus*	1

Paraisaria Samson & Brady
   P. dubia (Del.) Samson & Brady anamorph of *Cordyceps gracilis* which see.

Parapleurotheciopsis Kirk
   Saprophytes, *P. inaequiseptata* Kirk in the west.

Parapleurotheciopsis ilicina Kirk   In litter of *Quercus ilex*   1

Parasympodiella Ponappa
Lignicolous saprophyte, *P. clarkei* Sutton on *Pinus* in the west.

Passalora Fr.
Facultative parasite, compare also *Cercosporidium.*

Passalora bacilligera (Mont. & Fr.) Mont.   On leaves of *Alnus glutinosa*   1
& Fr.

Penicillium Link
   Predominantly saprophytes, anamorphs of *Eupenicillium* & *Talaromyces*, see also *Geosmithia*. Additional in Hampshire *P. terrestre* Jensen, in Berkshire *P. diversum* Raper & Fennell v. *aureum* R. & F., *P. luteum* Sopp, in Middlesex *P. alboaurantium* G. Smith, in E. Anglia *P. aculeatum* Raper & Fennell, *P. albidum* Sopp., *P. helicum* Raper & Fennell, *P. lilacinum* Thom, *P. piscarium* Westling, *P. roseopurpureum* Dierckx and *P. terlikowskii* Zal., in Dorset *P. griseolum* G.Smith & *P. janczewskii* Zal., others elsewhere.

Penicillium adametzi Zalewski	1		
P. asperosporum G.Smith	1		
P. atrovenetum G.Smith	"Sussex"		
P. avellaneum Thom & Turesson			4
P. brevicompactum Dierckx	1	2	
P. canescens Sopp.	2		
P. carneolutescens G.Smith			4
P. caseicolum Bainier	1		
P. charlesii G.Smith	1		
P. chrysogenum Thom	2		4
P. citreoviride Biourge	1		
P. citrinum Sopp.	1		
P. claviforme Bainier	1		
P. commune Thom			4
P. concentricum Sorrosa, Stolk & Hadlock	1		
P. corylophilum Dierckx			4
P. corymbiferum Westling	1		
P. crustosum Thom	1		4
P. cyclopium Westling	1	3	4
P. daleae Zalewski	1		
P. digitatum Sacc.	1		
P. duclauxii Del.	1		
P. expansum Link	1		4
P. fellutanum Biourge	1		
P. frequentans Westling	1	2	
P. funiculosum Thom	1		
P. gladioli Machacek			4
P. granulatum Bainier	1		
P. herquei Bainier & Sartory			4
P. hirsutum Dierckx			4
P. islandicum Sopp.	3		
P. italicum Wehmer	1		
P. janthinellum Biourge	1	3	
P. kewense G.Smith	1		
P. mali Novobranova			4
P. martensii Biourge			4
P. melinii Thom	1		4
P. miczynskii Zalewski	1		
P. mineoluteum Dierckx			4
P. nigricans Bainier	1	2	4

	1	2	3	4
P. *notatum* Westling	London			
P. *novaezealandiae* v. Beyma	1		3	
P. *olsoni* Bainier & Sartory				4
P. *oxalicum* Currie & Thom	1			
P. *paxilli* Bainier	London			
P. *piceum* Raper & Fennell	1			
P. *pulvillosum* Turfitt	London			
P. *purpurescens* (Sopp) Biourge				4
P. *purpurogenum* Stoll	1			
P. *pusillum* G.Smith		2		
P. *radulatum* G.Smith	1			
P. *raistrickii* G.Smith	1			
P. *restrictum* Gilman & Abbott	1			
P. *rocquefortii* Thom	1			4
P. *rolfsii* Thom	1			
P. *rugulosum* Thom	1			
P. *sclerotiorum* v. Beyma	1			4
P. *spinulosum* Thom				4
P. *steckii* Zalewski				4
P. *stoloniferum* Thom		2		
P. *thomii* Maire	1	2		4
P. *variabile* Sopp	1			
P. *verruculosum* Peyronel	1			4
P. *vinaceum* Gilman & Abbott	1			
P. *viridicatum* Westling	1	2		
P. *waksmanii* Zalewski	1			
P. *zonatum* Hodge & Perry		1		

Periconia Tode

Saprophytes, anamorph of *Didymosphaeria*, also *P. atra* Corda, *P. curta* (Berk) Mason & Ellis, *P. digitata* (Cooke) Sacc., *P. funerea* (Ces.) Mason & Ellis, *P. laminella* Mason & Ellis, *P. macrospinosa* Lefebvre & Johnson, *P. paludosa* Mason & Ellis & *P. typhicola* Mason & Ellis in E. Anglia, *P. circinata* (Mangin) Sacc. in Hertfordshire and *P. shyamala* Roy in the west.

	1	2	3	4
Periconia britannica Ellis	1			
P. *byssoides* Pers.	1	2	3	4
P. *cambrensis* Mason & Ellis		2		
P. *cookei* Mason & Ellis	1			4
P. *glyceriicola* Mason & Ellis	1			
P. *(Dematium) hispidula* (Pers.) Mason & Ellis	1		3	4
P. *minutissima* Corda	1			

Periconiella Sacc.
Saprophytes.

Periconiella ilicis Kirk	On *Ilex*	1

Pesotum Crane & Schoknecht
Anamorphs of *Ceratocystis*, which see.

Peyronelina Fisher et al.
Aquatic saprophyte, *P. glomerulata* Fisher et al. in the north.

Phaeococcomyces de Hoog
Saprophyte, *P. exophialeae* de Hoog in the west.

Phaeoisaria v. Höhn.
Saprophytes on fallen branches and dead woody stems.

Phaeoisaria clavulata (Grove) Mason & Hughes (*Graphium grovei* Sacc.)		1		
P. clematidis (Fuck.) Hughes (*Graphiopsis cornui* Bainier)		1		4

Phaeoramularia Muntañola
Facultative parasites, *P. antipus* (Ell. & Holw.) Deighton on *Lonicera* in the north.

Phaeostalagmus Gams
Saprophytes, segregate from *Verticillium*.

Phaeostalagmus cyclosporus (Grove) Gams	Ubiquitous on bark and wood	1	2	
P. peregrinus Minter & Hol.-Jech.		1		
P. tenuissimus (Corda) Gams	Widespread on bark and wood of many trees	1		3	4

Phialocephala Kendrick
Saprophytes, *P. truncata* Sutton on *Castanea* cupules in the west.

Phialocephala fumosa (Ell. & Ev.) Sutton	On *Castanea* cupules	1	3

Phialophora Medlar
Saprophytes and facultative parasites, *P. graminicola* Walker & *P. radicicola* Cain in East Anglia.

Phialophora (*Cephalosporium*) *asteris*	Wilt of *Aster* & *Callistephus*	1		
(Dowson) Burge & Isaac (*Verticillium vilmorinii* (Gueguin) Westerdijk & v.Luijk)				
P. bubakii (Laxa) Schol.-Schwarz	In wood	1		
P. cinerescens (Woll.) v.Beyma	Wilt of *Dianthus*		2	4
(*Verticillium cinerescens* Woll.)				
P. fastigiata (Lagerb. et al.) Conant	In soil and wood	1		4
P. hoffmanni (v.Beyma) Schol.-Schwarz		1		
P. malorum (Kidd & Beaumont) McColloch				4
P. melinii (Nannf.) Conant		1		
P. parasitica Ajello et al.		1		
P. rhodogena (Mangenot) Gams		1		

Phialophoropsis Batra
Symbionts of *Coleoptera* & facultative parasites, *P. cambrensis* Brady & Sutton in leaves of *Embothrium lanceolatum* in the west.

Phragmocephala Mason & Hughes = **Endophragmia**

Piricauda Bub.
Saprophyte, *P. arcticoceanum* R.T.Moore in Hampshire.

Pithomyces Berk. & Br.
Saprophytes and agent of animal mycosis.

Pithomyces chartarum (Berk. & Curt.) Ellis	On paper, dead grass &c.	1	
P. valparadisiacus (Speg.) Kirk	On *Calluna*	1	

Pityrosporum Sabour.
Agents of animal mycoses,

Pityrosporum pachydermatis Weidman	In cattle & dog	?1	

Pleiochaeta (Sacc.) Hughes
Facultative parasite of Leguminosae

Pleiochaeta setosa (Kirchn.) Hughes (*Ceratophorum setosum* Kirchn.)	On *Lupinus*	1			

Pleuropedium Marvanova & Iqbal
Aquatic saprophytes, *P. tricladioides* Marv. & Iqbal in the west.

Pleurophragmium Cost.
Saprophytes, *P. tritici* Ellis in East Anglia.

Pleurophragmium acutum (Grove) Ellis (*Acrotheca acuta* Grove)	On *Leptosphaeria acuta*	1			
P. parvisporum (Preuss) Hol.-Jech. (*P. simplex* (B. & Br.) Hughes)	On *Urtica* &c.	1			4

Pleurotheciopsis Sutton
Lignicolous saprophytes.

Pleurotheciopsis bramleyi Sutton	On *Ilex, Laurus nobilis*	1			
P. pusilla Sutton	On cupules of *Castanea sativa*	1			

Pleurothecium v.Höhn.
Lignicolous saprophyte.

Pleurothecium recurvatum (Morgan) v.Höhn.	On *Acer, Betula, Fagus, Quercus, Tilia*	1	2	3	4

Pollaccia Baldacci & Ciferri
Anamorphs of *Venturia*, which see.

Polycladium Ingold
Aquatic saprophyte, *P. equiseti* Ingold in the north.

Polypaecilum G.Smith
Saprophytes, *P. botryoides* (Brooks & Hansford) G.Smith on leaf litter in the north, also
P. insolitum G.Smith in human mycoses.

Polyscytalum Riess
Saprophytes and facultative parasites, *P. sericeum* Sacc. in the west.

Polyscytalum berkeleyi Ellis (*Dendryphium* *griseum* B. & Br.)	On *Urtica* &c.	1	2		4
P. fagicola Kirk	On *Fagus* leaf litter	1			
P. fecundissimum Riess	On leaves of *Betula, Fagus, Quercus, Salix*	1			
P. gracilisporum (Matsushima) Sutton & Hodges		1	2		4
P. hareae (Sutton) Kirk		1			
P. pini Kirk & Minter		1			
P. (Oospora) pustulans (Owen & Wakef.) Ellis	On *Solanum tuberosum* as skin spot				4
P. truncatum Sutton	On *Quercus ilex*	1			

Polyspora Lafferty = **Aureobasidium**

Polythrincium Kunze & Schmidt
Anamorph of *Mycosphaerella killiani*.

Porocladium Descals
Aquatic saprophyte, *P. aquaticum* Descals in the west.

Pseudaegerita Crane & Schoknecht
Saprophyte, *P. corticalis* (Peck) Crane & Schoknecht in the west.

		1		3	
Pseudaegerita viridis (Bayliss Elloitt) Abdullah & Webster	On wood	1		3	

Pseudoanguillospora Iqbal
Aquatic saprophytes, *P. prolifera* Iqbal & *P. stricta* Iqbal in the west.

Pseudobotrytis Krzemieniewska & Badura
Saprophyte, *P. (Spicularia) terrestris* (Timonin) Subramanian in East Anglia.

Pseudocercospora Speg.
Saprophytes or facultative parasites, *P. unicolor* (Sacc. & Penz.) Kirk on *Laurus* in the west.

		1	2
Pseudocercospora deightonii Minter	On needles of *Pinus*	1	
P. rubi (Sacc.) Deighton	On *Rubus*		2

Pseudocercosporella Deighton
Facultative parasites.

		1	2	3	4
Pseudocercosporella capsellae (Ell. & Ev.) Deighton	On *Brassica, Capsella, Raphanus*		2		
P. herpotrichoides (Fron) Deighton	Eyespot of *Agropyron, Avena, Bromus, Hordeum, Triticum*	1	2	3	4
P. scirpi (Moesz) Deighton	On culms of *Scirpus lacustris*				4

Pseudomicrodochium Sutton
Saprophytes, *P. lauri* Kirk in the west.

		1	2	3
Pseudomicrodochium aciculare Sutton	On cupules of *Castanea sativa*	1		3
P. candidum (Bres.) de Hoog	On debris of *Pinus*		2	
P. cylindricum Sutton	On cupules of *Castanea sativa*	1		

Pseudospiropes Ellis
Saprophytes, anamorph of *Melanomma*, also *P. rousselianum* (Mont. Ellis in Hertfordshire.

		1	2	3
Pseudospiropes hughesii Ellis	On *Fagus* and *Fraxinus*	1		
P. nodosus (Wallr.) Ellis	On *Acer, Betula, Corylus, Fagus, Fraxinus, Populus, Salix* &c.	1	2	3
P. obclavatus Ellis	On *Castanea, Hedera, Fagus*	1		
P. simplex (Kunze) Ellis (*Pleurophragmium cylindricum* Hughes)	On wood	1	2	
P. subuliferus (Corda) Ellis	On *Acer, Carpinus, Crataegus, Fagus, Fraxinus, Salix* &c.	1		

Pteroconium Sacc.
Saprophytes, *P. intermedium* Ellis on *Arundinaria* in the north.

Pterygosporopsis Kirk
Saprophytes, *P. rhododendri* Kirk in the north.

		4
Pterygosporopsis fragilis Kirk	On *Laurus nobilis*	4

Pycnostysanus Lindau
Facultative parasite and saprophytes, *P. medius* (Sacc.) Bat. & Peres in the north.

Pycnostysanus azaleae (Peck) Mason   On buds of *Rhododendron* spp.   1 2 3 4

Pyricularia Sacc.
Saprophyte, *P. lauri* Kirk in Essex, also *Massarina* anamorph.

Ramichloridium Stahel
Saprophytes & facultative parasites, *R. pini* de Hoog & Rahman in the north.

Ramularia Unger
Facultative foliar parasites, anamorphs of *Mycosphaerella*, see also *Didymaria*, *Hyalodendron* and *Ovularia*. Most herbaceous plants carry a *Ramularia* and the names are largely "host-species"; of these also *R. destructiva* Phill & Plowr. in Hampshire, *P. ludoviciana* Minter et al. on mites and *R. pastinacae* Bub. in Berkshire, *R. cardamines* in Buckingham, *R. coriandri* Moesz & Smarods and *R. lamiicola* Massal. in Hertfordshire, *R. alborosella* (Desm.) Gjaerum, *R. aromatica* (Sacc.) v.Höhn., *R. doronici* Grove, *R. erodii* Bres., *R. menthicola* Sacc., *R. obducens* Thüm., *R. ulmariae* Cooke, *R. valerianae* (Speg.) Sacc. & *R. winteri* Thüm. in E. Anglia, *R. gei* Lind. & *R. heraclei* (Oud.) Sacc. in the midlands, *R. anthrisci* v.Höhn., *R. centranthi* Brun., *R. cynoglossi* Lindr., *R. farinosa* (Bon.) Sacc., *R. mimuli* Ell. & Kell., *R. onobrychidis* All., *R. ovata* Fuck., *R. rutamurariae* Trotter, *R. schulzeri* Baum., *R. scolopendrii* Fautr., *R. senecionis* (B. & Br.) Sacc. & *R. sylvestris* Sacc. in the west, more in the north.

Ramularia acris Lindr.	On *Ranunculus acris*	1	2		
R. adoxae (Rab.) Karst.	On *Adoxa moschatellina*	1	2		
R. aequivoca (Ces.) Sacc.	On *Ranunculus repens* &c.			3	4
R. agrestis Sacc.	On *Viola tricolor*	1		3	4
R. ajugae (Niessl) Sacc.	On *Ajuga reptans*	1	2		
R. alba (Dowson) Nannf.	On *Lathyrus odoratus*	1			
R. alismatis Fautr.	On *Alisma plantago-aquatica*	1			
R. arenariae Smith & Rams.	On *Moehringia trinervia*	1		3	
R. ari Fautr.	On *Arum maculatum*	1	2	3	4
R. armoraciae Fuck.	On *Armoracia rusticana*	1	2		4
R. barbareae Peck	On *Barbarea praecox*	1			
R. beticola Fautr. & Lambotte	On *Beta vulgaris* & v. *cicla*	1	2		4
R. bistortae Fuck.	On *Polygonum bistorta*	1			
R. brunnea Peck	On *Tussilago farfara*	1	2		4
R. calcea (Desm.) Ces.	On *Glechoma hederacea*	1	2		
R. calthae Lindr.	On *Caltha palustris*		2		
R. cardui Karst.	On *Carduus* & *Cirsium*		2		
R. centaureae Lindr.	On *Centaurea nigra* & *C. scabiosa*	1		3	
R. circaeae All.	On *Circaea lutetiana*		2		
R. cirsii All.	On *Cirsium*	1			
R. cynarae Sacc.	On *Cynara scolymus*		2		
R. exilis Syd. & Syd.	On *Galeobdolon luteum*	1			
R. filaris Fres.	On *Picris echioides*	1	2		
R. filaris v. *lappae* Bres.	On *Arctium*	1	2		
R. geranii Fuck.	On *Geranium dissectum, G. molle, G. phaeum, G. pratense*	1			
R. hypochaeridis Magnus	On *Hypochaeris radicata*	1			
R. knautiae Bub.	On *Knautia arvensis*	1			
R. lactea (Desm.) Sacc.	On *Viola hirta, V. odorata, V. riviniana*	1	2		
R. lampsanae (Desm.) Sacc.	On *Lapsana communis*		2		
R. lysimachiarum Lindr.	On *Lysimachia nemorum* & *L. nummularia*	1			
R. macrospora Fres.	On *Campanula persicifolia* &c.	"Sussex"			4
R. meliloti Ell. & Ev.	On *Melilotus officinalis*				4
R. parietariae Pass.	On *Parietaria diffusa*		2		
R. picridis Fautr. & Roum.	On *Picris hieracioides*				4
R. plantaginis Ell. & Mart.	On *Plantago major* & *P. media*		2		

		1	2	3	4
R. pratensis Sacc.	On Rumex acetosa	1			
R. primulae Thüm.	On Primula vulgaris & cultivars	1	2	3	4
R. punctiformis (Schlecht.) v.Höhn.	On Chamaenerion & Epilobium		2		4
(R. montana Speg.)					
R. purpurescens Wint.	On Petasites fragrans			3	
R. rhabdospora (B. & Br.) Nannf.	On Plantago lanceolata	1		3	
(R. plantaginea Sacc. & Berl.)					
R. rhei All.	On Rheum rhaponticum	"Sussex"			
R. rubella (Bon.) Nannf. (Ovularia obliqua)	On Rumex spp.	1	2	3	4
(Cooke) Oud.					
R. sambucina Sacc.	On Sambucus nigra	1	2		
R. scrophulariae Fautr. & Roum	On Scrophularia	1		3	
R. sonchi-oleracea Fautr.	On Sonchus	1			
R. sphaeroidea Sacc.	On Lotus uliginosus	1	2	3	4
R. spiraeae Peck	On Filipendula ulmaria			3	
R. stachydis (Pass.) Massal.	On Stachys sylvatica	1			
R. tanaceti Lindr.	On Tanacetum vulgare	1			
R. taraxaci Karst.	On Taraxacum officinale	1	2	3	4
R. vallisumbrosae Cavara	On Narcissus spp. cult.	1			
R. variabilis Fuck.	On Digitalis purpurea	1	2		
R. veronicae Fuck.	On Veronica persica &c.		2		

Ramulaspera Liro
Facultative parasites, *R. holci-lanati* (Cav.) Lind. on *Holcus lanatus* in the west.

Refractohilum Hawks.
Hyperparasites, *R. galligenum* Hawks. on *Nephroma* in the west.

Rhexoampullifera Kirk
Saprophyte, *R. fagi* (Ellis) Kirk on leaves of *Fagus* in East Anglia.

Rhinocladiella Nannf.
Saprophytes, *R. atrovirens* Nannf. in the west.

		1			4
Rhinocladiella cellaris (Pers.) Ellis (*Racodium cellare* Pers.)		1			4

Rhinocladium Sacc. & Marchal
Saprophytes.

			1
Rhinocladium pulchrum Hughes & Hol.Jech.	On wood		1

Rhizoctonia DC.
Sterile mycelia of *Helicobasidium* & *Thanatephorus*, compare also *Sclerotium*.

Rhodotorula Harrison
Agents of animal mycoses, status as for *Candida*.

		1
Rhodotorula glutinis (Fres.) Harrison v. *rubescens* (Saito) Lodder		1
R. mucilaginosa (Jörg.) Harrison		1

Saccardaea Cavara
Saprophyte, *S. atra* (Desm.) Mason & Ellis (*Myrothecium atrum* (Desm.) Tulloch) in Hampshire.

Sarcinella Sacc.
Anamorphs of *Schiffenerula*, *S. heterospora* Sacc. in "Sooty mould" on trees and shrubs.

Sarcopodium Ehrenberg
   Saprophyte.

Sarcopodium circinatum Ehrenberg (*S. roseum* (Corda) Fr.)	On herbaceous stems	1

Sclerococcum Fr.
   Hyperparasites, *S. sphaerale* (Ach.) Fr. on *Pertusaria* in East Anglia.

Sclerotium Tode
   Sterile anamorphs of various basidiomycotina and ascomycotina, still unassigned are:

Sclerotium cepivorum Berk.	Parasitic on *Allium cepa*	1	2	4
S. glaucoalbidum Desm.		1		
S. muscorum Pers.	Saprophyte	1		
S. rhizodes Auerswald	On leaves of Gramineae	1		
S. rolfsii Sacc.	Casual alien			
S. tuliparum Kleb.	Grey bulb rot of *Tulipa* and other bulbous plants	1	2	4

Scolecobasidium Abbott
   Saprophytes, *S. arenarium* (Nicot) Ellis in maritime sands & immersed wood, *S. acanthacearum* (Cooke) Ellis casual alien.

Scolecobasidium constrictum Abbott		1		4
S. echinophilum (Massal.) Sutton	On cupules of *Castanea sativa*	1	3	4
S. salinum (Sutherland) Ellis	On *Laminaria* and immersed wood			4

Scolecosporium Lib.
   Saprophyte, *S. macrosporium* (Berk.) Sutton presumed anamorph of *Asteromassaria macrospora*.

Scopulariopsis Bainier
   Saprophytes = *Masoniella* G.Smith, anamorphs of Microascaceae; *S. acremonium* (Del.) Vuill. in East Anglia.

Scopulariopsis asperula (Sacc.) Hughes	1
S. brevicaulis (Sacc.) Bainier (*S. penicillioides* (Del.) Sm. & Rams.)	1
S. brumptii Salvanet-Duval (*Masonia grisea* G.Smith)	London
S. candida (Gueguin) Vuill.	London
S. chartarum (G.Smith) Morton & Smith	London
S. (Oospora) fimicola (Cost. & Matruchot) Vuill.	1 3 4
S. fusca Zach	London

Scorpiosporium Iqbal
   Aquatic saprophytes, *S. minutum* Iqbal & *S. rangiferinum* Descals in the north and west.

Scytilidium Pesante
   Lignicolous or terrestrial saprophytes, *S. thermophilum* (Cooney & Emerson) Austwick in the midlands.

Scytilidium lignicola Pesante	4

Selenosporella Arnaud
   Saprophytes.

Selenosporella curvispora MacGarvie	On cupules of *Fagus*	1

Sepedonium Link.
Hyperparasites, anamorphs of *Apiocrea*, also unassigned:

Sepedonium niveum Massee & Salmon		1	
S. simplex (Corda) Lindau		1	

Septofusidium W.Gams
Saprophytes or hyperparasites, *S. elegantulum* (Pidop.) Gams on *Laurus* in the west.

Septofusidium herbarum (Brown & Smith) Samson	On *Heracleum*	1	

Septonema Corda
Saprophytes, see also *Heteroconium*.

Septonema fasciculare (Corda) Hughes	On coniferous wood		2
S. secedens Corda	On coniferous wood	1	

Septosporium Corda
Lignicolous saprophytes.

Septosporium atrum Corda	Needs confirmation		3

Septotrullula v.Höhn.
Corticolous saprophytes.

Septotrulla bacilligera v.Höhn.	On *Betula, Fagus, Quercus*	1	

Sesquicillium W.Gams
Saprophytes in litter or soil, segregates from *Verticillium*.

Sesquicillium buxi (Schmidt) Gams	On *Buxus sempervirens*	1		
S. candelabrum (Bon.) Gams	On *Laurus* &c.	1	2	4

Sigmoidea Crane
Saprophytes, *S. marina* Haythorn & Jones in Hampshire, *S. aurantiaca* Descals in the north.

Spadicoides Hughes
Saprophytes or hyperparasites (anamorph of *Helminthosphaeria*).

Spadicoides atra (Corda) Hughes	On *Larix, Picea, Populus, Pseudotsuga, Quercus*	1		
S. bina (Corda) Hughes	On *Betula, Fagus, Pinus, Prunus, Quercus, Ulmus*	1		4
S. grovei Ellis	On *Corylus, Fagus* &c.	1	2	
S. xylogena (A.L.Smith) Hughes	On *Hypochnicium* spp.		2	

Spermospora Sprague
Facultative parasites of Gramineae, *S. lolii* MacGarvie & O'Rourke in Berkshire.

Sphaeridium Fres.
Saprophytes.

Sphaeridium candidum Fuck.	On cones and needles of *Juniperus* and *Pinus*	1	
S. citrinum Sacc.	On Fagus		2

Spilocaea Fr.
Facultative parasites, anamorphs of *Venturia* which see, also:

Spilocaea photinicola (McClain) Ellis	On *Photinia arbutifolia*	1	
S. pyracanthae (Otth) v.Arx	On *Eriobotrya japonica* & *Pyracantha*	1	

Spirosphaera van Beverwijk
Lignicolous saprophyte.

Spirosphaera floriforme van Beverwijk	On *Alnus, Calluna* &c.	1

Spondylocladiella Linder
Hyperparasite, *S. botrytioides* Linder on "*Corticium*" in the north.

Spondylocladiopsis Ellis
Saprophyte.

Spondylocladiopsis cupulicola Ellis	On *Fagus* cupules	1

Sporendonema Desm.
Saprophytes.

Sporendonema casei Desm. (*Oospora crustacea* auct.)	In cheese, bread, jam	1	4
S. purpurescens (Bon.) Mason & Hughes	In mushroom bed	2	

Sporidesmiella Kirk
Saprophytes, compare also *Endophragmia hyalosperma*; *S. coronata* (Sutton) Kirk in Berkshire.

Sporidesmiella claviformis Kirk	On *Rubus, Taxus, Ulex*	1	
S. parva (Ellis) Kirk	On *Buxus, Laurus, Rubus*		4

Sporidesmium Link
Saprophytes, mainly lignicolous, compare *Endophragmiella;* also *S. cladii* Ellis, *S. coronatum* Fuck., *S. ehrenbergii* Ellis & *S. paludosum* Ellis in E. Anglia, *S. anglicum* (Grove) Ellis, *S. cambrense* Ellis, *S. clarkei* Kirk, *S. ensiforme* Descals, *S. flexum* Matsushima, *S. parvissimum* Hughes, *S. pseudobambusae* Kirk, *S. rubi* Ellis, *S. vermiculatum* (Cooke) Ellis, *S. wroblewskii* (Bub.) Ellis in the west and *S. aturbinatum* (Hughes) Ellis in the north.

Sporidesmium adscendens Berk		1	
S. altum (Preuss) Ellis		1	
S. britannicum Sutton			2
S. cookei (Hughes) Ellis			4
S. cymbispermum Kirk	1		
S. densum (Sacc. & Roum.) Mason & Hughes	1		
S. eupatoriicolum Ellis	1		
S. folliculatum (Corda) Mason & Hughes	1	2	
S. goidanichii (Rambelli) Hughes	1		
S. hormiscioides Corda	1		
S. ilicinum Kirk	1		
S. larvatum Cooke & Ellis	1		
S. leptosporum (Sacc. & Roum.) Hughes	1	2	
S. pedunculatum (Peck) Ellis	1		

Sporoschisma Berk. & Br.
Predominantly lignicolous saprophytes.

Sporoschisma juvenile Boud.	Plurivorous	1
S. mirabile B. & Br.	Plurivorous	1

Sporothrix Hekt. & Perkins
Saprophytes and agents of animal mycoses, anamorphs of Eurotiales, *S. vizei* de Hoog in the west. & see *Engyodontium*.

Sporothrix schenckii Hektoen & Perkins	London

Sporotrichum Link
Saprophytes, anamorphs of Aphyllophorales but see also *Arthrographis* & *Tritirachium*.

Sporotrichum aureum Link	1 2 3

Stachybotrys Corda
Saprophytes, anamorph of *Melanopsamma*, also:

Stachybotrys chartarum (Ehrenb.) Hughes	Plurivorous	1	4
(*S. atra* Corda)			
S. cylindrospora Jensen		1	
S. dichroa Grove		1	

Stachylidium Link
Saprophyte, see also *Phaeostalagmus*.

Stachylidium bicolor Link	On *Heracleum, Oenanthe, Pteridium, Sambucus, Urtica*	1

Stemmaria Preuss
Saprophyte.

Stemmaria aeruginosa Massee	On bird dung	1

Stemphylium Wallr.
Saprophytes, anamorphs of *Pleospora*, also:

Stemphyllum ilicis Tengwall (*S. dendriticum* da Camara)		1	2	4
S. sarciniforme (Cav.) Wiltshire	On *Trifolium pratense*	1		
S. vesicarium (Wallr.) Simmons				4

Stenella Syd.
Facultative foliar parasites segregated from *Cercospora*, *S. lythri* (West.) Mulder in Hampshire.

Stephanoma Wallr.
Hyperparasite, *S. strigosa* Wallr. on hymenium of *Humaria hemispherica* in East Anglia.

Stephanosporium DalVesco
Saprophyte.

Stephanosporium cerealis (Thüm.) Swart	In soil and cereal flour	1

Sterigmatobotrys Oud.
Saprophyte.

Sterigmatobotrys macrocarpa (Corda) Hughes	On wood of *Abies, Picea, Taxus*	1

Stigmina Sacc.
Facultative parasites.

Stigmina carpophila (Lév.) Ellis	Shot-hole of *Prunus* spp.	1			
S. platani (Fuck.) Sacc.	On *Platanus.* Needs confirmation	1			
S. pulvinata (Kunze) Ellis	Especially on *Ulmus*	1			
S. tinea (Sacc.) Ellis	On *Viburnum davidii* & *V. tinus*	1			

Stilbella Lindau
Saprophytes, *S. pellucida* (Schrad.) Lindau in E. Anglia, *S. kervillei* (Quél.) Samson in caves.

Stilbella erythrocephala (Ditmar) Lindau (*Stilbum fimetarium* (Pers.) B & Br.)		1	2	3	4
S. fasciculata (B. & Br.) Mason & Grainger	On wood	1			
S. thermophila Fergus			2		
S. turbinata (Tode) Lindau	On wood	1			

Doubtful species perhaps belonging here:

Stilbum citrinellum Cooke & Massee	On *Lycopodium*	1	
S. sphaerocephalum Massee	On *Philodendron*	1	

Subramaniomyces Mani Verghese & Rao
Saprophyte, *S. fusisaprophyticum* (Matsushima) Kirk on *Laurus* & *Quercus ilex* in the west.

Subulispora Sutton
Saprophytes.

Subulispora britannica Sutton	On *Hedera, Ilex, Laurus* and *Quercus ilex*	1			
S. minima Kirk	On *Laurus nobilis*	1	2		4

Symphyosira Preuss
Anamorphs of *Symphyosirinia.*

Sympodiella Kendrick
Saprophytes.

Sympodiella acicola Kendrick	On needles of *Pinus*	1	
S. foliicola Kirk	On *Quercus*	1	

Sympodiocladium Descals
Aquatic saprophyte, *S. frondosum* Descals in the west.

Syngliocladium Petch
Entomogenous parasites, *C. cleoni* (Wize) Petch on wireworm larvae in Hertfordshire.

Syngliocladium aranearum Petch	On a spider	2	

Taeniolella Hughes
Saprophytes, mainly lignicolous, formerly referred to *Torula* or *Hormiscium*; *T. delicata* Christ. & Hawks. in Hampshire, *T. plantaginis* (Corda) Hughes in E. Anglia, *T. alta* (Ehrenb.) Hughes, *T. breviuscula* (B. & C.) Hughes, *T. faginea* (Fuck.) Hughes, *T. phaeophysciae* Hawks. and *T. pulvillus* (B. & Br.) Ellis in the west.

Taeniolella rudis (Sacc.) Hughes	On worked wood		4
T. scripta (Karst.) Hughes	On *Betula, Corylus, Fagus, Sorbus*	1	
T. stilbospora (Corda) Hughes	On *Alnus, Corylus* and especially *Salix*	3	

Taeniolina Ellis
Facultative parasite.

Taeniolina centaurii (Fuck.) Ellis On *Centaurium erythraea* 1 4

Taeniospora Marvanova
Aquatic saprophytes, *T. descalsii* Marv. & Stalp. and *T. gracilis* Marv. in the west.

Teratosperma Syd.
Facultative parasites & hyperparasite, *T. oligocladium* Decker et al. on *Sclerotinia* in the north.

Tetrachaetum Ingold
Aquatic saprophyte.

Tetrachaetum elegans Ingold 1 4

Tetracladium de Wild.
Aquatic saprophytes, others in the west.

Tetracladium marchalianum de Wild. 2 4
T. maxilliformis (Rostrup) Ingold 4
T. setigerum (Grove) Ingold 4

Tetraploa B. & Br.
Saprophyte.

Tetraploa aristata B. & Br. Especially on Cyperaceae, Gramineae & 1 4
Juncaceae

Tetraposporium Hughes
Saprophyte.

Tetraposporium ravenelii (Cooke) Hughes On *Laurus nobilis* 1

Thallomicrosporon Benedek
Agent of animal and human mycosis.

Thallomicrosporon kuehnii Benedek On *Canis* 1

Thedgonia Sutton
Facultative parasites.

Thedgonia ligustrina (Boerema) Sutton On *Ligustrum vulgare* in Hampshire

Thermoideum Miehe
Saprophyte, *T. sulphureum* Miehe in soil in the west midlands.

Thermomyces Tsiklinsky
Saprophytes in soil & fermenting vegetable matter, *T. (Humicola) stellatus* (Bunce) Apinis & *T. verrucosus* Pugh et al. in the midlands.

Thermomyces (Humicola) lanuginosus In soil, grain silos etc. 1
Tsiklinsky

Thielaviopsis Went.
Saprophyte and facultative parasite, especially of root systems of glasshouse crops.

Thielaviopsis basicola (B. & Br.) Ferraris (*Milowia nivea* Massee) 1 2 4

Thyrostromella v. Höhn.
 Saprophyte.

Thyrostromella myriana (Desm.) v. Höhn.   On *Ammophila arenaria*   4

Thysanophora Kendrick
 Saprophytes on tissues of Gymnosperms.

Thysanophora penicillioides (Roum.) Kendrick		3	
T. taxi (Schneider) Stolk & Hennebert	1	2	

Tilachlidium Preuss
 Saprophyte.

Tilachlidium (*Isaria*) *brachiatum* (Batsch)   On rotting Agaricales, *Helvella* etc.   1
 Petch

Titaea Sacc. see *Volucrispora*

Tolypocladium Gams
 Parasites, *T. cylindrosporium* W. Gams widespread on dipterous larvae, *T. geodes* Gams in the north.

Torula Pers.
 Saprophytes, see also *Scytilidium, Taeniolella* & *Xylohypha.*

Torula fusca (Bon.) Sacc.				4
T. herbarum (Pers.) Link	1	2	3	4
T. ligniperda (Willk.) Sacc.				4
T. ovalispora Berk	1			4
T. spongicola Dufour	1			

Torulopsis Berlese
 Animal mycoses, status as for *Candida* and see *Cryptococcus.*

Torulopsis candida (Saito) Lodder	In cattle	1
T. famata (Harrison) Lodder & van Rij	In cattle, hay etc.	1
T. glabrata (Anderson) Lodder & Vries	In swine	1

Triadelphia Shearer & Crane
 Saprophyte.

Triadelphia (*Dicoccum*) *uniseptatum*   On *Rubus fruticosus*   1
 (B. & Br.) Kirk

Tricellula van Beverwijk
 Saprophyte, *T. botryosa* Descals in the west.

Tricellula aquatica Webster   2

Trichocladium Harz
 Saprophytes, *T. macrosporum* Kirk & *T. pyriforme* Dixon in the west and north.

Trichocladium asperum Harz	1	
T. opacum (Corda) Hughes	1	4

Trichoconis Clem.
 Hyperparasite, *T. lichenicola* Hawks, on galled lichen thalli.

Trichoderma Pers.
Saprophytes and hyperparasites, anamorphs of *Hypocrea* which see, also:

Trichoderma hamatum (Bon.) Bainier (*Pachybasium hamatum* (Bon.) Sacc.)	1		
T. harzianum Rifai			4
T. koningii Oud.	1		4
T. polysporum (Link) Rifai	1	3	
T. pseudokoningii Rifai	1		

Trichophyton Malmsten
Saprophytes and agents of animal mycoses, anamorphs of *Arthroderma*, also:

Trichophyton equinum Gedeolst	1	4
T. gallina (Magnus) Silva & Banham	1	
T. mentagrophytes (Robin) Blanchard (*T. interdigitale* Priestley)	1	
T. verrucosum Bodin	1	
T. verrucosum Bodin v. *discoides* (Sabour.) Georg.	1	
Several species on man unlocalised.		

Trichosporon Behrend
Animal mycoses, *T. beiglii* (Kuchenm. & Rab.) Vuill. in man.

Trichosporon cutaneum (deBeurm et al.) Ota	1
T. capitatum Diddens & Lodder	1

Trichothecium Link
Saprophytes, but see *Duddingtonia*.

Trichothecium roseum (Pers.) Link	Ubiquitous & plurivorous	1	2	3	4

Trichurus Clem.
Saprophytes.

Trichurus spiralis Hasselbring	2

Tricladiopsis Descals
Aquatic saprophytes, *T. flagelliformis* Descals anamorph of *Coccomyces dentatus*, also *T. foliosa* Descals in the west.

Tricladium Ingold
Saprophytes, mainly aquatic, *T. gracile* Ingold in Berkshire, many more in the west.

Tricladium angulatum Ingold		2	4
T. castaneicola Sutton	On cupules of *Castanea, Fagus, Laurus* leaves etc.	2	4
T. splendens Ingold		2	4
T. terrestre Park			4

Tridentaria Preuss
Saprophytes or predaceous and see *Tetracladium setigerum*.

Tridentaria implicans Drechsler	1

Trimmatostroma Corda
Corticolous or lignicolous saprophytes, *T. scutellare* (B. & Br.) Ellis anamorph of *Eriosphaeria*.

Trimmatostroma betulinum (Corda) Hughes (*Coniothecium betulinum* Corda)		1	2	4
T. salicis Corda	On *Rosa, Salix*	1		4

Tripospermum Speg.
Saprophytes, *T. camelopardus* Ingold et al. & *T. prolongatus* (Sinclair & Jones in streams.

Tripospermum myrtii (Lind) Hughes	On *Myrtus*	4

Triposporium Corda
Saprophyte, see also *Actinocladium*.

Triposporium elegans Corda	On *Chamaenerion, Fagus, Filipendula, Rubus, Quercus* etc.	1	2	3	4

Triscelophorus Ingold
Aquatic saprophytes, *T. monosporus* Ingold in Berkshire.

Trisulcosporium Hudson & Sutton
Saprophyte, *T. acerinum* Hudson & Sutton on rotting leaves in the east midlands.

Tritirachium Limber
Saprophytes, see also *Acrodontium, Beauveria, Engyodontium*.

Tritirachium dependens Limber	On conifers	1	
T. isariae (Petch) de Hoog (*Sporotrichum isariae* Petch)	On *Paecilomyces farinosus*		2
T. oryzae (Vincens) de Hoog	In soap	1	

Troposporella Karst.
Saprophytes, *T. fumosa* Karst. in East Anglia.

Troposporella monospora (Kendrick) Ellis	On *Juniperus* and *Pinus sylvestris*	1

Tubercularia Tode
Anamorphs of *Nectria*.

Tuberculina Sacc.
Hyperparasite of uredineae, especially the aecidial state.

Tuberculina persicina (Ditm.) Sacc.	On *Cronartium, Puccinia caricina, P. coronata, P. dioicae, P. poarum, P. punctiformis, P. sessilis, P. vincae, Tranzschelia* etc.	1	2

Ulocladium Preuss
Saprophytes on plant products, damp paper &c., *U. consortiale* (Thüm.) Simmons in E. Anglia.

Ulocladium alternariae (Cooke) Simmons (*Stemphylium alternariae* (Cke.) Sacc.)	1	
U. atrum Preuss (*Stemphylium atrum* (Preuss) Sacc.	1	4
U. botrytis Preuss	1	
U. chartarum (Preuss) Simmons (*Alternaria chartarum* Preuss)	1	4
U. oudemansii Simmons	"London"	

Valdensia Peyronel
Facultative parasite, *V. heterodoxa* Peyrol on *Vaccinium* etc. in the north.

Vanrija Moore
Saprophyte, *V. aquatica* (Jones & Sloof) Moore in the west.

Vargamyces Tóth
Aquatic saprophyte, *V. aquaticus* (Dudka) Tóth in the west.

Varicosporium Negel
Aquatic saprophytes, *V. delicatum* Iqbal in the midlands and west.

Varicosporium elodeae Negel   4

Vermiculariopsella Bender
Saprophyte = *Oramasia* Urries.

Vermiculariopsella immersa (Desm.) Bender (*Oramasia hirsuta* Urries)   1

Veronaea Ciferri & Montemartini
Saprophytes or ?hyperparasites, *V. botryosa* Cif. & Mont. in the north, *V. caricis* Ellis in the west, *V. carlinae* Ellis, *V. parvispora* Ellis, *V. verrucosa* Geeson in East Anglia.

Verticicladium Preuss
Anamorph of *Desmazierella*, which see.

Verticillium Nees
Saprophytes, facultative parasites and hyperparasites, including *Acrostalagmus* Corda; see also *Phaeostalagmus*, *Phialophora* and *Sesquicillium*, a few are anamorphs of *Hypomyces* and *Nectria* which see.
Elsewhere *V. bulbillosum* Gams & Malla in the north, *V. effusum* Otth in the midlands, *V. tricorpus* Isaac in the west, *V. (Diheterospora) catenulatum* (Kamyschko) Gams and *V. menisporoides* Petch in East Anglia.

Verticillium alboatrum Reinke & Berthold	Ubiquitous as agent of vascular wilt	1	2	3	4
V. ampelinum Cooke & Massee	Needs clarification	1			
V. cf. *cephalosporum* Gams		1			
V. chlamydosporum Goddard		1			
V. dahliae Kleb.	Vascular wilt	1	2		4
V. fungicola (Preuss) Hassebr. (*V. malthousei* Ware)		1	2		4
V. fungicola v. *aleophilum* Gams & van Zaayen					4
V. fungicola v. *flavidum* Gams & van Zaayen					4
V. lecanii (Zimm.) Viegas		1	2		
V. nigrescens Pethybr.		1			4
V. nubilum Pethybr.					4
V. obovatum (Drechsler) Subramanian		1			
V. psalliotae Treschow		1			

Virgaria Nees
Saprophyte.

Virgaria nigra (Link) Nees	Plurivorous on bark and wood	1	2

Virgariella Hughes
Lignicolous saprophytes.

Virgariella atra Hughes	On *Fagus, Fraxinus, Quercus*	1
V. ovoidea Kirk	On *Quercus*	1

Volucrispora Haskins
Aquatic saprophytes.

Volucrispora aurantiaca Haskins			4
V. graminea Ingold et al.		2	4
V. ornithomorpha (Trotter) Haskins (*Titaea ornithomorpha* Trotter)	1		4

Volutella Tode
Saprophytes, anamorphs of *Pseudonectria* which see; *V. arundinis* (Desm.) Sacc. & *V. melaloma* B. & Br. on marsh plants in East Anglia.

Volutella ciliata (A. & S.) Fr.	Ubiquitous and plurivorous	1	4
V. colletotricoides Chilton	On *Zea mays*		4
V. setosa (Grev.) B. & Br.	On herbaceous stems		4

Wallemia Johan.-Olsen
Saprophytes.

Wallemia sebi (Fr.) v. Arx	Cosmopolitan, especially in foodstuffs	1	2

Wardomyces Brooks & Hansford
Saprophytes, especially in soil but also litter and foodstuffs, five species in the midlands.

Wiesneriomyces Koorders
Foliicolous saprophyte.

Wiesneriomyces laurinus (Tassi) Kirk (*W. javanicus* Koorders)	On *Laurus nobilis*	1	4

Xanthoriicola Hawks.
Lichenicolous hyperparasite, *X. physciae* (Kalchbr.) Hawksw. on *Xanthoria* in East Anglia.

Xylohypha (Fr.) Mason
Lignicolous saprophytes, segregates from *Torula*.

Xylohypha ferruginosa (Corda) Hughes	On *Carpinus, Fagus, Quercus, Sorbus*	1	
X. nigrescens (Pers.) Mason	On decorticated angiosperm wood of all kinds	1	3
X. ortmansiae Minter	On cones of *Pinus sylvestris*	1	2
X. pinicola Hawksw.	On cones of *Pinus sylvestris*		4

Zalerion Moore & Mayers
Lignicolous saprophytes, *Z. arboricola* Buczacki on *Larix* & *Picea* in the west midlands.

Zalerion maritima (Linder) Anastasiou	On *Fagus* & *Pinus* immersed in the sea	3	4

Zygophiala Mason
Facultative parasite, mainly in the tropics.

Zygophiala jamaicensis Mason	On *Dianthus* under glass	1	2

Zygosporium Mont.
Saprophytes, *Z. echinosporum* Bunting & Mason in Essex, *Z. minus* Hughes in the west.

Zygosporium gibbum (Sacc., Rouss. & Bomm.) Hughes (*Z. parasiticum* (Grove) Mason	1
Z. mycophilum (Vuill.) Sacc.	1
Z. oscheiodes Mont.	1

Index

Cymadothea 172
Cyphellopsis 27
Cyphellostereum 61
Cystochytrium 202
Cystodendron 239
Cystoderma 27
Cystolepiota 36
Cystopezizella 110
Cytidia 56
Cytoplacosphaeria 211
Cytospora 156, 211
Cytosporopsis 155

Dacampia 172
Dacrymyces 81
Dacryobolus 61
Dacryomitra 81
Dactylaria 239
Dactylariopsis 239
Dactylella 177, 239
Dactylellina 240
Dactylium 240
Dactylomyces 161
Dactylospora 130
Dactylosporium 240
Daedalea 61
Daedaleopsis 61
Daldinia 141
Dangeardia 202
Dangeardiella 172
Darluca 225
Daruvedia 173
Dasyscyphella 115
Dasyscyphus 115
Datronia 61
Debaryella 152
Debaryomyces 166
Deconica 27, 49
Deightoniella 240
Dekkera 166
Delicatula 27, 41
Delitschia 173
Delphinella 173
Dematioscypha 110
Dematium 257
Dematophora 240
Dencoeliopsis 111
Dendrodochium 240
Dendrophoma 211
Dendrospora 240
Dendrostilbella 110
Dendryphiella 240
Dendryphion 240
Dendryphiopsis 240

Dennisiella 173
Dennisiodiscus 111
Dermea 111
Dermocybe 25
Dermoloma 27
Desmazierella 99
Dialonectria 134
Diapleella 141
Diaporthe 153
Diaporthopsis 153
Diatrype 150
Diatrypella 151
Dibeloniella 111
Dicellomyces 81
Dichobotrys 105
Dichomera 211
Dichomitus 61
Dichomyces 167
Dichostereum 61
Dichotomophthora 240
Dicoccum 240
Dicranidion 240
Dicranophora 188
Dictyochaeta 241
Dictyopolyschema 241
Dictyosporium 241
Dictyotrichiella 171
Dictyuchus 192
Didymaria 241
Didymascella 127
Didymella 173
Didymochora 170
Didymopleella 173
Didymopsis 241
Didymosphaeria 173
Didymotrichiella 171
Diehliomyces 165
Digitodesmium 241
Diheterospora 241, 272
Dilophospora 177, 212
Dimargaris 188
Dimerium 173
Dimeromyces 167
Dimerosporium 170
Dimorphospora 241
Dinemasporium 144
Diplocarpa 111
Diplocarpon 111
Diplocladiella 241
Diplocladium 241
Diplococcium 241
Diplodia 212
Diplodina 152, 212
Diplophlyctis 202